ABOUT ISLAND PRESS

Island Press, a nonprofit organization, publishes, markets, and distributes the most advanced thinking on the conservation of our natural resources—books about soil, land, water, forests, wildlife, and hazardous and toxic wastes. These books are practical tools used by public officials, business and industry leaders, natural resource managers, and concerned citizens working to solve both local and global resource problems.

Founded in 1978, Island Press reorganized in 1984 to meet the increasing demand for substantive books on all resource-related issues. Island Press publishes and distributes under its own imprint and offers these services to other nonprofit organizations.

Support for Island Press is provided by Apple Computers, Inc., The Mary Reynolds Babcock Foundation, The Educational Foundation of America, The Charles Engelhard Foundation, The Ford Foundation, The George Gund Foundation, The William and Flora Hewlett Foundation, The Joyce Foundation, The J. M. Kaplan Fund, The John D. and Catherine T. MacArthur Foundation, The Andrew W. Mellon Foundation, The Joyce Mertz-Gilmore Foundation, The New-Land Foundation, Northwest Area Foundation, The Jessie Smith Noyes Foundation, The J. N. Pew, Jr., Charitable Trust, The Rockefeller Brothers Fund, The Florence and John Schumann Foundation, The Tides Foundation, and individual donors.

For additional information about Island Press publishing services and a catalog of current and forthcoming titles, contact Island Press, Box 7, Covelo, California 95428

FORESTS AND FORESTRY IN CHINA

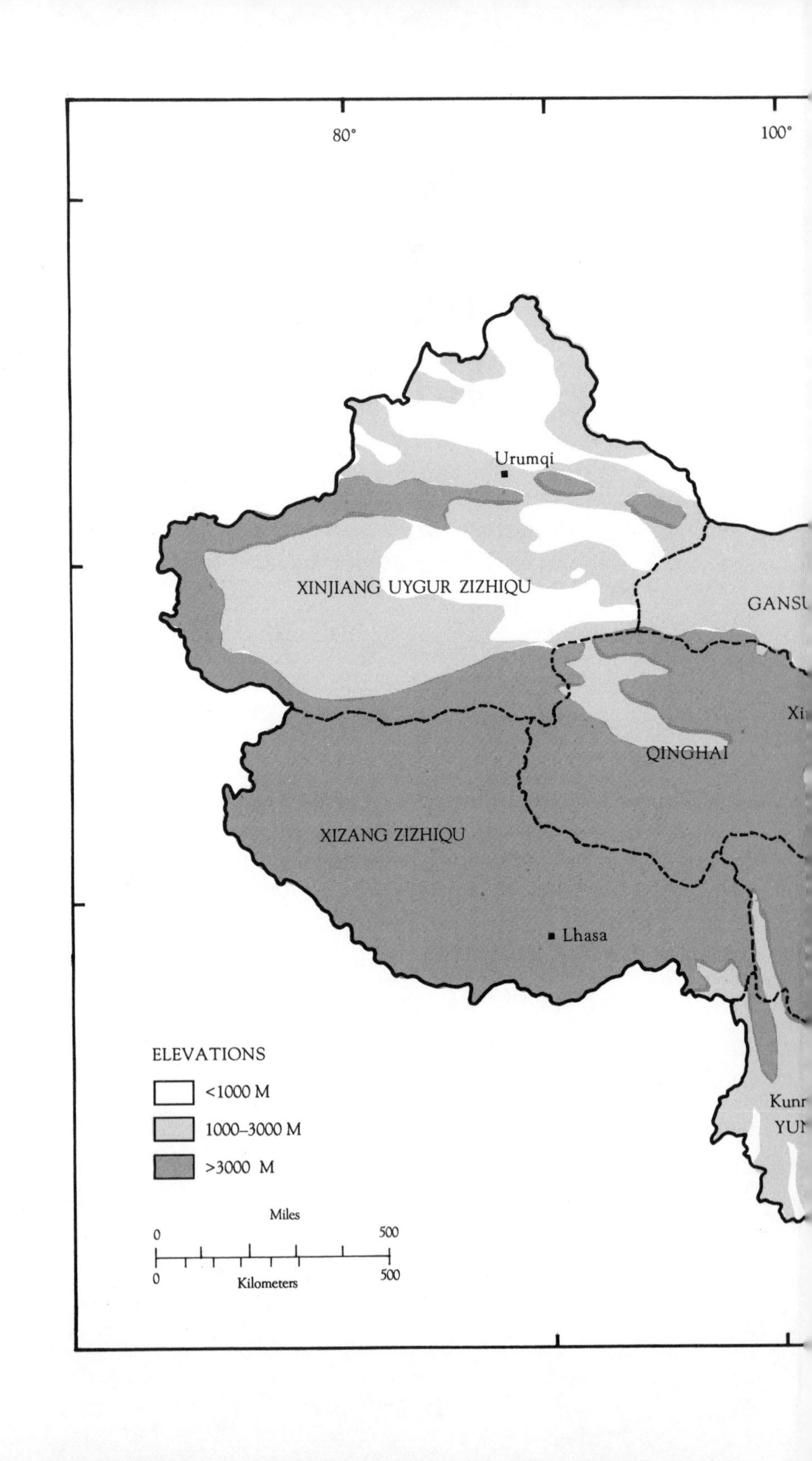

80°
100°
Urumqi
XINJIANG UYGUR ZIZHIQU
QINGHAI
XIZANG ZIZHIQU
Lhasa
ELEVATIONS
<1000 M
1000–3000 M
>3000 M
Miles
0
500
0
Kilometers
500

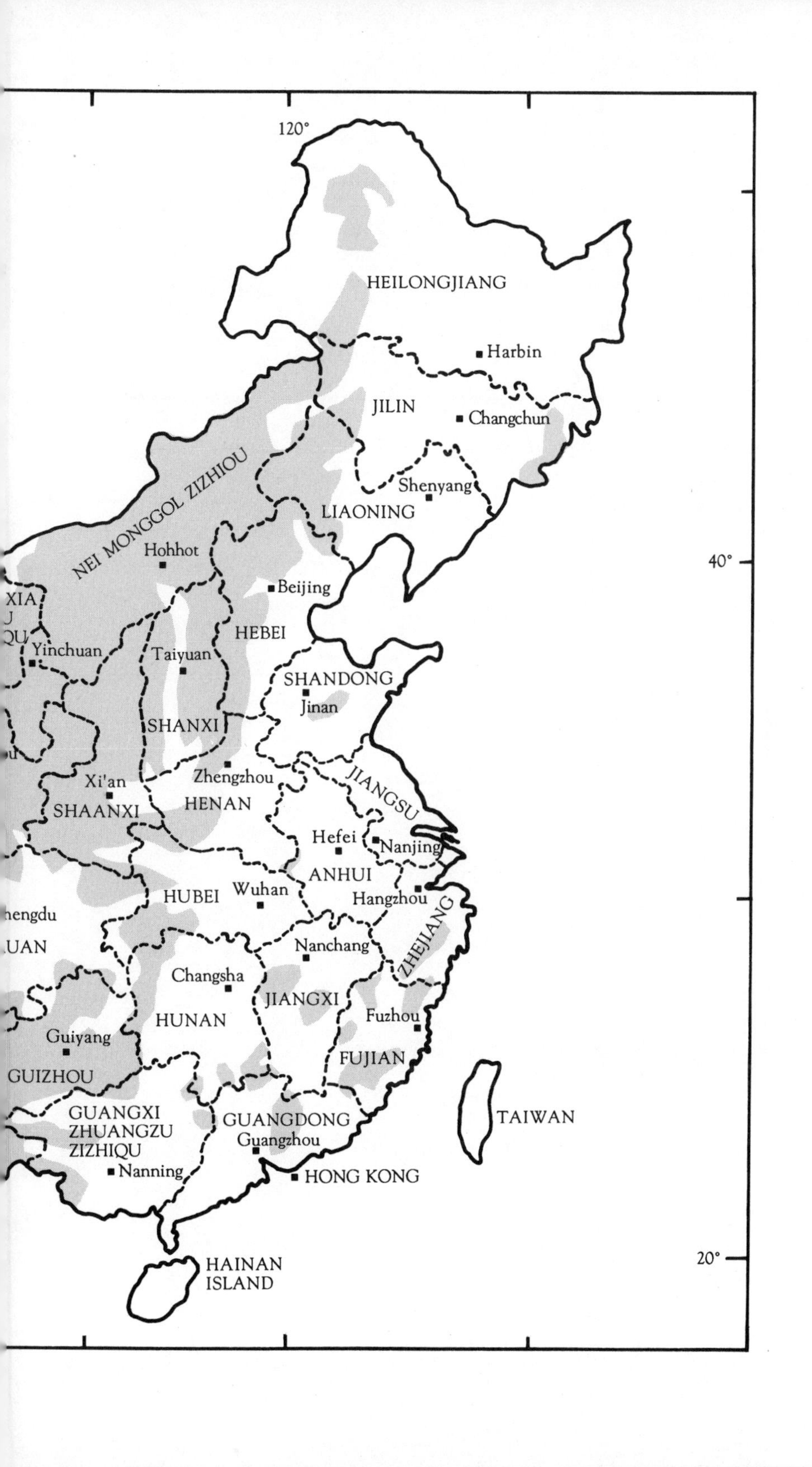
120°
40°
20°
HEILONGJIANG
Harbin
JILIN
Changchun
Shenyang
LIAONING
NEI MONGGOL ZIZHIOU
Hohhot
Beijing
HEBEI
Yinchuan
Taiyuan
SHANDONG
Jinan
SHANXI
Zhengzhou
Xi'an
JIANGSU
SHAANXI
HENAN
Hefei
Nanjing
ANHUI
Hangzhou
HUBEI
Wuhan
ZHEJIANG
Nanchang
Changsha
JIANGXI
Fuzhou
HUNAN
Guiyang
FUJIAN
GUIZHOU
GUANGXI
ZHUANGZU
ZIZHIQU
GUANGDONG
Guangzhou
TAIWAN
Nanning
HONG KONG
HAINAN
ISLAND

FORESTS AND FORESTRY IN CHINA

CHANGING PATTERNS OF RESOURCE DEVELOPMENT

S. D. RICHARDSON

Fellow, Environmental and Policy Institute
East-West Center

FOREWORD BY JEFF ROMM

ISLAND PRESS

Washington, D.C. ☐ *Covelo, California*

Portions of Chapters 1 and 2 were published previously in *Forestry in Communist China* by The Johns Hopkins University Press, 1966.

Library of Congress Cataloging-in-Publication Data

Richardson, S. D. (Stanley Dennis), 1925–
Forests and forestry in China: changing patterns in resource development / by S. D. Richardson ; foreword by Jeff Romm.
p. cm.
Includes bibliographical references.
ISBN 1-55963-023-X. — ISBN 1-55963-022-1 (pbk.)
1. Forests and forestry—Economic aspects—China. 2. Forest policy—China. 3. Forest management—China. I. Title.
HD9766.C52R53 1990
333.75′0951—dc20 89-24514
CIP

Photographs by the author

Printed on recycled, acid-free paper

Manufactured in the United States of America

10 9 8 7 6 5 4 3 2 1

To the memory of Cheng Wanchun,
who *achieved* the impossible

CONTENTS

FOREWORD

Although the many social values of forests have been recorded for at least 2,500 years, the social impacts of any one forest site have always expressed the special needs of the men and women who controlled its uses. People—hunters, gatherers, shifting cultivators, commercial loggers, politicians, peasant farmers, government foresters, city dwellers, soldiers, road and dam builders—decide what trees to cut and to grow, and where and when to do so. The forest landscape, the collective consequence of these actions, manifests the distribution of forest control among such groups and the motives and abilities that govern what they do. Whether viewed in terms of the forest's area, spatial distribution, productivity, species mixes, hydrologic effects, or spiritual significance, the qualities of the landscape are shaped by the social forces that distribute control of the forest and motivate its use.

These social forces have intensified in the past four decades. Nations throughout the world have tried to strengthen their sovereignty, independence, and livelihood by converting forests into timber, agricultural land, agrarian settlement, and administered territory. The process of national development, the sustained growth of social opportunity and well-being, has rapidly increased the size, mobility, hopes, and capacities of human populations, thus their needs for and access to land. Trade and communications networks are connecting every village with distant events and diverse views; democratic values and pluralistic politics have spread; people vote with their feet as never before to take their "fair share" of the land's bounties.

Such trends are transforming societies and their landscapes, for good and for ill. Measures of human well-being—life expectancy, security, income, opportunity, access—display encouraging rises in most parts of the world. On the other hand, the ecological slack that subsidized earlier surges in human welfare—and on which strategies of development had come to depend—is disappearing or is gone, with effects on people that rarely are beneficial or equitable. These trends have reached and penetrated the world's forests, which had survived protected by their inaccessibility. The trends have shaken the once-quiet domains of forests, forestry, and forest policy.

Typical forest policies until recently were fence lines of formal rights, authorities, and obligations around classes of land in which the state had some interest. In policy, the "forest" was a jurisdictional rather than a biophysical concept. It could include treeless land the state claimed; it could exclude tree-covered land outside a regime of state control—"wasteland," "jungle," "wilderness"—or within settlements and agricultural regions that defied the definitions of "forest" regimes. The "proper" use and ecological character of the forest, the focus of forestry, depended on the interests of those who created and enforced the fence; over time, forestry and forest policy became specialized to achieve the relatively narrow range of social and ecological possibilities that the state's vision of propriety required.

But actual forest use and character differed radically from the proper and professional vision when the pressures of other interests exceeded the enforceability or scope of law. By the 1970s, developing societies were projecting their energies across the landscape with such force that fences and the forests they protected were collapsing everywhere. Massive clearance occurred in some circumstances: the rates of tropical deforestation are familiar. Elsewhere, forests of new or unconventional kinds—private woods, village aggregations of forest gardens, temple forests—arose or simply were noticed by professionals for the first time as their vision expanded beyond the jurisdictional forest. People grew trees around their houses, in their fields, along roads and canals; they managed "natural" forests on a sustainable basis in forms and for purposes—fuel, food, fodder, fertilizer, medicines—never cited in forestry texts; they preserved and planted trees, species, and groves for religious reasons. The contributions of such forests to human well-being were found to be significant. Ninety percent of the wood supply of Bangladesh, for example, grows in home gardens—a fact not known until five years ago—and has no connection to the disappearing "forest" to which all professional and policy attention had been devoted.

Developmental processes were at once destroying conventional forests and creating or revealing new ones, usually in different places and of different kinds, by and for different people. The ecological and social consequences depended on how these changes modified the landscape as a whole. Although the welfare effects of landscape changes could be profound, the conventional formulations of development policy were not prepared to address them. Such policies had proven their worth in stimulating and steering the impacts of developmental process on other sources of human well-being—industrial, agricultural, educational, public health, and administrative capacities, for example—but they had yet to be conceived as means for shaping the environmental basis of life in more than specific places. When forestry issues exploded from their neat compartments into pervasive dynamic forces, they provoked a reconsideration that is just beginning to provide glimpses of the challenges ahead.

Over the past fifteen years, forestry analysts and professionals have been developing means to use forests—in their broadest sense—as active contributors in the growth of human opportunity and well-being over a wide range of social and ecological conditions. Some programmatic terms suggest the flavor and extent of the movement: social forestry, community forestry, village forestry, agroforestry, homestead forestry, participatory forest planning, stewardship agreements, cooperative forestry, forestry for rural or community development. These programs are still experimental, valuable perhaps more for their lessons than their outcomes, but each implies its own family of departures from historic conventions of forestry technique, organization, and control. Together, they have become the most influential, if not the predominant, features of forest policy in most nations.

A less obvious but equally profound trend has been occurring over the same period at the national level of development policy. Forestry analysts are demonstrating how standard instruments of policy—for trade, commodity prices, wages, savings, public and private investment, research and development, and intergovernmental, interagency, and intersectoral distributions of authority—determine where people go, what they do, how conservatively or destructively they do it, the effects on the landscape, and the well-being it might sustain. Such analyses promise some future convergent understanding that will enable nations to add "environmental" to the accepted economic and social objectives of development policy.

Dennis Richardson's *Forests and Forestry in China* is a unique contribution to our understanding of these developments in forestry and forest policy. Professor Richardson's vantage point is special: he has had the rare opportunity to witness the evolution of forestry and its roles in national development in China over a period of three decades. Since Richardson's 1966 publication of *Forestry in Communist China,* China's strategies of development have changed profoundly. Richardson's interpretation of the influence of these changes upon current forest policies, actions, and conditions creates a framework within which forestry becomes a window upon the development of the nation as a whole.

Professor Richardson's observations join a scholar's broad vision of contemporary China with a forester's sharp attention to the detail of how people use, control, and affect diverse forest ecosystems. He manages to convey at once the large forces that shape China's forests and an "on the ground" sense of the realities of China's forest resources, regional forest economies, and forestry practices. China may well possess the oldest and most diverse forestry experience in the world. As its circumstances—social, ecological, historical—have spawned inventions in forestry and forest policy that are not yet known but are possibly useful elsewhere, many of the technical and policy approaches that Professor Richardson presents advance ideas for possible trial in other situations. *Forests and Forestry in China* is a valuable source of such provocations as well as a novel perspective of developmental process and of China itself.

Jeff Romm
Professor
Department of Forestry and
Resource Management
University of California, Berkeley

PREFACE

IN 1963, at the invitation of the Academia Sinica via the Royal Society of London, I toured China's forests—from Yunnan to Manchuria. The visit resulted in the publication of *Forestry in Communist China* in 1966 and subsequently a Japanese translation by Norin Joho Chosakai. In 1986, the Chinese Academy of Forestry invited me back to China; I was able to retrace my earlier itinerary, to visit the same forest areas, and to meet some of the same people; the present book is the outcome. It follows the general outline of the 1966 publication but focuses on change and development.

China has changed considerably since 1963; so has this observer; and so, too, have global perceptions of forestry. The Cultural Revolution, like the Great Leap Forward before it, resulted in excess cutting of forests—to fuel backyard industries, to clear land for cereal grain, and, simply, to keep warm. Historically China has always been short of fuel and has a long history of forest cutting followed by flood and by drought. The observer in the 1960s saw little of the damage done to ecosystems and the eager disregard of simple conservation that post-Mao economic change has triggered. The new slogan "To Get Rich Is Glorious" is hardly compatible with sustainable, agriculture-supportive forestry—and, as in most countries where there is malpractice, the timber trade is in the forefront.

Another notable change since 1963 is in the availability—and reliability—of economic data. Following the extravagant (and subsequently withdrawn) productivity claims of the Great Leap Forward, Chinese statistics became general and guarded; indeed, analysts were reduced to what Alexander Eckstein has called "economic archaeology." It was not until 1979 that credible information became available. By then two things had become clear: First, forest production plays a crucial role in economic development; and, second, in China economic development reflects political change. The evaluation of future requirements for industrial wood, therefore, calls for discussion of political change and economic development in more detail than would be the case for most countries. By 1986, attempts to change the economic system from one of "command" planning to one that is both dispersed and market-driven had been in train long enough to enable one to evaluate the problems as well as the merits of "reform" policies. Moreover, the changes have involved a switch in priority from high investment in heavy industry to a brand of consumerism and investment favoring light industry and agriculture. Forestry is a component of both agricultural and industrial sectors; wood is both a consumer good and an industrial raw material. The role of forestry, therefore, can only grow in importance.

But with respect to forest products China faces a supply/demand crisis greater than ever before—despite controls, the shortfall will not be made up by new plantations until well into the twenty-first century. Until recently, it was assumed that the mass afforestation schemes launched with such fervor in the 1950s and 1960s would provide for China's raw material needs by 1990. The target date has now been revised to 2040. Moreover, the instrument of organizing mass participation in campaigns—the commune—is no more, and an alternative force for public works (including tree-planting, water conservation, and garbage collection) has yet to emerge. This book quantifies the crisis and discusses China's attempts "to design wood out of the economy" against a background of bureaucratic rigidity and conservative, but not conservationist, forest practices.

Forest industries are described in these pages, and a chapter is devoted to forest policy and law. China's industrial model followed Soviet guidelines and is based on large, nominally integrated wood-processing complexes. They are inefficient and wasteful, however, and lack the flexibility to cope with power shortages and disruptions to material supplies. In remote locations, they are centers for unemployment among middle-school leavers. China is coming to accept that its interests would be better served by smaller mills closer to consuming centers—which would absorb the waste wood that currently litters the woodyards. Consideration of the pulp and paper industry highlights problems of pollution in an "energy-constrained" society. The open economy has increased the demand for all forest products, with imports rising from less than 0.5 million cubic meters in 1963 to nearly 10 million in 1985.

The success story of China's forestry is, without doubt, the integration of wood production with agriculture—the planting of multiple-purpose species along roads and canals and around houses and fields, now called "Four Sides" forestry. Though regarded as a new development in many countries, it is the traditional forestry of China's northern plains and has been documented for hundreds of years. So, too, have the successful urban tree plantings. As early as the first unified empire of China under Qin Shi Huang (around 220 B.C.), there is evidence (annotated by Menzies, 1989) of a strategic barrier of elm trees planted along the Yellow River to prevent the barbarians from watering their horses. Commercial tree planting is recorded before the sixth century A.D., and the earliest known silvicultural monograph (on *Paulownia*) dates from the eleventh century. Despite these ancient traditions, industrial timber shortages have been documented for several hundred years.

Forests and Forestry in China updates developments in protection forestry (which are bedeviled by political and bureaucratic demarcation disputes) and notes the principal changes in forestry research and education. Research has become more competitive and more imitative, but the administrative problems are recognized. Forestry education and training, as indeed all disciplines, suffered from the loss of an entire decade through the Cultural Revolution, and the restricted vision of the cadres and managers is a legacy that may take much longer to overcome.

The present generation of foresters (the world over) has come to accept that forestry must concern itself with people as much as with trees. *Forests and Forestry in China* underlines that lesson: The successes (and the failures) are man-made. The book ends with a chapter on "Learning from China," which concludes that the country with the most to learn from China is China.

This book provides a personal and professional evaluation of forestry over the past quarter of a century in the world's most populous, but least known, country. My return to China was made possible by the award of a fellowship by the East-West Center in Honolulu and a grant from the New Zealand Ministry of Foreign Affairs. I am indebted to them both and to many colleagues in China (some, like myself, older and perhaps wiser survivors from the earlier tour).

In acknowledging the major role of the East-West Center in providing facilities and encouragement for me to write this book, I want to thank particularly Helen Takeuchi, Senior Editor, who assembled the manuscript, ensured that it was acceptable for publication, and nursed it through infancy. I am also indebted to Drs. Lawrence Hamilton (who sponsored my fellowship), John Dixon, and Norton Ginsburg for their contributions in reviewing early drafts.

Finally, I have been continually reminded of the extent to which my own perceptions and priorities have changed over a quarter of a century and am aware, too, of the influence of generations of students and colleagues and my own family. I owe them much.

My first book on forestry in China was dedicated to the "memory of King Canute, who also attempted the impossible." This one recalls my earlier guide in China, Cheng Wanchun—a great dendrologist, poet, philosopher, and mentor and the only forester ever to achieve the highest ranks in the Academia Sinica.

FORESTS AND FORESTRY IN CHINA

1

ECONOMIC BACKGROUND

THE PEOPLE'S REPUBLIC OF CHINA covers an area of 9.6 million square kilometers spanning 50 degrees of latitude (including the islands within Chinese territorial waters) and 62 degrees of longitude. This vast section of the Eurasian landmass amounts to one-fourteenth of the total land area of the globe and supports nearly one-quarter of the world's population. It is characterized by immense variation in topography, climate, soils, ecology, and ethnic features. In altitude it extends from almost 300 m below sea level in the Turpan depression of northern Xinjiang to over 8800 m on the peak of Qomolungma in the Himalayas. The climate ranges from the humid tropics of Guangdong and Taiwan, with a mean annual rainfall of over 250 cm and mean annual temperatures in excess of 25° C, through periglacial conditions on the Tibetan Plateau, to the arid deserts of Inner Mongolia, northern Gansu, and Xinjiang, where the mean annual precipitation may be less than 1 cm, the mean annual temperature below 5°C, and the annual temperature range greater than 45°C. Most of the major soil types of the world are represented—from chernozems and chestnut soils of the Manchurian steppes to tropical laterites and terra rossa. China's flora is among the richest in the world, including more than 5000 woody species in almost 700 genera; the vegetation pattern is unique in Eurasia in that natural forest extends in an unbroken sequence of communities from tropical rain forest in the south to montane-boreal coniferous forest in the northeast. The native fauna, though much depleted, is almost as varied.

The inhabitants of present-day China are as diverse as their country. Made up of more than fifty nationalities, they differ widely in ethnic type, culture, language, and religion. Muslims, Lamaist and Southern Buddhists, and Taoists coexist with nominal Christians of all denominations; the still nomadic Mongols of the north are aligned with peoples who had developed a settled agriculture by 10,000 B.C., who cultivated silkworms on a commercial scale 3000 years ago, and who in April 1970 launched around a startled globe the first satellite from a Third World country (and which played the raucous folk tune "The East Is Red").

On 1 October 1949, after forty years of more or less continuous strife and civil war, the People's Republic of China was established—the third Communist state in the world and the second (after Outer Mongolia) in Asia. The country has defied many predictions of economic and political failure to emerge in the 1980s confident, powerful, and resourceful. To appreciate the pervasive roles of forestry in the development of China, it is necessary to understand their economic background.

PHYSICAL GEOGRAPHY AND CLIMATE

China is a country of mountains, high plateaus, deserts, and river plains. On a topographic map, it is the high country that is most conspicuous. In the west, and covering nearly half the land area, lie the sparsely populated plateaus, depressions, and lofty mountains of Tibet, Xinjiang, and Qinghai—an area largely without external drainage and accessible only via high mountains or, in the north, through the Gansu corridor. From the Pamirs, in the extreme northwest, a series of mountain ranges radiates to the eastern edge of the Qinghai–Xinjiang–Tibetan Plateau; one such—the Qilian range—continues across Central China, separating the major rivers and increasing the latitudinal climatic differences between the north and south of "China Proper."[1] A second conspicuous mountain chain runs southwest from the Greater Hinggan range in northwestern Manchuria to the Yunnan–Guizhou Plateau. Somewhat lower ranges form the north, northeast, and southeast borders of Manchuria and run through the southeast coastal provinces. The major rivers run west–east from the interior highlands and include three of the longest in the world—the Chang (Yangtze), the Huang (Yellow), and the Heilong (Amur).

The general landform is what Ren Mei'e et al. (1985) describe as a topographical staircase. The Qinghai–Tibetan Plateau has an average height of more than 4000 m (and rises to over 8800 m—amply justifying the sobriquet "Roof of the World"). The second stair runs from the outer rim of the Qinghai–Tibetan Plateau to the Greater Hinggan, Taihang, Wushan, and Xuefeng mountains, forming a vast area of plateaus and basins (among them the Inner Mongolian and loess plateaus, the Sichuan basin and Yunnan–Guizhou Plateau, and, to the northwest, the basins of Tarim—between the Kunlun and Tian mountains—and Junggar, between the Tian and Altay mountains). The lowest step of the staircase is formed from the plains and hills of eastern China. The five major plains that stretch generally from north to south—the Sanjiang, the Northeast, the North China, and the Middle and Lower Chang plains—cover an area of nearly 100 million hectares and make up nearly 10 percent of China's total land area. They support some 350 million people—one-third of China's population—and their cultivated farmland amounts to 40 percent of the national total.

All three steps of the staircase slope gently in the west and more steeply in the east. The predominance of high country is illustrated by the fact that 65 percent of China's territory is over 1000 m above sea level and nearly 20 percent above 5000 m (see map). The northern part of the Greater Hinggan range merges with the Qinling and Taihang ranges to act as a barrier between the green humid ricelands to the south and the dry loess hills and sand deserts farther north. It also forms the limit of the monsoon airflow system of moist air during early summer from the Pacific region and the prevailing dry and dust-laden winds from the west, which by the time they reach China have dropped their moisture.

China's landform provides enormous variety—from the magnificence of the world's highest peak (Mount Qomolungma, Sagarmatha, or Everest) through the almost lunar terrain of the eroded loess plateaus, the rolling grasslands of Nei Monggol, the extraordinary karst formations of Yunnan and Guangxi, the largest sand desert in China (the Taklamakan), to the rich ricelands of the river floodplains. Offshore, China lays claim to more than 5000 coastal islands predominantly in the South China Sea south of Hangzhou. The most extensive is Taiwan, but Hainan (with an area of more than 32,000 km^2) is almost as big.

Again, the landform varies greatly from coral atolls to the mountain ranges of Hainan and Taiwan.

In general, the climate of China is affected by interactions between the arid north and the northwestern interior, the monsoon zone of the south and southwest, and the Tibetan Plateau. Western China is cold and life-styles are affected by the pervasive tongues of glaciers that may drop to 3000 m in some valleys. On the high plateaus there is permafrost (covering over 20 percent of the total land surface), but massive fluctuations in winter temperature can result in accelerated weathering and violent mudflows during the spring thaw. In the arid basins of northwestern China, where annual precipitation may be less than 200 mm, seasonal temperature changes may also be dramatic and wind erosion and redeposition are widespread. China's desert lands exceed 1.25 million square kilometers—nearly 15 percent of the land surface. There is historical evidence, however, of climatic change in the valleys and river terraces of the semiarid regions, while even the Qinghai–Tibetan Plateau shows once-tropical limestone formations at altitudes exceeding 5000 m.

Surface geology is exemplified by loess in the north and karst in the southwest. The best-known formation is the loess plateau of central and eastern Gansu, northern Shaanxi, and Shanxi, which averages over 1000 m in elevation and covers some 300,000 km^2: It is the largest formation of its kind in the world and may be as much as 100 to 200 m thick. Most of the silt in the Huang River comes from this plateau. Limestone covers 1.3 million square kilometers ranging from the tropical karst pinnacles of Guangdong and Guangxi to the temperate springs and dry valleys of northern China.

China's climate is as complex as its landforms. The chief characteristic of the temperature is its mild variation between north and south in summer and great fluctuations in winter (see Table 1). Rainfall is heavy in the southeast; it decreases in intensity to the north and becomes increasingly unreliable. The provinces south of the Chang River are always subject to summer typhoons, and the abrupt changes in topography that characterize the plateaus and basins prevent any simple classification of climatic effects. Ren Mei'e et al. (1985) divide eastern China into five zones: cold temperate, temperate, warm temperate, subtropical, and tropical. Moving west, with an increasing distance from the ocean, macroclimates change from humid, through semihumid and semiarid, to arid; mountainous areas, of course, are vertically zoned.

In the present context, the importance of physical geography is

TABLE 1
TEMPERATURE RANGES:
1951–1970

Location	Mean Temperature in Hottest Month (°C)	Extreme High (°C)	Mean Temperature in Coldest Month (°C)	Extreme Low (°C)
Nenjiang	20.4	37.4	−25.8	−47.3
Changchun	22.9	38.0	−17.0	−36.5
Hohhot	21.8	37.3	−13.5	−31.2
Beijing	26.0	40.6	− 4.7	−27.4
Zhengzhou	27.5	43.0	− 0.2	−17.9
Nanchang	29.7	40.6	4.8	− 7.7
Guangzhou	28.3	38.7	13.4	0.0
Yulin	23.5	38.6	−10.0	−32.7
Urumqi	24.5	40.9	−15.6	−41.5
Lanzhou	22.4	39.1	− 7.4	−21.7
Chongqing	28.6	42.2	7.5	− 1.8
Kunming	19.9	31.5	7.9	− 5.4
Guiyang	24.0	37.5	4.9	− 7.8
Lhasa	15.5	29.4	− 2.3	−16.5

SOURCE: Ren Mei'e et al. (1985).

threefold. First, the relative scarcity of productive agricultural land in relation to the size and distribution of China's population necessitates the close integration of cultivation, pastoral agriculture, and forestry. Oversimplification (and excessive optimism on the part of doctrinaire planners) have—as in many countries—led to expensive mistakes in land use. They have been compounded by the second key characteristic of China's physical geography: widespread erosion. The uneven seasonal and regional distribution of rainfall, and its great variation from year to year, is often characterized by prolonged torrential precipitation and results in extensive soil erosion by wind and water. China's susceptibility to "natural calamities"—as well as the country's ability to cope with them—are legendary. Finally, both landform and climate highlight the economic importance of drainage systems and water conservation.

There are reportedly more than 50,000 rivers in China draining areas of more than 100 km^2. The exterior drainage basins linked to the Pacific Ocean are the biggest and include the Chang, the Huang, the Heilong, and the Zhujiang. The Indian Ocean drainage basin is not large, but it takes in tributaries of such famous rivers as the Ganges and the Irrawaddy. The Arctic drainage basin has only one river, the Irtysh; after leaving China, it flows across Kazakhstan and western Siberia to the Kara Sea.

Interior drainage basins cover 3.5 million square kilometers. Very low rainfall and excessive summer temperatures pose enormous problems of salinity and soil instability. Again, climatic inconsistency can lead to abrupt changes in sedimentation and shifts in location. There is active marsh formation in the drier areas of northwestern China and at high altitudes on the Qinghai–Tibetan Plateau. In northeastern China, where there is permafrost, peat moss formation is widespread; marshes in the valleys and gullies of the Hinggan and Changbai ranges show layers of peat over 1 m thick.

China's rivers, most notably the Huang, carry high silt loads (see Table 2) and those with low silt levels tend to have high runoff. Although silt is deposited as alluvial sediment along the lower reaches of the rivers, the Huang Delta (for example) is advancing seaward at some 50 m per year. Moreover, even where silt is dropped before it reaches the ocean, the water-soluble nutrients leached out of it are not.

The Chang and the Huang rivers are among the world's most intractable. There are records of the Chang flooding in 1871 to a height of 85 m above normal in the gorges downstream from Chongqing, and floods reportedly caused the evacuation of 10 million people from the lower valleys in 1954. The Huang ("China's Sorrow") has allegedly killed more people than any other feature of the earth's surface (Wijkman and Timberlake, 1984). The first major work on it began in the third millennium B.C.; in 1887 between 900,000 and 2 million people were lost by drowning and starvation (due to loss of food crops) during devastating floods; and in 1931 between 1.0 and 3.7 million people reportedly died. In 1938, the banks were deliberately breached to hold back the Japanese army; this stratagem was successful, but half a million people were killed in the process.

The Huang travels 4000 km, much of it through loess, and has a 40 percent silt load by the time it reaches the floodplain, some 800 km from the sea. For much of this distance it flows some 8 m above the level of the floodplain.

Deforestation is blamed as a major cause of flooding in China. In recent times, both the Great Leap Forward and the policy of "grain first" were responsible for accelerated deforestation. Even on Hainan Island, the vegetation cover has dropped from 25 to 7 percent during the past three decades.

TABLE 2
SILT LOADS OF CHINA'S EXTERNAL DRAINAGE SYSTEMS

River Basin	Average Silt Content (kg/m^3)	Silt Discharge Million Tons	Silt Discharge Percentage of Total
Rivers in NE China	0.51	86	3.2
Rivers in N China	8.72	150	5.7
Huang He (Shaanxi)	37.70	1593	60.6
Huai He, Yi He, and She He	0.25	15	0.6
Chang (Datong)	0.54	502	19.1
Rivers in Zhejiang and Fujian coastal areas	0.11	26	1.0
Zhu and other rivers in S China	0.20	95	3.6
Rivers in SW China	0.08	162	6.2
Total	6.01	2629	100

SOURCE: Ren Mei'e et al. (1985).

PHYTOGEOGRAPHY AND SOILS

Until the 1950s, descriptive material in the English language on the natural vegetation of China rested heavily on reports contributed by the late-nineteenth- and early-twentieth-century professional plant hunters from Europe and America. Their efforts have been outlined by Cox (1945). From the prolific (if somewhat humorless) Robert Fortune—who left England in 1843 equipped "with fowling piece and pistols and a Chinese vocabulary"—to the tireless Frank Kingdon Ward, the catalog of these pioneers is impressive. It includes men like Pratt, Wilson, Forrest, Farrer, and the French missionaries David, Delavy, Soulie, and Ducloux. The collectors operated mainly in the southern and southwestern provinces, and it is through their efforts that these areas became better known, from the standpoint of their flora, than the much simpler regions of the north and northeast. Certainly the subtropical and high-altitude areas of the southwest provided a rich source of horticultural and garden material that could be propagated in Europe. Wang (1961) and, more recently, Menzies (1989) point out that in Chinese literature many of the major plant communities were recognized as long ago as the fifth century A.D., and there are many descriptions of plants and habitats providing valuable records of land-use patterns and vegetational changes. Menzies cites monographs on species and groups of species—

including bamboos—dating from the Chin dynasty (between A.D. 317 and 419) and a compendium of lore and techniques relating to the cultivation and utilization of *Paulownia*, which appeared in 1049 (see also Deng, 1927).

There have been several recent attempts to describe and interpret the broad vegetational patterns of the landmass. These attempts have been both physiognomic and floristic: Given that China has 2700 genera of seed plants and 30,000 species, including 2800 tree species (embracing 95 percent of the woody angiosperms in the world), complexity has to be expected. Within the three major forest formations, Wang (1961) recognizes nine forest types; in a series of regional treatises he provides detailed floristic descriptions of types and subtypes and discusses the origin and segregation of the forest communities. Wang deliberately avoids any attempt to relate vegetation type to soil or climate: "One should make inquiries that are independent of presumed causal relations," he believes. Nor does he consider historical land use, which has brought about major changes in the distribution and composition of China's forests.

As background to an understanding of land use in China, the outline published by Hou et al. (1956) and refined by Hou (1983) is more relevant. It includes a map showing the climatic regions of China (based on temperature and rainfall distribution), a soil map (which recognizes nineteen soil regions), and a vegetation map indicating thirty-nine natural vegetation types and thirteen cultural types. The forest components of this map and certain land-use implications are discussed more fully in Chapter 2. Broadly, in southern and eastern China the natural forest distribution is determined by soil type; in the arid west and northwest, climate-induced steppe, semidesert, and desert dominate the vegetation types; in the mountain and plateau regions, vertical zonation of soils and vegetation is marked.

The greater part of China's natural forest has been destroyed over the years, resulting in acute erosion problems and a severe shortage of forest produce. Significant virgin forest remains only in the northeastern horseshoe formed by the Greater and Lesser Hinggan mountains and the Changbai range; in the remoter parts of Guangdong, Yunnan, and Sichuan; and in the river valleys of the Qinghai–Tibetan Plateau. Over most of China, the primary forest vegetation is indicated only by groves surrounding temples and shrines.

Climatic change has been a key factor in the development of China's vegetation and soils. When the global climate was warmer (during

Tertiary times), most of China was subtropical or tropical. When the Himalayas and the Tibetan Plateau were uplifted and the Sea of Tethys disappeared, northwestern China became dry and monsoon circulation developed in East Asia. During the Quaternary period, the world climate became colder again, but there was no continental cover of glaciers during the ice ages in China and plant forms that were destroyed by glaciation in other parts of the world retreated up the mountains and survived. As a result, China preserves to an extent some relict plants from the Cretaceous and Tertiary periods. In the subtropical regions of eastern China such species include *Metasequoia* in western Hubei and *Gingko* in Zhejiang. The desert vegetation and soils in the huge inland basins of northwestern China also reflect climatic change over a long period of time.

In northeastern China there are fringes of the great Northern Coniferous Forest formation, which stretches around the world from Scandinavia to Alaska. They are pure forests carrying species similar to those in Europe and North America—larch, pine, and spruce primary, with the secondary succession to birch, alder, aspen, and the like. The tree line is at about 1100 m. Based on the findings of an FAO mission to Outer Mongolia in 1984–1985 (Richardson, 1987b)—where this forest type is much more extensive than in China—the average stand volume is of the order of 100 m^3 per hectare and the overall increment less than 1 m^3 per hectare per year. Climatically the area has little agricultural potential and the forests are intact. From the production viewpoint, however, they are overmature and, as was shown in June 1987, can be devastated by fire.

Most of the northeastern forests are classified within the Mixed Coniferous and Deciduous Broad-Leaved formation—similar with respect to genera (though not species) to those of northern Europe and North America, but much richer in species than elsewhere. This formation contains valuable resources of ash, oak, sycamore, walnut, and *Phellodendron* (a cork tree)—as well as the prized conifers Korean pine and true fir. The climate is milder than in the far north, and forest areas have been given up to agriculture at lower elevations (below 600 m). Nonetheless, this forest type supplies 40 percent of China's industrial timber needs. The average stand volume is higher than that of the Northern Coniferous Forest—perhaps 150 to 160 m^3 per hectare—and the annual increment is 1.5 to 2.5 m^3 per hectare. Secondary forest in this zone tends to be richer in broad-leaved species than the primary forest, but regeneration is of birch, poplar, and the like rather than the more valuable oak, ash, and walnut.

The natural forests of the southwest cover a much wider range of climatic (and topographic) environments than do the northeastern forests. The Evergreen Broad-Leaved Forest formation, which extends across the southern provinces of China, is one of the most important in Asia; it characterizes large areas of Khmer, Vietnam, Laos, Thailand, Myanmar, and Japan as well as China. The better-known tree genera are *Castanopsis, Liquidambar, Beilschmiedia, Lithocarpus, Cyclobalanopsis, Lindera,* and *Cinnamomum,* together with the coniferous genera *Cryptomeria, Keteleeria, Cunninghamia* (Chinese fir), *Tsuga, Pinus,* and *Podocarpus.* In contrast with the northeastern forests, secondary forests of the Evergreen Broad-Leaved Forest formation are richer in coniferous species than the primary communities. In the west, more forest has survived at high altitudes—where spruce, fir, and larch are common. The secondary forest includes a number of interesting tropical pines (*Pinus yunnanensis, P. kesiya*). At lower altitudes there are broad-leaved species (including teak, rosewood, and *Shorea*) and rich bamboo resources, as well as a variety of *Camellia* and other flowering shrubs. The latter represent a resource that China is beginning to develop deliberately in international commerce. The mountains of the southeastern Qinghai–Tibetan Plateau form a zone some 300,000 km^2 in area, some of which is heavily wooded. The altitude averages 4000 m, but the plateau is much incised by tributaries of the great rivers of Southeast Asia—the Chang, Mekong, Irrawaddy, Salween, and Brahmaputra. The primary forests comprise many species of spruce and fir in admixture with larch and juniper; rhododendron, *Ribes,* and the like form an understory. Again, the shrubs represent gene pools that are beginning to be tapped commercially.

DEMOGRAPHY

Despite the small proportion of the land surface that is cultivable, China remains a predominantly agricultural country. The industrialized eastern provinces are to some extent urbanized, but the most densely settled regions of China remain the central parts of the North China Plain, the delta of the Chang River, the east coast, the middle Chang River basin around Wuhan (Hubei province), and the Red Basin of

Sichuan. In parts of the lower Chang River area, there is extreme congestion, with population densities of up to 2500 per square mile (965 per square kilometer). In the border provinces, densities fall below 25 per square mile (10 per square kilometer) and large tracts of Qinghai, Xinjiang, and Nei Monggol are uninhabited, as are parts of Yunnan, Sichuan, Gansu, Heilongjiang, and Jilin. Some 90 percent of China's population live on little more than 15 percent of the land surface.

The 1982 census (see Banister, 1986) shows a change in the proportion of urban population from 10.6 percent in 1949 to 23.5 percent in 1983. The growth has not been uniform, however, and there have been a number of reductions in urban population as a result of the various Xiafeng movements from urban to rural areas. There have also been deliberate attempts to settle the remoter areas—the provinces of Nei Monggol, Yunnan, and Xinjiang show a higher than average growth rate—while the provinces supporting minority nations have attracted immigrants.

There are fifty-five national minorities numbering 67 million people, but it is difficult to gauge their growth because of differences in census method between 1964 and 1982. Over most of China ethnic patterns are fairly uncomplicated, but in the southwest (Yunnan, Guizhou, Guangxi, and Guangdong) and in the northwest there are many minorities and attendant problems. The so-called autonomous regions are self-governing only to the extent of internal control, and while the nominal party head is usually a member of the national minority, administrative power remains concentrated in the hands of the ethnic majority Han Chinese. The minorities have their own life-styles—some are pastoral, some traditional shifting cultivators, some hunters and fishermen. Influxes of Han Chinese have created problems and, for political reasons, restrictive legislation is being applied rather more gently in the minority areas than in much of China (with respect to population growth controls, environmental legislation, and religious observance, for example). Administrative difficulties are not made easier by language differences and the population's general immobility. Excluding Hong Kong, there are also some 20 million overseas Chinese—generally scattered throughout the world but predominantly in Southeast Asia. Their link with the mother country tends to be closer (culturally and economically) than is the case with many ethnic minorities. Appendix A discusses the aspects of population distribution and growth that are likely to influence demand for forest products.

There is no evidence that, until recently, China has ever considered population size to constitute an overwhelming problem. Ideologically, it has been difficult to accept overpopulation; indeed, despite lip service to birth control in the late 1960s, it was not until the demise of Mao Zedong that China seriously addressed the issue. The draconian measures adopted in the 1980s have been widely publicized; some of their less desirable side effects are discussed in Aird (1986).

Freeberne (1971) describes the Chinese as "obsessed with the distinction between town and countryside," a preoccupation that found its ultimate expression in the thesis of the late Lin Piao that the world divides into two parts—the metropoles (urban islands) of wealthy, industrialized North America, Western Europe, Australia, and Japan; and the underdeveloped rural areas of Asia, Africa, and Latin America, surrounding these islands in a sea of discontent. This is doubtless an oversimplification. But China's feverish—albeit sporadic—economic growth from a strictly limited resource base—and one vulnerable to catastrophe and cataclysm—has been possible only because of support from peasant farmers. According to Needham (1964) the moral stature of the Communist Party in China came from the fact that it had "lived in the wilderness" with the peasants. More recently, the success of reform owes more to agriculture and rural-based small industry (see Crook, 1986; Travers, 1986) than to urban activities.

DEVELOPMENT OF RESOURCES

In 1963 I argued that forestry is of greater importance in China than in any other country of the world. My argument rested on three features of China's national economy: first, the extent to which agricultural productivity depends on water conservation and, hence, on protective reforestation; second, the importance of timber in the development of industry; and third, the existence of an acute shortage of forest products throughout most of China. Nothing has happened in the last two decades to change that assessment, though in view of China's present willingness to import large volumes of logs, the third feature might be regarded as somewhat less crucial. The key to an understanding of these arguments lies in an appreciation of China's economic development.

On 1 October 1949, in Beijing, Mao Zedong proclaimed the birth of the People's Republic of China. His immediate objectives were land reform, the dismantling of a corrupt, bureaucratic establishment, and—with the help and guidance of the USSR—the transformation of the national economy and society. The major phases of postrevolutionary development were as follows. From 1949 to 1953 China recovered from victory: Attempts at currency stabilization and the revitalization of moribund industry, involving stringent economic and political regimentation, placed severe strains on the rural economy. Widespread floods in the latter half of 1949, the outbreak of the Korean War, and the prohibition in 1950 of imports of rice and wheat (reversing a trade flow of more than fifty years' standing) provided the incentive for collectivization of agriculture and the formulation of the First Five-Year Plan (FYP) in 1952. The first plan (1953–1957) was based on the conventional socialist priority of developing heavy industry and, as a result, agriculture was neglected.

During the First FYP, 58 percent of all state investment was earmarked for industry, with heavy industry receiving much the greater part. It was this doctrinaire stance that led to the establishment (under Russian tutelage) of the large, so-called integrated wood complexes in the natural forest areas of northeastern China, projects that remain today monuments of inefficiency and waste. (See Chapter 4.) Investment in heavy industry remained at the center of the Second FYP, but it soon became clear that the economy was in severe difficulty—especially with respect to food production. For the first time since 1949, there was a threat of widespread famine and it became apparent that the level of agricultural production simply could not sustain the target for industrialization. The winter of 1960–1961 was particularly harsh, and substantial imports of wheat from Canada, Australia, and other countries were begun. At least 16 million metric tons were purchased between 1960 and 1963, and purchases continued at about 5 million metric tons annually until the 1980s.

Traditionally in China, agriculture does much more than feed the population: It is a source of raw material for major textile industries; it provides locally used products such as timber, fish, soap, cooking oils, tobacco, and handicrafts; and, most important, until very recently China relied overwhelmingly on agricultural products to earn foreign exchange. Yet the total cultivated area amounts to only 11 percent of the country and (even including multiple cropping) it is doubtful whether it is equivalent to more than 15 percent of the land area. Agriculture has always been intensive and strongly dependent on irrigation. Indeed, irrigation works were built during the second century B.C. (at Dujiangyan, in Sichuan province, for example).

The need for industrial retrenchment was realized as early as 1956 and reflected in the first specific plan for agriculture issued in that year. This twelve-year plan emphasized the need for higher yields per hectare of crops, livestock, and fish. The country was divided into three "output" zones according to grain yield potential, and the 1967 targets indicate that the aim was to raise the yield per hectare by at least 100 percent.

Somewhat paradoxically, the plan also heralded a reduction in the cultivated area, with the realization that ambitious schemes for the reclamation of marginal land would not succeed. Even in 1956, it had been suggested that "our country will not in the near future . . . be in a position to carry out reclamation of wasteland to any great extent, or rapidly to end flood and drought" (NCNA, Suppl. 248, 10 January 1956). In 1957–1958, the cultivated area of Shanxi was reduced by one-third and again by a further third in 1958–1959; half the newly cultivated area of Nei Monggol was abandoned. Overall, the cultivated area declined by 4 million hectares in 1958 and by a further million in 1959. Increased productivity was to be achieved by concentrating resources of capital and labor on areas known to be fertile—"higher output on a smaller quantity of land." Ma Yinchu's eight-point charter for agriculture (deep plowing, fertilizers, water conservation, seed improvement, close planting, protection from pathogens, improved management, and improved tools) was revived and adopted by Mao Zedong. (Foresters adopted their own charter based on Ma Yinchu, which led to wasteful practices such as ultra-close spacing—see Chapter 5).

Justification for the partial abandonment of a land reclamation policy is not hard to find in China; it lies in the marked increase in soil erosion triggered by indiscriminate cultivation of unstable soil. Of all countries in the world, China has had most reason historically to be aware of erosion hazards (see Sowerby, 1924a; Lowdermilk, 1924; Deng, 1927; Cressey, 1955); yet many of the development projects of the early years of Communist control were put into effect without regard for the consequences. Changes in land use included the expansion of winter wheat in the Gansu corridor, Ningxia plain, and northern Xinjiang; wheat, sugarbeet, and rice in Manchuria; tropical cash crops (palm oil, rubber) in the south; and the extension of cultivation in the pastoral areas. Since the early days, there have also been attempts to extend livestock breeding in the settled agricultural areas as well as the development of inland fisheries. Many of these innovations are now being reversed—for example, a shift of cotton production back to the traditional producing regions of the North China Plain. Environmental damage resulted from the emphasis on maximization of grain production, which led to the plowing of grassland to the detriment of both crop production and livestock. Walker (1984) records that in thirty years some 6.7 million hectares of pasture was plowed in the major livestock provinces, but the decline in grass production (30 to 50 percent between the 1950s and 1981) was not offset by increasing supplies of grain. Grain output per head of population also fell—particularly between 1965 and 1976. And there was a steady fall in the number of livestock grazed in both Xinjiang and Nei Monggol during the late 1960s and early 1970s, not compensated by any significant increase in arable yields until the 1980s (primarily because of poor infrastructure and a lack of incentives).

By 1958, then, a significant shift in national economic policy was apparent: Agricultural productivity was to be encouraged by all possible means. The withdrawal of Russian technical assistance and financial aid (beginning in 1957) also doubtless influenced the pace of industrialization and brought into focus, all the more sharply, the need to develop agricultural output—since only by providing an exportable surplus of agricultural products could the Chinese pay for the industrial equipment they were forced to purchase abroad.

The year of the Great Leap Forward (1958) demonstrated the impracticality of combining agriculture and backyard industry in the communes; and, from 1959 on, the two sectors of the economy were more or less segregated. Major industrial areas were centered on a number of iron and steel "complexes" (including one at Paotow, in the heart of a sparsely settled pastoral region of Nei Monggol), for which some 24,000 communes (averaging 4200 ha, 5000 households, and 10,000 laborers) provided the agricultural base. Further decentralization of agricultural organization occurred in 1959, with the production brigade (averaging 250 households) forming the key unit.

Inevitably, attempts to increase production demanded increased irrigation works and properly coordinated water conservation measures, particularly in the two northern output zones. As noted already, however, China's rivers do not flow peacefully; the major catchments are enormous and for the most part steep, barren, and located in areas of uncertain and unevenly distributed rainfall. The task of creating storage reservoirs, building dams, stopbanks, new canals, and irrigation channels on the scale required was truly gargantuan. Some measure of the importance given to water conservation from 1956 on can be obtained from Chinese claims of increases in irrigated areas. From 1949 to 1955, the area is said to have increased by over 2 million hectares annually; from 1956 to 1959, by 11 million hectares annually; and from 1960 to 1962, by 5 million hectares annually. The reduced rate reported from 1960 to 1962 was allegedly due to "incomplete statistics" and to concentration on large-capital, multipurpose projects.

Changes in objective are also apparent. Thus the principal projects during the period of the First FYP were designed to regulate river flow onto the North China Plain, while from 1958 to 1962 the main emphasis was on long-term development of the northwest. Projects included forty-six large dams on the upper reaches of the Huang, an 800-mile irrigation canal in eastern Jiangsu, major irrigation projects in the Tian Mountains of Xinjiang, and the transfer of water from the headwaters of the Chang (on the eastern Tibetan Plateau) to Jiangsu and Nei Monggol. To be successful, these ambitious projects required extensive catchment afforestation schemes and shelterbelt establishment. There is abundant evidence that, by 1957, China had begun to appreciate the folly of attempting water conservation projects and land reclamation schemes without proper regard for the revegetation of areas subject to virtually constant wind or water erosion.

The withdrawal of land from cultivation was also accompanied by ambitious plans for protection forestry. Between 1956 and 1967, some 100 million hectares of "wasteland and desert" were to be afforested. And this was to be only a start—the minister of forestry announced that "to plant trees on 100 million hectares of land in twelve years is only a beginning. There will still be 70 million or 100 million, or even 200 million, hectares of barren mountains for us to afforest." Following these pronouncements, the project for establishing the "Great Green Wall"—a shelterbelt covering 1.6 million hectares intended to serve as a protective barrier against the northern deserts—was accelerated. Moreover, a 600-km belt skirting the Tengger Desert was initiated, and in 1958 there was a call for one-third of the entire cultivated area of China to be planted in trees. Recent communiqués continue to highlight shelter planting and attacks upon the "sand dragon." There can be no doubt that the Chinese now appreciate the dependence of agricultural development on water; the necessity, in water conservation works, for catchment stability; and the value of plantation forestry in revegetation programs. (These topics are discussed further in Chapter 7.)

Industrial retrenchment was ushered in by the "hundred flowers" speech of Mao Zedong in February 1957, when criticisms of the administration were invited and expressions of liberal political opinion encouraged. ("Let a hundred flowers bloom and a hundred schools of thought contend.") A major demand made by the workers was for a reduction in unproductive staff at all levels of administration and was followed by a reduction in national expenditure and investment and by a decentralization policy. Regulations on industrial control published in November 1957 placed many of the enterprises controlled by the economic ministries under the aegis of provincial and local bodies, with profit-sharing in the ratio of 20 percent to the local authority and 80 percent to the central government. By the following year, some 80 percent of former government enterprises had been handed over to local bodies; these authorities were empowered to raise capital by bond issues and to retain certain taxes and excess profits. These sweeping changes heralded the Great Leap Forward, the formation of the communes, and the startling proliferation of backyard industries manufacturing everything from ball bearings to bootlaces.

Several indices of production have been published that demonstrate the effects of the Great Leap and its aftermath. There was, without doubt, an overall increase in output in both agricultural and industrial sectors during 1958. But while industrial output may have continued to rise in 1959 and 1960, food crop production declined drastically in these years and only began to approach the 1957 level again by 1962. The nonfood agricultural sector declined through 1961 and even by 1965 had not achieved its 1957 level. Field (1982) argues that post-Leap industrial growth was achieved by the reemployment of plant installed prior to the Great Leap, rather than by the installation of new capacity. The Great Leap also demonstrated—if evidence were needed—the inadequacy of China's transportation system. Before 1949, significant industrial development was limited to the eastern seaboard and (under Japanese control) the central lowlands of Manchuria. With few exceptions, industrial centers were remote from their raw materials, which it was often cheaper to import than to supply from the hinterlands. When the Communists came to power, their first concern was to rehabilitate existing industrial centers and then, under the First FYP, to establish new centers in northern China based on mineral resources. Vigorous expansion occurred in Manchuria (utilizing known resources of coal, power, and iron); in North China at Taiyuan, Shijiazhuang, and Paotow, based on coal and iron; and in the northwest (Xinjiang, Gansu, Shaanxi, and Qinghai) to exploit nonferrous metals and oil resources. These developments (coupled with an effective ban on major food imports) demanded a vastly increased interchange of the products of industry and agriculture, as well as the stimulation of internal trade among the various regions of China.

Political considerations, also, conditioned the government's determination to promote interregional economic integration. One reason why previous regimes in China had lost effective national control is that local self-sufficiency, encouraged by poor communications, allowed the authority of local officials and warlords to become virtual independence. According to Hughes and Luard (1959:115): "The Communist Government have been anxious from the start to restore and to maintain the unity of the country. It was partly in order to obviate the fissiparous tendencies which had endangered previous regimes that they have made such efforts to ensure that every part of the country should be firmly linked with all others and thus merged, economically as well as politically, in a single integrated and organic unit." In this connection, developments in the northwest take on a special significance, since Chinese control of that area had for many decades been precarious, and the location there of a substantial force of nonindigenous Chinese enabled both effective political control and the discovery and development of natural resources. A pervasive role has been played by the People's Liberation Army (PLA), which, since it was founded in 1927, has made specific contributions to the economy, especially in the border regions. It has operated logging and sawmilling units and undertaken reclamation and reforestation projects on a massive scale.

The period 1966–1969 reflects the turmoil of the Great Proletarian Cultural Revolution—which has been described as "one of the most enormous psychological experiments ever made by man" (Malraux, 1968). The first stirrings of hostility to "capitalist roaders" and to the hubristic attitudes of bureaucrats could be sensed by visitors to China in the early 1960s. On my arrival in Yunnan in May 1963, I was made immediately and acutely aware of the cultural and social rift between my suave interpreter (dressed in a silk suit and wearing patent leather shoes) and the local forestry fraternity. The first political murmurs were heard in 1964, following the publication in *Red Flag* of an editorial on 30 June entitled "A Great Cultural Revolution." They grew until 1966 when, in May, right and left within the central party polarized and the famous poster attack on the principal of Beijing University and his subsequent dismissal publicly launched the Cultural Revolution (and party purges).

It is not my intention here to discuss the ideology and politics of the Cultural Revolution. Its economic impact, however, is indeed relevant. It was undoubtedly far-reaching, triggering what Deng Xiaoping has described as the "Decade of Catastrophe" (*Beijing Review,* 11 August 1986). One of its principal effects was the disruption of transport by the Red Guard movement at the end of 1966. (The Red Guards were granted free rail travel facilities and so cluttered up the railways—especially the main lines—that the movement of industrial raw materials and produce was seriously hampered.) The economy slowed down but perhaps more selectively than during the Great Leap. The end of 1968 saw major changes in economic strategy: Increased resources were devoted to agriculture and to rural capital reconstruction, increased emphasis was placed on raw materials procurement (especially fuels), decentralization was urged yet again (with stress laid on the development of small and medium-scale enterprises rather than large complexes), and there was a call for the mass movement of urban labor (and, more particularly, intellectuals and cadres) into the rural areas.

The pattern of economic growth during these early years has been described by Ishikawa (1983) with respect to various gross parameters (national income, total values of production for heavy and light industry and agriculture, and per capita availability of food grains). It shows a scale of fluctuations in the "trough" periods more characteristic of a war economy than one of peacetime. It also illustrates the *relative* insensitivity of agriculture to change. Werner Klatt (1983) points out that in agriculture, because of environmental vagaries (soil, climate), economies of scale are less significant than in other sectors (a fact not appreciated by Marx or Lenin); nor is agricultural production as amenable to centralized controls as are industry and commerce. In the Soviet Union and China (in Western countries too), problems become manifest when farmers are treated as if they were machine operators and when "key links" are emphasized at the expense of other factors (arable crops versus livestock, nitrogenous fertilizer versus potash, and so forth). (A corollary is that rural subsectors are less amenable to command planning than other components of the economy; indeed, for all practical purposes certain operations are unplannable and thus beyond the reach of the statistician.) It is perhaps significant that increased yields in Chinese agriculture have come not from increases in capital investment but from transferring production and market decision-making from the commune level to brigade and production teams (now, in effect, households).

There is increasing evidence (see Smil, 1984) of ecological damage during the Great Leap from woodland destruction in order to fuel the backyard furnaces and the degradation of pasture in order to "take grain as the key link." Memoirs of the Cultural Revolution period are now being published (Nien Cheng, 1985; Yang Chiang, 1986; Liang Heng and Shapiro, 1983) describing the acute fuelwood shortages in many rural areas that sapped the will of entire communities. In southern Sichuan, the World Food Program (Brown, 1985) is assisting in afforestation (Project China 2606). In one county, Gongxian, it is recorded that forest cover was reduced from 27.3 percent to 9.6 percent of the land area between 1958 and the mid-1970s; in another, Gulin, it was reduced from 48 percent to 10.7 percent, most of it on steep slopes subsequently abandoned because of erosion. These features are not revealed by forest sector statistics (recording "industrial roundwood removals"), but their effect on productivity and the quality of life can be devastating.

The U.S. Joint Economic Committee's 1975 review ("China: A Reassessment of the Economy") indicated a resumption of growth in most economic parameters beginning in 1969. In fact, the assessment was misleading. China was still in the turmoil of the Cultural Revolution and, despite real increases in production, *factor productivity* was declining. Yeh (1984) records a long-term negative trend in productivity (which averaged *minus* 1.5 percent a year since 1957); growth in production had been due to increasing factor inputs (especially labor) rather than improvements in productivity. There had been no increase in per capita grain yield for two decades (Walker, 1984). Moreover, *quality* of both inputs and outputs left much to be desired. Prybla (1986a) observes that in some years more than 50 percent of steel production was unusable—because of poor strength or being in the wrong place; China used three times more energy per unit of national income than any other country in the world; production was "at an enormous expenditure of waste, monumental waste." There is little doubt that the leadership was seriously concerned about the quality as well as the quantity of economic growth at the end of 1974.

Nor had the effects of the Cultural Revolution yet dissipated. Indeed, their aftermath will remain for many years (see Chapter 8). Hu Yaobang, former general secretary of the Central Committee of the Communist Party, claimed that 160 million young people who ranged in age from 8 to 18 in 1966 had received a "poisonous" education. Mao Zedong had reviled intellectuals as "the stinking ninth" category of bad elements; even as late as 1976, wall posters queried the value of study; and Garside (1981) has described the deadness of the universities during that time—for ten years, no full-length university course was completed in any field. There is, therefore, an entire generation with totally inadequate training and covertly envious of the educated.

The mass movement of urban youth into rural areas, a feature of the Cultural Revolution, also had its aftermath. The lost generation—and their reluctant hosts—became disillusioned and resentful, especially when the visitors were not permitted to return home. And those who avoided rural reeducation acquired no skills and few scruples. The problems facing China's leaders in the final year of the Fourth FYP were thus not only economic.

In mid-1975, Deng Xiaoping (then the senior deputy to Zhou Enlai) directed the preparation of a major report on "Some Problems in Accelerating Industrial Development." (At the National People's Congress in the previous January, Zhou had singled out the Fifth FYP—1976–1981—as crucial to China's achievement of "front rank" status as an industrial nation by the year 2000.) The report was a realistic appraisal, and its recommendations reinforced Deng's reputation for pragmatism. (He had earlier been castigated by leftists for his observation that it does not matter whether a cat is black or white so long as it catches mice!) The report advocated closer economic controls, substantial pay increases for workers, and massive imports of technology. It was to provide the basis for the "Four Modernizations" (Industry, Science and Technology, Agriculture, and the Military). Perhaps its most important feature, however, was its recognition of the need for new incentives. Stressing improvements in quality of life from the days before 1949 had lost much of its impact in China as fewer and fewer people remembered those early times; the new incentives were to be consumer goods.

The year 1976 was eventful, but it did little for the economy. The death of Zhou Enlai and the "Spring Rites for the Departed" in April revealed a grief for him which led to rioting when mourning was officially curtailed. In September, both Mao Zedong and the revered General Zhu De died, and the disaster of Tangshan (an earthquake in which some 750,000 died) was interpreted as a traditional sign of dynastic change.

After Mao, the speed of political change was breathtaking. Within two years, twenty-six of twenty-nine provincial leaders had been replaced and officials dismissed during the Cultural Revolution were reappointed. Science—and the Academia Sinica—was restored to respectability and the need for "world standard" technology was emphasized. New problems called for new solutions ("Stalin never wore Dacron," according to Deng), and thousands of young students were sent abroad for training—most to the United States and Japan.

The year 1977 was a time of fence-mending and ideological spring-cleaning. The Cultural Revolution was officially declared over; there began the era of the Four Modernizations and the eclipse of doctrinaire ideologues (who, in the earthy imagery of Deng, "occupy the lavatory, but only fart"). The economy, however, was unstable and the tuning sporadic as the leadership prepared for the period of "adjustment and reform."

Development in China since the rehabilitation of Deng Xiaoping divides into two periods. From 1978 to 1980 was the time of "readjustment," when it was considered that changes in the government's "style of work," in education, and in the use of capitalist techniques would enable economic modernization without significantly altering the institutional framework. And from 1980, a consensus—albeit fragile—developed that more fundamental changes of the system itself were needed ("reform"). A chronology of major events during the period 1978–1984 has been published in *China Quarterly* (see nos. 51 and 54). More recent developments are annotated in *CELT 2000* and summarized (with a degree of skepticism) by Prybla (1986b).

The key event was the third plenum of the Eleventh Central Committee in December 1978, which vindicated Deng Xiaoping's pragmatism. ("Practice is the sole criterion of truth": In other words, theory—even Marxist-Leninist theory—must be judged by its results.) The communiqué called for a shift in decision-making from party to state and from state to economic enterprise and urged a resolve to increase investment in agriculture at the expense of heavy industry. The next year saw the opening of China to foreign investment, the establishment of the Special Economic Zones and the encouragement of joint ventures, membership in the IMF and the World Bank, and—perhaps most important—significant increases in agricultural procurement prices (to the point where rural-sector goods increased relative to industrial products by 25 percent, according to Yahuda, 1986).

What necessitated (and enabled) increases in real prices was the contract system of net output delivery by households—the *Baogan Daohu.*[2] Instead of receiving work points (the value of which was determined by the production of the entire collective or commune), the production team was able to contract to produce at specified procurement prices. Underproduction is penalized but overproduction remains the property of the family to use itself, offer on the open market, or sell to the state. The contract system led to devolution of decision-making with respect to choice of crop and farming method, specialization, and work scheduling; it also led to de facto tenant farming under secure landholding. Land can now be leased for up to fifteen years (thirty to eighty years in the case of grazing land and woodland); labor can be hired; machinery and draft animals—and planted trees—can now be owned and inherited.

There are local variations in the system—some necessitated by its very success. At a forest nursery in Jilin, for example, seedling production expanded beyond the forest planting capacity and the nursery workers turned to the production of ginseng. Within two years, they totally abandoned seedling production and the management had to make participation contingent on the production of minimum tree seedling quotas. Aspects of the contract responsibility system as they apply to forestry are discussed in Chapter 3.

The effects of contract responsibility systems have been palpable. Per capita agricultural production increased to the point that China became a net exporter of maize, cotton, and soybeans. The administrative and managerial functions of the communes have become separate; moves toward market socialism have been accepted enthusiastically by farmers (if not by all cadres). The next phase was the application of market mechanisms to rural industry, which benefited from unused capacity within the now discredited commune system. Results were even more impressive: While the total output value of agriculture in 1984 grew by 9.9 percent (helped by good weather and more readily available fertilizer), that of rural industries leaped by 44.5 percent. In 1984 and 1985, the acreage under grain crops actually decreased as farmers turned to higher-value crops and rural manufactures—yet the yield still exceeded 400 million metric tons (USDA, 1985; FEER, 16 January 1986). In 1985, too, the state monopoly purchase of agricultural products ended and production quotas on nonstaple foodstuffs (such as livestock and vegetables) were abolished (He Kang, 1985).

Changes in agricultural production are documented in Walker (1984). The major developments include the formulation of a national land-use plan, which urges more rational distribution of production based on ecological, scientific, and economic criteria. Many changes appear to involve a return to traditional cropping patterns. As well as grain yield, there have been increases in the areas under the principal industrial crops (cotton, oilseed, sugar, tobacco) and even more impressive increases in output—enabling spectacular reductions in imports. (The most startling change is the rise in cotton production, which, according to Walker, shows an increase in average yield per hectare of 56 percent between 1979 and 1983.) The emphasis on increasing yields is underlined in the report "China in the Year 2000" prepared by the State Council and the Chinese Academy of Social Sciences in 1985. This report envisages a further decline in area under grain by some 20 million hectares by the year 2000 but an increase in output from the record 407 million metric tons in 1984 to more than 500 million metric tons. This increase in output will involve an increase in yield from 2.74 tons per hectare (1980) to 5.3 tons.

Major problems envisaged in the future relate to the maintenance of agricultural investment and, in particular, long-term investment in such subsectors as forestry and water conservation (which declined by 36 percent from 1978 to 1982). A major focus of "Document No. 1, 1984" is on how to mobilize long-term levels of investment within the agricultural sector. The second problem relates to prices and the need for both incentives and infrastructure—not only to increase production but also to reduce what a Chinese source (SWB, 24 January 1984) refers to as "astonishing waste of rotting farm products." Market signals, it seems, will not provide the universal panacea if there are no hearing aids.

Agricultural policies have relevance for forestry for two reasons. First, developments in intensive arable farming increase the demand for wood and, at the same time, extend the range of tree species that can be raised under conditions of more intensive cultivation, fertilization, and the like. Agroforestry, for example, has greatly extended the geographic range of such species as *Paulownia* and Chinese fir (*Cunninghamia lanceolata*) as well as many multiple-purpose "industrial" species. The highly fertile soils made available by "Four Sides" forestry (linear tree planting around houses, villages, roads, and rivers) and the use of irrigation and fertilizers are changing the patterns of China's landscape. Second, there are implications for changes in livestock farming, of which three components are important in the present context.

The first component is increased meat production from domestic animals (which are 95 percent pigs)—a measure that will require grain feeding or an alternative. In the early days of the regime, pigs were regarded as producers of fertilizer rather than meat and were not intensively managed. The fattening period in China (according to Walker, 1984) was 19 months compared with 5.5 months in Japan and 7.7 in the United States. Carcass weights were also lower. Rises in pork production since 1978 follow the increasing use of grain, which must be either imported or grown; the production of green manure and fodder crops from trees can thus enable a reduction in the allocations of feed grain.

The second aspect relates to draft animals—which account for 50 percent of all large animals reared in China. Despite some mechanization, there is still a need for draft animals and shortages are being reported in the Chinese press. The abandonment of large-scale mechanized state farming in the pastoral areas has raised demands for draft animals far beyond the short-run supply situation, and prices during the 1980s have more than doubled. Again, the use of fodder trees enables an increase in the number of draft animals (especially bullocks) that can be raised on small farms.

The third component is pastoral production and the need to repair the damage caused through the destruction of grazing lands. There is indeed a potential for increased productivity in the traditional pastoral areas, but their vulnerability to climatic hazards makes it essential that stall feeding in combination with natural grazing be contemplated. Given improvements in infrastructure (transport, marketing, weather forecasting), there is no reason why the productivity of pastoral farming should not increase. The intensive utilization of grassland and forest achieved in Europe is not impossible in parts of China, nor is the application of integrated agroforestry learned in the plains to the foothills of the mountain ranges and plateaus. Such developments will involve a combination of planning and market-driven decision-making, as well as commitments to long-term policies and investment. But they could markedly affect land use and productivity in China.

Reporting on the Seventh FYP to the National People's Congress (NPC) on 25 March 1986, Premier Zhao Ziyang identified highlights of the Sixth FYP—citing growth in agricultural production, coal and power generation, foreign trade, and living standards. (Per capita net income of rural residents rose 13.7 percent and that of urban dwellers, 6.9 percent; house construction reached 3.2 billion square meters in rural areas and 0.63 billion in towns.) He declared "class struggle" to represent "erroneous theory" and to be inappropriate in a socialist society. This charge heralded the first overt criticism of Mao Zedong by the CCP general secretary, Hu Yaobang, in June 1986.

Reform of the planning system has endeavored to substitute "guidance planning" for command planning and to establish a system of "macrocontrol and microautonomy." In agriculture and rural industry, the initial successes are undoubted. Attempts to reform the urban industrial sector—the traditional domain of the cumbersome state enterprises—have been less successful, leading to the reassertion of certain economic controls (notably over foreign exchange and international trade as well as the allocation of certain strategic raw materials) and the articulation of ideological doubts by Ye Jianying and the veteran Chen Yun (FEER, 26 March 1986).

In the forest industry subsector, reform is experimental but augurs well for the development of competitive markets (and, in turn, much needed product quality control and the acceptance of managerial responsibility). The first step was to allow the southern provinces (those south of the Chang River—Hubei and Sichuan) to opt out of mandatory supply quotas for timber (roundwood) to be allocated—at fixed prices—under the state plan.[3] They have restrictions on their annual production (which in theory must not exceed the annual increment), and they must sell the bulk of their products through provincial authorities (forestry departments and local timber corporations) at "negotiated" prices; but surplus wood can be sold at floating prices—subject to state-imposed ceilings. As a result, log and lumber prices have more than doubled and the stage has been set to enforce managerial fiscal responsibility (and accountability) at enterprise level. The second component of decentralization relates to timber distribution—formerly a strict monopoly of the State Timber Corporation through its local subsidiaries. The latter are now subject to guidance planning only. Priority consumers are still assured of minimal supplies, but requirements over and above their quotas are subject to local contract and negotiated prices.

Although this is not a free market in a capitalist sense, prices have been permitted to reflect more closely scarcity and production values.

Timber imports are strictly controlled, but some are being channeled into "freer" retail markets. There is also illegal leakage from the system (to take advantage of high prices and lower taxes), which is leading to complaints about the poor quality of allocated domestic wood. According to NFPA (1986), the central government is having to subsidize official purchases from collectives in order to fulfill priority needs. Developments in the forest industry are outlined in Chapter 4.

PROBLEMS AND PROGNOSES

In fairness to Mao Zedong, it should be acknowledged that both decentralization and the introduction of incentives were tried at various stages in socialist China's economic experiment. In particular, the Cultural Revolution policy of creating industrial systems in each province on the basis of "self-reliance" began the transfer of resources from central to local control; it failed because it was administrative control only that was transferred—not decision-making capacity, which at all levels remained with governmental agencies rather than production enterprises. Indeed, following the Cultural Revolution, there was an increase in the number of bureaucrats—accompanied by a decline in work efficiency, which Deng Xiaoping ascribes to three factors: overcentralization of power; the lack of promotion available to "younger, better educated, more technically competent" personnel because of an aging leadership; and, finally, the absence of rules and regulations and personnel assessment criteria resulting from the antibureaucratic ethic of the Cultural Revolution. Bureaucratic deficiencies remain the biggest constraint in implementing reform in the industrial sector.

Some problems of reform stem from its very success. Early on, urban industry came under pressure for its failure to meet the burgeoning demand for machines and equipment from the rural sector, while inadequacies in the marketing infrastructure (including transportation and storage facilities) soon became apparent. Even partial decontrol of procurement prices led to sharp price rises in the cities, necessitating subsidies and other controls. Bureaucrats exposed to a market environment have behaved (according to Prybla) "more like black marketeers than capitalists"; even honest ones feel threatened inasmuch as the market—if it works—either makes many of them redundant or removes their privileges.

Numerous examples of corruption and nepotism may be cited from the Chinese press. Apart from major scandals such as the widely publicized Hainan Island affair (when senior officials abused their import powers to purchase huge numbers of cars, TV sets, and video recorders through over 900 companies for resale on the mainland), many less damaging "economic crimes" have been reported and denounced (see Fewsmith, 1985; FEER, 20 March 1986). They include illicit timber dealing by logging units of the PLA in Tibet (FBIS, 8 February 1982) and the setting up of companies by party and government organizations to buy timber from farmers (now that there is no longer a state purchasing monopoly)—but using their mandate to issue felling licenses to force low buying prices and give themselves high profits (*China Daily*, 3 June 1985). Currency black markets, smuggling, bribery, and tax evasion; sheer inefficiency in management (especially in the featherbeds of state enterprises, of which there can be no more glaring example than the timber industry);[4] bureaucratic indifference to joint-venture problems; a lack of realistic quality/price differentials in all kinds of goods; the inability of rudimentary laws to cope effectively with malpractice—all have contributed to the checks and reverses faced by the leadership.

An interesting study by Gold (1985) not only documents numerous cases of corruption on the part of cadres but also points to radical changes in personal relationships within communities because of intensifying interpersonal competition. Similarly, Zweig (1986) analyzes the inequalities that developed in a commune in which specialized households raised tree seedlings (very profitably) for urban planting. When supplies became short, seedlings were stolen, trees already sold to the township were mutilated for cuttings, families looted each other's holdings, cadres imposed illegal sanctions, and conflict erupted at all levels. Such conflicts are exacerbated in China by the difficulties people have in changing their domiciles.

The sheer size of China, of course, poses problems of governability faced by no other country in the world. It is easy to call for devolution of authority, and there is ample evidence that the economy has in the past suffered from excessive control. It is less simple to prescribe the optimal degree of decentralization, given the variety of outcomes that may emerge (see Naughton, 1986; Wong, 1986). The division into three economic regions under the Seventh FYP with different development priorities—and different degrees of central control—is acknowledgment perhaps that China is not a single economy (*Beijing Review,* 24 April 1986).

The distinguishing features of later "reform" policies have been the attempts to stimulate lateral rather than hierarchical material flows and the decision (set forth in Document No. 1 for 1985) to "gradually abolish" state monopolies over procurement and sale of products. Where products are also raw materials (as is the case with timber), the picture is further complicated, especially where controls are shared. It is not unknown for both local and central government each to claim exclusive control. A recent joint venture to manufacture panel products in Heilongjiang, for example, took four years to negotiate because different departments claimed control of the wood resource.

The progressive decentralization of raw material allocations has distorted prices and created problems in the rural sector. For example, the attraction of higher net revenues has led to investment in subsectors (such as sliced-veneer production and furniture-making), which may not be sustainable by material supplies; once in place, enterprises have sometimes been protected by local interests—in defiance of central government attempts to exert control through material supply. With multiple pricing in operation, quotas tend to be underfulfilled and ex-plan production (for sale on the free market) repeatedly in excess of targets.

State enterprises in a dual system are at a competitive disadvantage vis-à-vis local operations, which can more readily take advantage of higher free-market prices and outlets. Wong (1986) points out the adverse efficiency implications likely to be shown up when small-scale production displaces that from large-scale enterprises. There may be little consequence at the village level, but there are likely to be adverse effects on capital-intensive industries amenable to economies of scale—such as pulp and paper mills and plywood plants. And where the only means of control is through raw material and capital allocations, state intervention is likely to hurt the state enterprises more than those under local control, which may be able to obtain materials through extrasectoral barter deals or political influence.

Apart from price reforms, a shift for state enterprises from the remittance of net profit to the payment of income tax is designed to redress some of these effects. Delfs (1989) notes the difficulties of implementation of a tax system under conditions of "fiscal feudalism." It is also government policy to strengthen managerial control of enterprises (reducing the powers of party committees) and, in return, to demand greater accountability.

There have been several recent studies of industrial management in China, and the papers in *CELT 2000* are relevant. Fischer (1986) examined industrial management effectiveness under four heads—decentralization of management authority, improvement of accountability, the nature of economic incentives (for organizations and individuals), and organizational flexibility. He concluded that although China has made "substantial progress," deficiencies in numbers and quality of managers will continue. The Chinese would accept this latter observation. According to Ma Hong, president of the Chinese Academy of Social Sciences, "Less than one-third of the leading cadres in our country's large and medium-sized enterprises are well versed and know how to run them."

Decentralization of decision-making has gone a long way in production enterprises and, in light industry particularly, there is considerably more autonomy than existed even five years ago. Managers can decide in some cases what they will produce, how much, to whom they will sell it, and how they will allocate profits after taxes. Other enterprises have a limited autonomy, but in some there is no change at all; these tend to be the larger state-controlled enterprises working within the plan on allocated raw materials and set production targets. The integrated wood enterprises fall within this category.

Problems arising from increased decentralization of decision-making relate to the development of overcapacity with respect to products with a higher than average profit margin, excessive competition for limited resources that can lead to inefficiency in resource use, and problems that can arise from what Fischer describes as the "considerable skill and experience that Chinese managers have in finessing both manufacturing and supply frustrations."

Accountability is assessed in China by the "Eight Great Standards"; these are production volume, quality, profit, cost, labor efficiency, consumption of resources, capital utilization, and production values. As well, a large number of social considerations may enter into the assessment of managerial competence—pollution abatement, family planning among the workers, performance of schools and clinics, and the like. Work units are responsible not only to the administration but also to party committees and the executive council of the enterprise "workers congress." In theory, the party committee and the workers executive play advisory functions only with respect to management accountability except where ideological issues are involved. This, however, is clearly a gray area of responsibility and is influenced by the personal strength of the figures involved. Fischer makes the point that "politics is no longer in command at the level of operational or even strategic decision-making within the enterprise. Increasingly, the Chinese manager is expected to perform as an economically efficient and effective manager or be replaced."

Incentives are considered at two levels—that of the enterprise and that of the individual. There is considerable variation with respect to the retention of earnings by enterprises. And since performance is measured from a base year (and the taxes to be paid as well as the earnings retained are calculated on the basis of improvements over that base year), there is little element of competition between enterprises in the measurement of performance.

Similarly with respect to individual incentives, there is a range of "carrots" but virtually no "sticks." Bonus systems tend to be uniform—indeed there have been pressures to equalize bonus payments (or for individual bonus awards to be shared among all workers). Moreover, the fringe benefits that enterprises have to provide their workers may be distributed not on the basis of performance but seniority—and at the behest of the party committee rather than the enterprise manager. Nonetheless, living standards are increasingly being linked to the economic performance of the enterprise. Field and Noyes (1981) show that while productivity remained constant in the state sector between 1979 and 1980, that of collective enterprises rose by 16.6 percent—a fact attributable to increased managerial flexibility.

Organizational flexibility in the socialist system is limited by the commitment of enterprises to the supply of social amenities—housing, medical care, child care, subsidized meals and utilities, transportation, access to cultural entertainment and sporting events, and more. Dismissing a worker, therefore, is to condemn him (and his extended family) to a standard of life below the poverty line. (Conversely, employing an additional worker involves a significant increase in overheads.)

There is perhaps more organizational flexibility in cooperative enterprises. Fischer cites an example from Liaoning province where thirty-one scientific research and design academies and institutes, six universities and specialized colleges, and seventy-six plants and enterprises combined to form seventeen "research-production" units. Institutional affiliations remained unchanged in these units, but the ventures were enabled to respond realistically to market demands. The increasing involvement of research institutes in consulting and in ad hoc projects provides other examples.

Decentralization has not everywhere been successful. Where there is local involvement in materials or product allocation there tends to be a proliferation of local bodies demanding a voice in enterprise management—which, again depending on the strength and personality of the individual, can be less efficient than central "command" planning. (See Naughton, 1986.)

China has made progress in reform of middle-sized and small enterprises, but much has to be done at the level of the larger state enterprises. Moreover, the reforms have been structural, not functional, and it remains questionable whether there is a sufficient pool of competent executives from which to draw managers (after the devastation of the Cultural Revolution). There remain problems with the failure of bureaucrats to adapt to a competitive environment, and Clarke (1986) makes the interesting point that prosperity in rural areas has reduced incentives to join the bureaucracy. Similarly, managerial responsibility without power will only frustrate the able cadres (see Walder, 1983; Naughton, 1983). It remains to be seen how inevitable tensions between the traditional and modern managerial generations now being trained in the West are resolved.

Experience in Eastern Europe has shown that decentralization of responsibility does not always result in acceptance of it: There may be a continuing "paternalistic" relationship that prevents the state from strictly enforcing regulations and enabling the market to drive. Walder (1986:644) concludes that "no reform scheme for enlarging enterprise autonomy and increasing financial accountability is likely to succeed without simultaneous attention to the difficult problem of enforcement at the local level."

China's intentions with respect to industrial reform are clear enough. A State Council regulation of May 1984 stipulates the devolution of authority over: (1) production and sales of non-FYP output, subject only to price "guidance"; (2) the use of retained profits with no restriction on bonus payments; (3) enterprise personnel—except the appointment of the manager and the leading party secretary; (4) the right to dispose of fixed assets and to promote a proportion of workers in accordance with profitability. These are "Temporary Regulations on Expanded Autonomy." They make it clear that the authority of the plant manager with respect to the plant's operation overrides that of the party secretary. This, too, is an expression of intent rather than achievement.

Decentralization is likely to continue. Projects financed by the World Bank have revealed dissension between the provinces and the central government over loan servicing. Beijing decided that provincial governments should assume responsibility for a proportion of the project loan. In one case, the province of Heilongjiang objected to having to assume responsibility for a forestry loan—recording its dissatisfaction that for many years it had supplied timber for state allocation at well below world market prices. (Such complaints are familiar in all countries with primary producing—and relatively undeveloped—regions, even if the product is only rainwater.) Possibly in response, the largest remaining primary forest area in the northeast (Yakashih) has been transferred to the provincial control of Nei Monggol.

Despite setbacks, China's leaders are confident. The Seventh FYP reveals a determination to continue reform. ("Mistakes are inevitable but the reform must go on.") The outward orientation of the economy is being underwritten politically by the globe-trotting of the leadership and detente (albeit falling short of entente) with the USSR (Nations, 1986).

Economists (socialist and capitalist) recognize that China's experiment in market socialism has no successful precedent. Prybla (1986b) argues that it must conflict with both the Marxist code of "the right to work" and Leninist ethics of party control. Klatt (1983), on the other hand, sees no reason why "precarious coexistence" of public corporations and private enterprise should not be accommodated in China as it is in the West. On present evidence, if China's experiment fails the reasons will be technical as well as political. The key problems relate to China's failure to mobilize resources—notably energy—efficiently enough (CIA, 1986; Fingar, 1986) and to the inflexibility of the entrenched bureaucracy. Both sets of problems are exemplified by the forestry sector.

NOTES

1. Traditionally there were two Chinas—"China Proper," which comprised the eighteen provinces within the Great Wall, and "Outer China," made up of dependencies beyond the wall (Manchuria, Mongolia, Xinjiang, and Tibet). Outer China covered almost two-thirds of the land surface of the country but contained only 5 percent of the population. The distinction is

no longer officially acceptable in China, but in news releases aimed at overseas readers, reference is occasionally made to "China Proper" and to its former subdivisions of North, Central, and South China. It must be remembered, too, that even in the nineteenth century, the Chinese empire included territories now independent or found in parts of other countries. Liu Peihua (1954) lists "Chinese territories taken by imperialism" as follows: the Great Northwest (parts of present-day Kazakh, Kirghiz, and Tadzhik SSRs); the Pamirs; Nepal; Sikkim; Bhutan; Assam; Burma; Andaman; Malaya; Thailand; Indochina; Sulu; the Ryukyus; the Great Northeast; and Sakhalin.

2. A model contract under the incentive system is analyzed in Crook (1985).
3. Three provinces—Fujian, Sichuan, and Yunnan—retained the option to supply some roundwood under the old system in return for state allocations of similar volumes; this enables them, in effect, to exchange part of their own production for northeastern softwoods and veneer species.
4. Field (1982) has published indices of industrial performance in state-operated enterprises by branch of industry; the timber industry is the only one of eleven production sectors to show a consistent decline in gross value of industrial output (GVIO) per worker since 1952.

2

ENVIRONMENT AND LAND USE

CHINA'S NEW POLICIES of involvement with the outside world have released a flood of publications on economic geography that are informative and objective—in contrast to the hubristic presentations of the pre–Cultural Revolution era. They include a series on natural resources and their management, descriptions of new provincial floras, and texts on land-use classification. A standard work is *An Outline of China's Physical Geography* (Ren Mei'e et al., 1985), which provides the basis for this discussion of physical regions in China and their subdivision. Ren Mei'e and colleagues recognize eight physical regions: Northeast China, North China, Central China, South China, Southwest China, Inner Mongolia, Northwest China, and Qinghai–Tibet. The four eastern regions—landforms of plains and low mountains—are differentiated on the basis of heat zones. The primary index is the annual total daily average temperature over the period when the daily average is above 10° C. Broadly speaking, Northeast China is a *temperate* zone, most of the North China Plain region is in the *warm temperate* zone, the Central China region occupies a *subtropical* zone, and the South China region is approximately equivalent to a *tropical* zone.

To the northwest of a line from the Greater Hinggan Mountains to northern Yunnan, moisture is more influential than temperature. The Inner Mongolia region is mainly a grassland steppe and the Northwest region a desert. The regional boundaries tend to run from north to south rather than west to east, but there are landscape variations from west to east according to differences in moisture. The Qinghai–Tibetan Plateau is 3000 to 4000 m above sea level and comprises desert, alpine meadow, and steppe. The Southwest region includes Yunnan and the southwestern part of Sichuan—which is mountainous (1500–2000 m above sea level), subtropical, and has a climate described as "spring in all seasons." Generally the Southwest region is a tropical mountainous plateau.

Throughout Southwest China, climate is greatly influenced by the tropical continental airmass and its reaction to altitude. Thus the natural environment is determined by three sets of factors: the three-step staircase system of topography outlined in Chapter 1; humidity and seasonality (which decreases from south to north and from east to west); and temperature, which shows a fairly clear-cut latitudinal zonation.

Within the eight major regions, Ren Mei'e and colleagues recognize twenty-six subregions and fifty-eight "areas." Physical descriptions of each area, together with notes on vegetation and land use, range over some 300 pages and are too detailed for present purposes. Information on natural vegetation derives from the work of Academia Sinica botanists working under the guidance of Hou Xueyu, whose classification issued in 1956 was used as the basis for the chapter on "The Natural Vegetation, Soils, and Land Use" published in *Forestry in Communist China* (Richardson, 1966). That classification has been refined and reissued by Hou (1983). An editorial board has been working on China's vegetation for some years and published a general description of vegetation in 1980. There have also been several provincial descriptions (see Sichuan, 1980; Xinjiang, 1978; Guangdong, 1976) as well as more detailed management-oriented studies of particular vegetation types in the principal forest areas. The natural vegetation and soils have not of course changed significantly since 1966, but there is a greater awareness of what Hou designates "cultural vegetation" in characterizing the environment. His vegetation map distinguishes six natural vegetation types (divided into thirty-nine subtypes according to landform, climate, and soil) together with five cultural vegetation types (divided into thirteen crop subtypes).

Changes between Hou et al. (1956) and Hou (1983) are conceptual rather than fundamental. The six categories broadly labeled the "forests of the east" in 1956 are in 1983 divided into twelve vegetation types;

there are six desert and xerophytic shrub types and four steppe and savanna types that replace the former categories of "steppes and deserts of the northwest and northeast" and "mountains and plateaus of the west and southwest." In addition, the 1983 classification includes four meadow and swamp vegetation types.

The map reproduced in Richardson (1966) was a combined vegetation/soil map published by Hou et al. (1956). It was backed with a table summarizing the main features of twelve phytogeographic regions—reproduced here in amended form (Table 3) to emphasize forest components of the ecosystems. The 1983 classification is too elaborate, however, for a general text on forestry in China. The outline presented here, therefore, follows the earlier system but incorporates some species revision and elements of Hou's (1983) "cultural vegetation" types. In detailing the specific composition of climax forest communities, I have relied on Wang (1961), results of Chinese forest inventories (principally that completed in 1981), and personal observations, as well as papers on forest management prepared for various regional forestry and ecological workshops. (See Petrov, 1979; Xu et al., 1986; Jiang, 1986.) Papers discussing regional characteristics of vegetation with summaries in English are also available (Chang, 1973; Wu, 1979; Wang, 1977).

FORESTS OF THE EAST

In general terms, that part of China which lies east of an imaginary line drawn between Dongchuan (Yunnan), Beijing, and Hailar (Heilongjiang)—but excluding the Manchurian Plains—has a climax vegetation that is essentially high forest. Natural forest once formed an unbroken sequence of communities from tropical monsoon rain forest in the south to montane coniferous forest in the north.

NORTHERN CONIFEROUS FOREST

Located in the extreme north of Manchuria, the Northern Coniferous Forest region includes the northern part of the Hinggan range (Greater Khingan Mountains)—the long and narrow ranges of gneiss and granite that form the uplifted eastern margin of the Mongolian Plateau. Ad-

TABLE 3

SOILS, CLIMATE, AND CHARACTERISTIC TREE SPECIES

Phytogeographic Region	Soils	Mean Annual Rainfall (in.)	Growing Season (frost-free days)	Major Tree Species	Minor Tree Species
Northern Coniferous Forest	podzolic	14–24	100–125	*Larix dahurica* *Pinus sylvestris* var. *mongolica* *Picea obovata*	*Betula* spp. *Alnus mandshurica*
Mixed Coniferous and Deciduous Broad-Leaved Forest	podzolic, brown forest, gleys	25–40	120–130	*Pinus koraiensis* *Abies nephrolepis* *Picea jezoensis* *Larix dahurica* *Betula* spp. *Acer* spp. *Tilia amurensis* *Quercus* spp.	*Abies holophylla* *Pinus densiflora* *Fraxinus mandshurica* *Populus koreana* *Corylus* spp.
Deciduous Broad-Leaved Forest	variable: brown forest, korichnevie, solonchaks	20–28	150–240 north–south	*Quercus* spp. *Ulmus* spp. *Acer* spp. *Betula* spp. *Tilia* spp.	*Populus* spp. *Salix* spp. *Pinus tabulaeformis* *P. bungeana*
Mixed Deciduous and Evergreen Broad-Leaved Forest	brown forest, yellow podzolic, yellow korichnevie	ca. 40	230–280	*Acer* spp. *Tilia* spp. *Carpinus* spp. *Fraxinus* spp. *Pinus massoniana* *Cunninghamia lanceolata* *Cryptomeria japonica* *Cupressus funebris*	*Quercus* spp. *Zelkova* spp. *Dalbergia hupeana* Bamboos

Evergreen Broad-Leaved Forest	yellow podzolic, rendzina (eastern), red podzolic, terra rossa (western)	40–75 (E) 36–60 (W)	250–300	*Castanopsis hystrix* *Schima superba* *Elaeocarpus japonicus* *Cinnamomum* spp. *Machilus pingii* *Phoebe bournei* *Photinia serrulata*, etc.	*Pinus massoniana* *Cunninghamia lanceolata* *Cyclobalanopsis glauca* *Castanopsis* spp. *Cupressus* spp., etc.
Tropical Monsoon Rain Forest	yellow lateritic	48–110	350–365	Many species of Myrtaceae, Moraceae, Lauraceae, Annonaceae, Leguminosae, Sapindaceae, Euphorbiaceae, Melastomaceae, Rubiaceae, Palmae, Theaceae, Fagaceae (see Wang, 1961)	
Forest Steppe	NE: gray forest, chernozems, solonchaks NW: siero-korichnevie (loess)	19–26 (NE) 12–20 (NW)	140–160 (NE) 90–100 (NW)	NE: *Quercus mongolica* *Betula platyphylla* *Populus davidiana* *Salix* spp. NW: *Quercus liaotungensis* *Betula japonica* *Salix matsudana* *Populus simonii*	NE: *Tilia amurensis* *Rosa dahurica* NW:*Pinus tabulaeformis* *Biota orientalis*
Steppe	chestnut, sands, solonchaks	8–15	145–155	*Salix* spp., *Populus cathayana*, *P. euphratica*, *Juniperus chinensis*, *Ulmus pumila* (along watercourses only)	
Desert and Semidesert	zierozem, solonchaks, solonetz, highly saline	< 4	160–180	Virtually nil	

TABLE 3 (*Continued*)

SOILS, CLIMATE, AND CHARACTERISTIC TREE SPECIES

Phytogeographic Region	Soils	Mean Annual Rainfall (in.)	Growing Season (frost-free days)	Major Tree Species	Minor Tree Species
Northwest Mountains	zierozem, chestnut, brown forest, meadow (altitudinal zonation)	8–12	80–100	*Juniperus* spp. *Larix sibirica* *Pinus sibirica* *Abies sibirica* *Picea obovata* *Betula* spp.	*Juglans fallax* *Pyrus malus* *Acer* spp. *Picea schrenkiana* *Pinus tabulaeformis*
East Tibetan Mountains and Plateaus	yellow podzolic red podzolic, brown forest, alpine, high mountain desert (altitudinal zonation)	15–35	80–120	*Abies squamata* *A. faxoniana* *Picea purpurea* *P. neoveitchii* *Larix potaninii* *Picea likiangensis* *Pinus yunnanensis*	*Betula* spp. *Populus* spp. *Pinus armandi* *Abies* spp. *Tsuga chinensis* *Picea brachytila*
Tibetan Plateau	high mountain desert, alpine meadow, alpine semidesert	10–45	45–130	*Juniperus* spp., *Salix* spp., *Ulmus pumila*, *Populus* spp. (along watercourses only)	

ministratively, the area covers some 200,000 km^2 and embraces parts of Heilongjiang province, Nei Monggol, and the autonomous district of Oroqen.

Coniferous deciduous forest (light taiga) is larch forest found on drier sites than either spruce or fir. In the northeast of China the dominant species is *Larix gmelini* (syn. *L. dahurica* and *L. kamtchatica*) with an understory of shrubs—*Vaccinium vitis-idaea* or *Ledum palustre*, depending on soil moisture. (The northwestern extension of the subtype in the Altay Mountains is dominated by *Larix sibirica* in mixture with *Pinus sibirica* and an understory of *Rhododendron dahuricum.*) These forests are contiguous with the coniferous forests that run along the northern border of Outer Mongolia and southern USSR. In elevation the Northern Coniferous Forest ranges from 450 to 1100 m, with peaks reaching 1500 m. The region is very cold, with a mean annual temperature from −1.2 to −3.2°C and an extreme minimum of −50.1°C. January temperatures are around −25°C and those for July about 18°C, giving a mean annual range of 43°C. Precipitation varies from 35 to 60 cm at the higher altitudes and, as might be expected, the soils are acid podzols, the degree of podzolization varying with the relief.

The major tree species apart from *Larix dahurica* is *Pinus sylvestris*, with *Picea obovata*, *P. microsperma*, *Pinus pumila*, and *Juniperus dahurica* locally dominant. Secondary species include *Betula platyphylla*, *B. dahurica*, and *Populus davidiana*, while in the understory *Alnus mandshurica*, *Vaccinium uliginosum*, *Rhododendron dahuricum*, and *Linnaea borealis* are frequent. The herbs of the forest floor include several *Pyrola* species, *Trientalis europaea*, *Nitella nuda*, and *Goodyera repens*. This vegetation type is strongly reminiscent of the northern coniferous forests of Europe and North America, which have many common or closely related species. It is distinguished, particularly at the lower elevations, by the large size of the trees and their stocking density.

The region is almost entirely forest, with the tree line at about 1100 m. Occasional meadows are characterized by such species as *Carex schmidtii*, *Calamagrostis hirsuta*, *Sanguisorba parviflora*, *Eriophorum vaginatum*, *Iris sibirica*, *Lilium dahuricum*, *L. pulchellum*, *Veratrum dahuricum*, *Geranium* and *Pedicularis* spp., and wet swamps of *Populus suaveolens*, *Betula fruticosa*, *Chosenia macrolepis*, *Salix rorida*, *S. brachypoda*, *S. sibirica*, and *S. myrtilloides*. The black, organic soils of the meadows are less acid and very fertile. Though many cultivated crops and fruit trees do not ripen in the area, potatoes, barley, and cabbage offer possibilities for cultivation. There are small groups of Oroqons who live by raising reindeer, and there are hunters and fur trappers active in the region.

The larch forests are being exploited for timber in the northeast, but in the Altay extension transport is too much of a problem to enable industrial use of the forests. Adjoining the larch forests are evergreen coniferous types (dark taiga) dominated in the northeast of China by spruce (*Picea jezoensis* and *P. koraiensis*), *Abies nephrolepis*, *A. holophylla*, and *Pinus silvestris* var. *mongolica*. (In the northwest, the species are *Picea obovata*, *Abies sibirica*, and *Pinus sibirica*.) Small areas of this forest type occur in drier parts (in the central part of the arid desert region, for example), but spruce only (not fir) is found. Secondary growth, where it occurs, is of *Betula platyphylla*.

On the western slopes of the Hinggan Mountains there is a forest grassland zone below the larch with scattered birch and aspen (*Populus davidiana*). This zone is transitional to the Inner Mongolian steppes and currently supports a very low population. It is regarded as available for "reclamation" and agricultural development. It is also not far from the area subject to disastrous fires in May and June 1987.

Mixed Coniferous and Deciduous Broad-Leaved Forest

This type forms a transition between the coniferous forest in the north of Manchuria and the deciduous broad-leaved forest in the south. It covers much of the Siao Hinggan Ling (Lesser Khingan Mountains), the Changkwansai Mountains, and the Changbaishan massif north of Korea. The latter consists of several parallel ridges, generally between 450 and 1100 m, but with peaks rising to almost 2750 m. In the Siao Hinggan Ling, the topography is rolling and generally below 600 m. The river valleys and alluvial floodplains of the Sungari and the Ussuri rivers (tributaries of the Heilong) form an extensive area of low-lying land to the east. Some 300,000 km^2 in area, the type covers parts of Heilongjiang and Jilin provinces and the Korean autonomous districts of Yenpien and Changbai.

The climate is both milder and moister than that of the far north. Annual precipitation ranges from 50 to 75 cm in the Siao Hinggan Ling to 100 cm in the eastern plains. Below 600 m, the mean annual temperature is from 0.5 to 5.5°C, with mean temperatures in January from −16 to −25°C and in July from 20 to 24°C. The extreme winter minimum ranges from −32 to −45°C, but the length of the growing season (average number of frost-free days) is more than 125 days.

The soils include moderately leached podzols derived from pre-Cambrian metamorphic rocks in the mountains, brown forest soils of low-base status on the lower slopes, and deeper and richer alluvial soils in the plains, with local areas of gleyed peat. The pattern again resembles that in Europe and North America, often showing local catenas with sharply differentiated topographic components. I have examined such a sequence in an area northeast of Dailing in Heilongjiang province and was struck by the similarity to that in, for example, parts of Upper Deeside in Scotland. Over a distance of perhaps 200 m and an altitude range of 50 m, the soil type changed from a shallow podzol at the top of the slope to a deep brown forest soil halfway down and then to a strongly gleyed alluvium in an area of impeded drainage at the bottom.

The relative richness of the flora distinguishes the Mixed Coniferous and Broad-Leaved Deciduous zone from similar European forest types. At higher elevations, *Pinus koraiensis* is the characteristic species, but *Abies nephrolepis, A. holophylla, Picea jezoensis, P. koyamai* var. *koraiensis,* and *Larix olgensis* are frequent among the conifers. The broad-leaved species include *Betula ermanii, B. platyphylla, B. costata, Sorbus pohuashanensis, Acer tegmentosum, A. mono, A. pseudo-sieboldianum, Tilia amurensis, Fraxinus mandshurica, Syringa amurensis, Phellodendron amurense, Juglans mandshurica, Populus koreana, P. ussuriensis,* and *Maackia amurensis.* Shrub layers are characterized by *Sorbaria sorbifolia, Spiraea salicifolia, Corylus mandshurica, Crataegus maximowiczii, Viburnum* spp., *Ribes* spp., *Vitis amurensis, Schizandra chinensis, Actinidia kolomicta,* and *Celastrus flagellaris.*

On the brown forest soils at lower elevation, the dominant *Pinus koraiensis* is supplemented by *P. densiflora* and *Abies holophylla; Acer, Populus,* and *Betula* spp., *Corylus heterophylla, Lespedeza bicolor, Quercus mongolica,* and *Q. liaotungensis* are common. The broad-leaved trees predominate on the well-drained alluvial soils, while in areas of impeded drainage and a fluctuating water table, the sedges *Carex rhynchophysa* and *C. meyeriana, Calamagrostis* spp., *Iris ensata, Trollius ledebourii,* and *Pogonia japonica* assume dominance; *Larix olgensis* is frequent.

To a greater extent than in the coniferous forest region (where the limiting factors are climatic), vegetation and soil patterns determine cultural practices in this zone. Production forestry is by far the most important land use since, in terms of quantity, quality, and accessibility, China's richest timber resources are located here. As long ago as 1913, Sowerby visited Manchuria and was impressed by the seemingly endless tree cover (Sowerby, 1923) and, in spite of later settlement and occupation by the Japanese, large forests remain that provide some 30 percent of China's present-day timber requirements. The "three precious trees" of Manchuria—walnut, ash, and cork (*Phellodendron amurense*)—are renowned throughout China.

Settlement and agricultural development have been mainly confined to the river valleys and plains, where crops of potatoes, soybeans, spring wheat, maize, barley, kaoliang, millet, rice, and cabbage are raised. The growing season is also adequate for such cucurbits as pumpkin, watermelon, and cucumbers. In addition, some plantation forestry (using mainly *Pinus koraiensis, P. sylvestris* var. *mongolica,* and *Larix dahurica* but also the hardwoods *Tilia amurensis* and *Acer mono*) has recently been started, and cultivation of ginseng (and the mass of small industries spawned by it) is becoming a major economic component. As well as three precious trees, Manchuria has "three treasures"—ginseng, deer antlers, and the fur of the marten. Wild ginger and medicinal plants are gathered from the deciduous forests, and there are deer and mushroom farms.

The lower-lying land of the northeast region has seen significant development during the last twenty years. On the Sanjiang plain, some 1.3 million hectares of land have been reclaimed and there are still large areas remaining to be developed. Heilongjiang province has become the largest marketable grain source in China. There are, however, flood problems that may be associated with the removal of the forests in the mountains and with earlier attempts at mechanized farming.

Deciduous Broad-Leaved Forest

The Deciduous Broad-Leaved Forest zone skirts the southwestern end of the Manchurian Plain to take in the peninsulas of Shandong and Liaoning, the North China Plain, the northern slopes of the Qinling Mountains, and the western half of the Shanxi highlands. It covers almost 1 million square kilometers in the provinces of Jilin, Liaoning, Hebei, Shanxi, Shandong, Jiangsu, Anhui, Henan, and Shaanxi, including the autonomous districts of Fuxin, Kolaqin, and Beijing. The plains are low-lying (below 150 m), but the Shanxi highlands and the

Qinling Mountains range generally between 450 and 2500 m, with ridges and peaks over 4000 m. Local areas rising to 900 m are also found in Shandong and Liaoning.

The climate is uniformly subcontinental, with hot, wet summers and cold, dry winters, and is noted for the large fluctuations from year to year in amount of precipitation. In general, however, the rainfall is between 50 and 85 cm, falling over periods of about 120 days in the north and 65 days in the south. The mean annual temperature ranges from 7°C in the north to 16°C in the south, with extreme minima varying from −33 to −15°C. The January mean is from −13 to 0°C and the July mean from 22 to 29°C. The length of the frost-free season varies from 150 days in the north to 240 days in the south. This mildness, together with dry-farming practices, enables double-cropping in many southeastern parts of the region.

Not surprisingly, parent materials and soils show considerable variation through the area. The soil-forming materials on the intensively cultivated plains are ancient alluvial deposits sometimes overlain with loess and giving rise to the so-called korichnevie soils; they are alkaline or neutral. Calcareous korichnevie soils (often leached) are also found on the limestone mountains and loess-covered hills of Shanxi. The granite and gneiss hills of Shandong and Liaoning, on the other hand, form acidic brown forest soils sometimes slightly podzolized. Solonchaks containing easily soluble chlorides occur along the eastern seaboard and, in local inland depressions, soils containing mixed chlorides and sulfates affect the vegetation.

The primary vegetation on the brown forest soils comprises several species of *Quercus*, *Ulmus*, and Acer (including *Quercus liaotungensis*, *Q. dentata*, *Q. variabilis*, *Q. aliena*, *Q. acutissima*, and *Acer truncatum*), *Tilia amurensis*, *Pterocarya stenoptera*, *Lindera obtusiloba*, *Sorbus alnifolia*, and *Fraxinus rhynchophylla*. The shrubs include *Callicarpa japonica*, *Clerodendron trichotomum*, *Grewia biloba* var. *parviflora*, *Styrax obassia*, *Celastrus orbiculatus*, *Rhododendron micranthum*, and *Lespedeza bicolor*. Destruction of the forest may give rise to dense thicket growth of *Corylus heterophylla*, *Crataegus pinnatifida*, *Zizyphus sativa* var. *spinosa*, *Vitex chinensis*, *Prunus humilis*, *Pyrus betulaefolia*, *Symplocos paniculata*, and *Periploca sepium*. On dry sites, *Pinus tabulaeformis* is common, together with *Biota orientalis*, *Pinus densiflora*, and *P. bungeana*. According to Wang (1961), the first two of these conifers may represent climax species and not simply seral stages.

The korichnevie soils of the plains are almost entirely cultivated, and the specific composition of the natural vegetation is somewhat conjectural. Remnants of forest occur on surrounding hilly country and in protected reserves around villages and temples. These vestiges have been studied intensively by Yang (1937), who recognized four types of oak forest, two birch associations, and a *Tilia–Betula* type. Following Wang (1961), the components of the oak forests are listed in Table 4, together with those of the higher elevation *Tilia–Betula* and *Pinus tabulaeformis* types. The *Betula* types (comprising *Betula fruticosa, B. japonica* var. *mandshurica, B. alba* var. *chinensis,* and *B. chinensis,* with admixtures of *Populus tremula* var. *davidiana, Sorbus, Prunus,* and *Salix* spp.) are found at higher altitudes still. The original cover of the North China Plain itself may have been grassland, wooded steppe, or forest. Wang (1961) concludes that it was forest, on the grounds that the existing gallery forest of *Salix matsudana, S. babylonica, Populus simonii, P. suaveolens,* and *P. tomentosa* is not confined to watercourses and that many species now growing in plantations or in seminatural stands have been present on the plain since the third century A.D. Because the status of these species may be important in evaluating the effectiveness of China's artificial protection forests, they are listed here in Table 5.

The limestones of the high mountains and the loess-covered hills are also very rich in woody species. *Quercus liaotungensis, Q. variabilis, Tilia mongolica, Fraxinus chinensis, Acer truncatum,* and *Biota orientalis* of the deciduous oak forests are enriched with, among other species, *Quercus acutissima, Prunus armeniaca, Oxytropis davidiana, Carpinus turczaninowii, Acer pictum* var. *parviflorum, Cotinus coggygria* var. *cinerea, Pinus bungeana, P. tabulaeformis, Juglans regia,* and *Crataegus pinnatifida.* Calcicolous shrubs are also found, including *Gleditschia heterophylla, Wikstroemia chamaedaphne, Indigofera bungeana, Prinsepia uniflora, Periploca sepium, Lycium halimifolium, Ulmus pumila, Lespedeza floribunda, L. dahurica,* and *Vitex chinensis.*

The coastal vegetation is strongly influenced by the sea. In the littoral zone, *Carex kobomugi, Calystegia soldanella, Ixeris repens, Phellopterus littoralis,* and *Lathyrus maritima* are examples of halophytic psammophytes. On the heavier solonchaks, *Suaeda ussuriensis* is commonly found in pure communities. *Statice bicolor, Artemisia scoparia, Scorzonera mongolica* var. *putiatae, Puccinellia distans, Tamarix juniperina, Chenopodium glaucum, Atriplex littoralis, Aster tripolium, Polygonum sibiricum,* and *Suaeda glauca* may also be found. In the inland depressions, where the solonchaks contain sulfates as well as chlorides, the common halo-

TABLE 4

COMPONENTS OF THE DECIDUOUS OAK FORESTS

	Forest Type and Altitudinal Distribution					
	1[a]	2	3	4	5	6
Plant Name	(Below 700 m)	(Below 500 m)	(400–1000 m)	(700–1000 m)	(1100–1400 m)	(Over 1400 m)
Quercus aliena	1**		1 r			
Q. dentata	1 f		1**			
Q. variabilis		1**				
Q. liaotungensis				1**	1 f	11 f
Pinus tabulaeformis						1**
Fraxinus chinensis	1**	1 r	1 o	1*		
Biota orientalis		1 r				
Cornus walteri	1 r					
Evodia danielii	1 f					
Populus tremula var. *davidiana*				1 o	1 f	1 f
Ulmus macrocarpa			1 r	1 r	1 o	
Betula fruticosa				1 f		
B. dahurica					11 o	
B. japonica var. *mandshurica*				11 r	1 f	
B. chinensis				11 r		
Carpinus turczaninowii	11**	11 r	1 r			
Acer mono	11 f			11 f	11 r	11 f
A. truncatum		11 r				
Tilia mandshurica	11 o			11 f	1*	
T. mongolica	11 o			11 f	1*	11 o
Malus baccata				11 r		
Prunus sibirica				11 o		
Salix phylicifolia					11 f	
Sorbus pohuashanensis					11 r	
S. alnifolia	11 o					
Pistacia chinensis		1 r				
Celtis bungeana	11 r					
C. koraiensis	11 r					
Chionanthus retusa	11 r					
Hovenia dulcis	11 o					
Picrasma quassioides	11 r					
Pyrus ussuriensis	11 r					
Ulmus japonica	11 r					

SOURCE: Wang (1961).

NOTE: 1 = trees of the crown layer; 11 = second tree layer; ** = dominant; * = abundant; f = frequent; o = occasional; r = rare.

[a] 1. *Quercus aliena–Fraxinus chinensis*; 2. *Q. variabilis*; 3. *Q. dentata*; 4. *Q. liaotungensis–Fraxinus chinensis*; 5. *Tilia–Betula*; 6. *Pinus tabulaeformis*.

TABLE 5
SPECIES CURRENTLY GROWING ON THE NORTH CHINA PLAIN

Ailanthus altissima	*Picrasma quassioides*
Albizzia julibrissin	*Pinus tabulaeformis*
Biota orientalis	*Pistacia chinensis*
Broussonetia papyrifera	*Populus simonii*
Castanea mollissima	*P. suaveolens*
Catalpa bungei	*P. tomentosa*
Cedrela sinensis	*Pterocarya stenoptera*
Celtis bungeana	*Pyrus betulaefolia*
C. koraiensis	*Quercus mongolica*
Diospyros lotus	*Q. serrata*
Fraxinus chinensis	*Q. variabilis*
Gleditschia heterophylla	*Salix babylonica*
G. sinensis	*S. cheilophila*
Hemiptelea davidii	*S. matsudana*
Hovenia dulcis	*Sophora japonica*
Juglans regia	*Ulmus japonica*
Juniperus chinensis	*U. parvifolia*
Koelreuteria paniculata	*U. pumila*
Morus mongolica	*Zelkova sinica*
Paulownia tomentosa	*Zizyphus vulgaris*

SOURCE: Wang (1961).

phytes include *Triglochin palustre, Crypsis aculeata, Halerpestes sarmentosa, Atriplex sibiricum,* and *Statice aurea.*

Most of the accessible forest in the deciduous broad-leaved region has been logged, and the further potential for timber production from natural forest is minimal. Nevertheless, the region is still important for forestry—partly because of the role that trees have to play in water and soil conservation and partly because of the long tradition of farm forestry and shelter planting in North China.

North China was the cultural center of ancient China—one of the so-called cradles of civilization. Flood control and land reclamation date back thousands of years. But many problems remain. A distinguishing feature is the extensive occurrence of loess, which is constantly relocated by wind. The loess regions cover 440,000 km^2 and are concentrated north of the Qinling, Qilian, and Kunlun mountains. The most extensive area, the so-called loess plateau, is in the middle reaches of the Huang River and extends over an area of some 300,000 km^2.

There is a more or less continuous deposit some 50 to 200 m thick, and its erodibility creates its own peculiar features in the landscape. It is calcareous material and hence gives rise to sinkholes, gullies, and vertical cliffs. The rivers carry tremendous silt loads, some 30 percent of which is estimated to travel out to sea. The steady rise in the riverbeds results in the river flowing up to 10 m above the floodplains. As well as the risk of overflow, immense problems arise because of the inability of the rivers to serve a drainage function.

Yet there is no doubt that if flood control can be achieved, North China still has potential for agricultural production—it has a good climate and the soils are not infertile. There is scope, as well, for plantation forestry in the foothills of the mountains and on the borders of the loess plateau.

The Liaodong and Jiaoding peninsulas in the east present few problems, and there are still forest remnants in the Qian Mountains (*Pinus tabulaeformis* var. *mukdenensis*, as well as the commoner oaks and mixed deciduous broad-leaved trees). Fruit growing has been developed with pears, apples, and grapes as the most important products, while tussah is produced by moths feeding on the oak trees. Arable crops are kaoliang (sorghum), maize, peanuts, wheat, paddy rice, and cotton.

The North China Plains are alluvial and of complex and variable structure. Soils are not infertile but can be highly saline. Most of the rivers (not just the Huang) have changed course frequently in the past, and there are high sand dunes that support only scrub growth. Over the region generally, the major problems of agricultural development are drought, waterlogging, salinization and alkalinization, and soil erosion. Flood control and water conservation are essential (see Chapter 7).

There is also further scope for the development of farm forestry and protective afforestation. Lowdermilk (1932) found "extensive farm woodlots of *Populus simonii* . . . cultivated and irrigated for construction material on farm land" and, in Henan, noted that "forest management is worked out to greater detail in this region than the writer found at any time in Germany." He was impressed by the fact that the North China Plain was self-supporting in timber and produced poplar and *Paulownia* logs for export. It is no coincidence that foresters visiting China are invariably shown integrated forestry development on the North China Plain—nor that the favorable conclusions contained in recent visitors' reports (including the 1979 and 1982 FAO papers) owe much to the examples provided from traditional agroforestry areas.

Hou (1983) notes that the broad-leaved deciduous forests in China differ from those in Japan and in eastern North America by the absence of beeches (*Fagus* spp.). He also includes within this broad classification two categories of microphyllous deciduous forest—that of the temperate and subtropical zones in which species of *Betula* and *Populus* are dominant (*Betula platyphylla* in the cold temperate zone and *Betula albo-sinensis* and *B. platyphylla* var. *szechuanica* in the subtropics) and some remnant pure stands of *B. ermanii* at the upper limit of the montane evergreen coniferous forests. Those of the temperate zone are dominated by *Populus euphratica* (*P. diversifolia*), which will grow in somewhat saline soil, along with *Elaeagnus angustifolia.*

Mixed Deciduous and Evergreen Broad-Leaved Forest

The Mixed Deciduous and Evergreen Broad-Leaved zone is a transition between the deciduous and the evergreen broad-leaved forest zones that forms a belt some 300,000 km^2 in extent along the Han Chiang River in the west (a tributary of the Chang) and the lower reaches of the Chang in the east. The western part is very rugged and ranges from 200 to 3500 m above sea level; the eastern region comprises an alluvial plain with hills seldom rising above 1300 m.

Hou distinguishes two categories—one on limestone soils in the subtropics and including calcium-adapted, xerophilous, deciduous trees; the other, the montane forest, in the western humid subtropics. In general the climate is milder and more humid than farther north, with mean annual temperatures from 15 to 17°C and a mean annual range of 23 to 27°C. Mean temperatures for January range from 2 to 5°C and, for July, from 26 to 30°C. The extreme minimum, however, may be as low as −14°C. The annual rainfall varies from 120 to 200 cm, only 10 percent of it falling during winter. The average frost-free season is from 230 to 280 days.

Soils are also to some extent transitional. In the hill country to the west, between the Qinlin and Daba ranges, brown forest soils of low-base status have developed from the acidic parent materials, while yellow podzolic soils are found in the mountain foothills and on the hills of the lower Chang plains. Over calcareous rocks, the resultant soils are neutral yellow-korichnevie types. Areas of alluvial soils border the lower reaches of the rivers.

Wang (1961) emphasizes two characteristics of the mixed mesophytic forest formation of this region. First, it is composed of many species, representing a large number of plant families that are not closely related. Second, no species or group of species is predominant—and in this respect the forest differs from all other deciduous forest types, resembling, rather, the tropical rain forest of the extreme south of China. The northern boundary of the region coincides with the northern distribution boundary of many evergreen species, including members of the Rutaceae, Theaceae, Ericaceae, Lauraceae, Fagaceae, and Euphorbiaceae. Together with *Pinus massoniana, Cunninghamia lanceolata, Cryptomeria japonica, Cupressus funebris,* and *Dicranopteris linearis,* they show the close affinity of the region with the evergreen broad-leaved types.

The richness of the forest flora is illustrated by the fact that within the canopy more than fifty broad-leaved and twelve coniferous genera are represented, only a few of which are localized or rare. Of those known in Europe and North America there are, in this region, over fifty species of *Acer,* nine of *Tilia,* eleven of *Carpinus,* ten of *Fraxinus,* eight of *Ulmus,* and thirty-two of *Sorbus.*

Wang (1961) points out that many genera of the deciduous broad-leaved forest type are found as small trees in the understory of the Mixed Deciduous and Evergreen Broad-Leaved zone. He lists the following:

Acer
Aesculus
Albizzia
Alnus
Betula
Carpinus
Castanea
Celtis
Diospyros
Evodia
Fraxinus
Gleditschia
Hovenia
Juglans
Kalopanax
Maackia
Magnolia
Malus
Morus
Ostrya
Paulownia
Phellodendron
Pistacia
Populus
Prunus
Pterocarya
Pyrus
Quercus
Salix
Sorbus
Tilia
Zelkova

In addition, the following broad-leaved genera are found:

Alniphyllum
Aphananthe
Camptotheca
Carya
Cercidiphyllum
Cladrastis
Daphniphyllum
Ehretia
Elaeocarpus

Emmenopteryx	*Hamamelis*	*Nyssa*
Eucommia	*Idesia*	*Platycarya*
Euptelea	*Liquidambar*	*Pterostyrax*
Fagus	*Liriodendron*	*Rhus*
Firmiana	*Litsea*	*Sassafras*
Gymnocladus	*Mallotus*	*Styrax*
Halesia	*Meliosma*	*Tetracentron*

The twelve coniferous genera are:

Cephalotaxus	*Juniperus*	*Pseudotsuga*
Cryptomeria	*Nothotaxus*	*Taxus*
Cunninghamia	*Pinus*	*Torreya*
Cupressus	*Pseudolarix*	*Tsuga*

Several coniferous species are of more than passing botanical interest, including the monotypic *Pseudolasia,* the rare *Larix leptolepis* var. *louchanensis,* and the relict *Ginkgo, Metasequoia,* and *Taiwania.* The complexity of the region is indicated by the fact that in spite of relatively intensive early botanical exploration in western parts (David, 1872–1874; Franchet, 1883–1888; Bretschneider, 1898; Potanin, 1899; Wilson, 1913; see also Cox, 1945), two of these conifers (*Metasequoia* and *Taiwania*) are recent discoveries. Indigenous conifers that are being cultivated include *Cryptomeria japonica, Torreya grandis, Cunninghamia lanceolata,* and *Cupressus funebris.* Of these, the third (Chinese fir) is the most important timber-producing species of the central and southern provinces of China. It is also a species with potential in other countries and produces a high-quality timber akin to the once renowned New Zealand kauri.

The Mixed Deciduous and Evergreen Broad-Leaved Forest type is characterized by understories, also very rich in species. Chen (1936) distinguishes five strata, including many climbers and epiphytes. The bamboos, *Phyllostachys* and *Arundinaria* spp., are important constituents now widely cultivated.

The existing vegetation, as in all accessible parts of China, has over a long period been modified by humans. The alluvial plains and the yellow-korichnevie soils of the lower Chang River are intensively cultivated, and most of the natural vegetation has disappeared. Densely populated, they form part of the traditional "ricebowl" of China. In addition to rice (double-cropped) are grown barley, wheat, cotton, maize, hemp, sweet potatoes, tobacco, tea, citrus, sugarcane (in the Sichuan basin and areas south of the river), mulberries, and bamboos. Only the protected areas around the secluded temples in the hills give an indication of the natural vegetation. On limestone, this includes the calcicoles *Zelkova schneideriana* and *Pteroceltis tatarinowii,* and such species as *Platycarya strobilacea, Dalbergia hupeana, Pistacia chinensis, Styrax japonica, Celtis sinensis, Quercus acutissima, Q. variabilis, Ulmus parvifolia, Sophora japonica, Gleditschia sinensis,* and *Zizyphus sativa.* On the yellow-podzolic soils of the Qinlin and Daba foothills, forests of *Pinus massoniana* and *Cunninghamia lanceolata* are common; also frequent are the broad-leaved species *Cinnamomum septentrionale, C. camphora, Photinia serrulata, Lindera glauca, Phyllostachys edulis, Thea sinensis, Myrica rubra, Aleurites montana,* and *Liquidambar formosana.* Brown forest soils of the western part of the area now support, among other species, *Pinus armandi, P. tabulaeformis* var. *acutiserrata, Quercus liaotungensis,* and *Q. variabilis.* This western part of the zone is growing in importance for timber production from the natural forests, and there is an almost unlimited choice of species for plantation forestry on the easier country. In addition to economic species mentioned earlier, there are lacquer trees, Chinese *Sapium,* tea, tea oil, tung oil, red bayberries, and loquats, as well as warm temperate fruit trees such as persimmon, Chinese chestnut, pear, peach, and apricot.

The area of the Chengdu Plain is of interest and had its first irrigation project in 250 B.C. It is a rich production area for silk, grain, and tobacco. The Chuanzhong basin is also noteworthy because of its sophisticated multiple cropping, which includes wheat or rape planted in winter and (before harvest) interplanted with maize, which in turn is interplanted with sweet potatoes. On the western rim of the basin is Wanfo Peak rising more than 3000 m above sea level and 2500 m above the plain. On it are preserved a number of Tertiary relict plants including *Gingko biloba, Davidia involucrata,* and *Larix potanini.*

The Chang River has considerable potential for further development of electrical power generation, navigation, and irrigation. It has a huge drainage area of 1.8 million square kilometers and accounts for nearly 40 percent of the total runoff in China. In terms of average flow, it has a water resource reserve of 230 million kilowatts—more than 40 percent of the whole country's reserve. The river and its tributaries have a navigation mileage of more than 70,000 km—70 percent of the national total. The lakes along the Chang River play an important role in regulating floodwaters (notably Dongling Lake, which is silting badly) and in supporting rich freshwater fish resources.

Ren Mei'e et al. (1985) divide Central China into two subregions and eight "areas" according to terrain. South of the western hills are yellow-brown earths best suited to tree cropping or plantation forestry. *Eucalyptus robusta* has been planted on the southern slopes of the Qinlin and Daba hills and does well, but the traditional species *Cunninghamia lanceolata* and *Paulownia fortunei* have higher value. Throughout the region, there is abundant degraded hill forest in need of rehabilitation—especially on the red earths in the east and south. The subtropical climate is suitable for oil trees (*Aleurites* and *Thea*) as well as timber.

Evergreen Broad-Leaved Forest

The Evergreen Broad-Leaved Forest formation is among the most important of eastern Asia, covering large areas of China, Vietnam, Laos, Thailand, Myanma, and Japan. In China it extends in a broad belt some 1,750,000 km^2 in area across the southern provinces and includes parts of Zhejiang, Fujian, Anhui, Jianxsi, Guangdong, Hubei, Hunan, Guizhou, Sichuan, and Yunnan provinces, the Guangxi Zhuangzu autonomous region, and most of the autonomous districts within these provinces. It even reaches the eastern counties of Tibet.

It is convenient to subdivide the formation into eastern and western subregions since, climatically and edaphically, they are somewhat different. Floristically they are very similar at the generic level and have many common species; in some genera, however, the species differ. Hou et al. (1956) list the following examples:

Eastern Subregion	**Western Subregion**
Myrica rubra	*Myrica nana*
Castanopsis hystrix	*Castanopsis concolor*
Machilus pingii	*Machilus yunnanensis*
Rhododendron simsii	*Rhododendron microphytum*
Cyclobalanopsis glauca	*Cyclobalanopsis glaucoides*
Schima superba	*Schima wallichii*
Keteleeria davidiana	*Keteleeria evelyniana*
Pinus massoniana	*Pinus yunnanensis*
Cupressus funebris	*Cupressus duclouxiana*
Cryptomeria japonica	*Cryptomeria kawai*

EASTERN EVERGREEN BROAD-LEAVED FOREST This subregion includes the great Red Basin of Sichuan (one of the most fertile areas in the world), the highlands of Guizhou (one of the *least* fertile areas in the world), the south Changjiang hills, the southeastern coast, and the highlands of Guangdong and Taiwan. The Red Basin fluctuates between 200 and 1000 m in elevation and is surrounded by mountains of over 5000 m. The Guizhou highlands range from 600 to 3000 m and are much dissected by tributaries of the Chang and Hsi rivers, but there are level areas between about 800 and 1000 m. To the south and southeast, the country is rolling and generally below 700 m in altitude, but coastal hills rise to over 1000 m.

The climate is favorable for plant growth—the mean annual rainfall of 120 to 220 cm is well distributed. Mean annual temperatures range from 15 to 21°C, with the coldest month between 3 and 6°C and the warmest 21 to 31°C. The extreme minimum seldom goes below −7°C and at least ten months of the year average more than 10°C.

The hills and mountains are mainly acidic sandstones, shales, or granite, giving rise to yellow podzolic soils merging into red podzolic soils ("red earth") in the south and east. In the Red Basin and the Guizhou highlands, extensive areas of calcareous parent materials occur, forming rendzinas, or alkaline purple soils.

The typical genera of the Eastern Evergreen Broad-Leaved Forests are the cupuliferous *Quercus, Castanopsis,* and *Pasania;* more than 150 species have been recorded in this belt and, including those of Vietnam, Thailand, and Myanmar, there are well over 250 species (Wang, 1961). The evergreen oaks occur in mixture with members of the Theaceae (*Schima, Hartia*), the Lauraceae (*Machilus, Phoebe, Beilschmiedia, Cinnamomum*), the Magnoliaceae (*Magnolia, Maglietia, Illicium, Michelia*), the Hamamelidaceae (*Altingia, Bucklandia, Liquidambar*), and others. Although dominated by a few genera, the forests are very rich in species (more so, in fact, than the mixed deciduous and evergreen broad-leaved type); nearly a thousand species have been reported as associates of the community, while the total flora of the region reaches several thousand, including members of the following coniferous genera (Wang, 1961):

Amentotaxus	*Cryptomeria*	*Fokienia*
Cathaya	*Cunninghamia*	*Juniperus*
Cephalotaxus	*Cupressus*	*Keteleeria*

Pinus	*Pseudolarix*	*Torreya*
Podocarpus	*Pseudotsuga*	*Tsuga*
	Taxus	

Cathaya is a conifer genus only recently described (Chun and Kuang, 1958). Coniferous species of economic importance associated with the Eastern Evergreen Broad-Leaved Forest include *Pinus massoniana, P. kwangtungensis, Cunninghamia lanceolata, Podocarpus javanicus, P. neriifolius, Fokienia hodginsii, Taxus chinensis, Cephalotaxus fortunei, Cryptomeria japonica, Keteleeria fortunei,* and *Tsuga longibracteata.*

Many deciduous species also occur, but there is no marked stratification in the forests. In open stands, and in secondary mixed pine and oak stands, xerophytic trees and shrubs (*Engelhardtia, Albizzia, Phyllanthus,* and others) are common over a ground cover of grasses. In the dense sclerophyllous forests there is no obvious understory, though lianas and epiphytes are frequent. Typical of the subsidiary vegetation in the lower-altitude forests of southern Hunan are *Rhododendron, Litsea, Ilex, Photinia, Pittosporum, Lindera, Sapium, Hydrangea, Actinidia,* and the bamboo *Phyllostachys.*

As in the more northern regions, much of the natural forest of this area has been cleared for crop cultivation, but secondary forest (often maintained by uncontrolled burning) occurs in patches, particularly in the west. In southern Guizhou, for example, less than 20 percent of the natural vegetation remains and the land surface is either bare or covered with sparse secondary forest. According to Yang (1962:8) it is maintained in this condition by forest fires, "unreasonable practices of tree-cutting and irrational methods of cultivation and reclamation." The coniferous component is significantly increased in the secondary forests, which comprise mixed *Pinus* species, *Cunninghamia lanceolata,* and hardwoods. *Pinus massoniana* is characteristic, with scattered white pines (*P. kwangtungensis, P. fenzeliana,* and others).

The evergreen broad-leaved forest type is among the most productive agricultural regions of China, and a variety of temperate and subtropical crops thrive. Most important is paddy rice (double-cropped), but a complete list would include tobacco, cotton, jute, ramie, mulberry and tung trees, rapeseed, peanuts, tea, oats, maize, sweet potatoes, kaoliang, sugarcane, millet, soybeans, barley, citrus fruits, and wheat. Of these, tobacco, jute, ramie, sweet potatoes, and barley are the most extensive. The culture of freshwater fish is also important, and the area is of considerable significance in Chinese plans for plantation establishment—particularly short-rotation, multiple-purpose tree crops. Apart from mulberry and tung, these crops include in this zone *Cinnamomum camphora, C. cassia, Thea sinensis, T. oleosa, Sapium sebiferum, Pterocarya stenoptera, Quercus acutissima, Castanea henryi, Aesculus chinensis, Sassafras tsuma, Cupressus funebris, Citrus sinensis, C. deliciosa, Bischoffia* spp., *Litchi chinensis, Canarium album, Camellia oleifera,* and the bamboos *Bambusa stenostachys, Dendrocalamus latiflorus, D. giganteus, D. strictus, Sinocalamus affinis, Arundinaria* spp., and *Phyllostachys* spp.

There are of course extensive bamboo forests in China, and their cultivation is highly developed in the evergreen broad-leaved vegetation zone. Among them, *Phyllostachys pubescens* is an edible bamboo that is more commonly planted on acid soils than, for example, *Sinocalamus affinis,* which is calcicolous. On marshy soils, *Phyllostachys congesta* is frequent.

South of the Chang River there are widespread deposits of red earth. Since it is acidic and low in organic matter, it is easily erodible and large expanses of wasteland remain to be reclaimed. The low hill slopes are suitable for growing tea as well as tea oil, tung oil, and citrus, while the steeper slopes are earmarked for reforestation.

Pinus massoniana is widely planted (or sown) as a production species, as is *Cunninghamia lanceolata* to an even greater extent. On a lesser scale, *Cedrela sinensis, Melia azedarach, Cupressus funebris, Cryptomeria japonica, Quercus acutissima, Acacia confusa,* and the exotic *Eucalyptus globulus, E. citriodora,* and *E. exserta* occur in plantations. Growth rates are good—*Cunninghamia lanceolata,* for example, can show mean annual increments (at age twenty years) of 25 m^3/ha, and *Eucalyptus globulus* as much as 30 m^3/ha at the same age. *Cryptomeria japonica, Cedrela sinensis, Eucalyptus citriodora,* and *Melia azedarach* are also impressive. It is not surprising, then, that the Chinese have decided to concentrate some 80 percent of their production reforestation in the region south of the Chang River.

WESTERN EVERGREEN BROAD-LEAVED FOREST The western subregion comprises most of the province of Yunnan and its many autonomous districts, as well as parts of southwestern Sichuan and western Guizhou. It is variable in topography, climate, and soils. In elevation it ranges from 600 to over 5000 m, with many high-altitude lakes and swift-flowing rivers. Much of it is difficult of access and, as a result, the natural vegetation has in many areas survived the predations of man. Significant forest resources (some of them newly discovered) are located here.

Climatically, it is generally drier than the eastern subregion, with a mean annual rainfall of 50 to 150 cm. This difference is reflected in the natural vegetation. Yang (1962) points out that the vegetation of southern Guizhou is sharply differentiated between east and west, with a more xerophytic flora (including *Schima argentea* and *Alnus nepalensis*) in the east.

The mean annual temperatures range from 8 to 23°C, with the coldest month averaging 4 to 8°C and the warmest 18 to 22°C. Eight to ten months of the year have temperatures above 10°C (except at very high altitudes). The soils include alpine meadow soils (and lithosols and tundra soils) at high altitudes on the edge of the Tibetan Plateau, purple-brown forest soils in central Yunnan, and acidic red podzolic soils interspersed with gray-brown podzols; in the east of the subregion, yellow earths predominate, while terra rossas are common over limestone.

The evergreen broad-leaved forest is again dominated by the cupuliferous genera *Castanopsis*, *Pasania*, and *Quercus*. In the north, it forms a transition to montane conifers (*Tsuga*, *Picea*, and *Pseudotsuga*); farther south, it reaches altitudes ranging from 1200 to 3000 m and is characterized by the predominance of *Quercus* species. In central and southern Yunnan, the elevation decreases and *Castanopsis*, *Pasania*, and *Cyclobalanopsis* spp. become more common. In the vicinity of Kunming (central Yunnan), seven associations have been recognized: *Castanopsis delavayi*, *Quercus schottkyana*, *Lithocarpus (Pasania) mairei*, *Quercus franchettii*, *Pinus armandi*, *Keteleeria evelyniana*, and *Pinus yunnanensis*. The conifer associations are secondary. Minor constituents of the first four associations include *Machilus yunnanensis*, *Castanopsis concolor*, *Lindera communis*, *Schima argentea*, *Pistacia* spp., *Albizzia mollis*, *Celtis bungeana*, and *Juniperus formosana*. Climbers are represented by *Hedera sinensis* and *Smilax* and *Rosa* spp.; at high elevations in central and southern Yunnan, epiphytic mosses and ferns are abundant. The forests of Yunnan are well described by Wang (1961), who, before the war, carried out considerable fieldwork in the region. Because of the presence of national minorities, the area also attracts students of ethnobotany.

Most of the evergreen cupuliferous species are useful timber trees and, where readily accessible, they have been extensively logged. The secondary forest types are coniferous (*Pinus armandi, P. yunnanensis, Keteleeria* spp.) and, particularly in the east of the region (western Guizhou), they now dominate the forest landscape. There is abundant evidence, however, that if left without interference the forests would quickly revert to evergreen broad-leaved associations—one of the problems in coniferous plantation management in Yunnan, for instance, arises from dense regrowth of the hardwoods.

The conifers are of interest. In the general area of the Western Evergreen Broad-Leaved Forest type, the following coniferous timber species are exploited:

North	**East**	**South**	**West**
Abies fabri	*Pinus armandi*	*Abies delavayi*	*Abies delavayi*
A. forrestii	*P. yunnanensis*	*Pinus armandi*	*Pinus armandi*
Picea brachytila	*Keteleeria davidiana*	*P. yunnanensis*	*P. yunnanensis*
P. likiangensis		*P. simaoensis*	*Keteleeria evelyniana*
Tsuga chinensis		*Keteleeria evelyniana*	*Torreya fargesii*
T. yunnanensis			*Taiwania cryptomerioides*
Larix potaninii			
Pinus armandi			
P. yunnanensis			

Notwithstanding logging operations and land clearance for agriculture, the western subregion of the Evergreen Broad-Leaved Forest type still contains extensive resources, which have been significantly extended by border changes agreed with Myanmar. If problems of access can be solved and communication to the north further developed, they could—together with the forests of Sichuan—play an important part in the development of northwestern China.

Apart from forestry, the accessible areas of the region are cultivated and grow many of the same crop species as the eastern subregion. With an adequate water supply for irrigation, rice can be double-cropped, and the cultivation of tropical fruits holds promise. The population of the western subregion, however, is very different ethnically from that of the east. Swidden agriculture is still practiced, and needless destruction of the forest has yet to be fully controlled. For political reasons the national minorities are less subject to control from Beijing than Han Chinese. Moreover, populations have been increased in recent years by in-migration, including retired PLA members resettling in the rural areas. Not surprisingly, agriculture is more primitive than in the east and less productive; the principal cash crops are rice, peanuts, tobacco, rape, cotton, maize, sweet potatoes, kaoliang, sugarcane, millet, soybean, and barley. Tibetan sheep and goats are grazed extensively in the high country, while cattle raising is a feature of the lower-altitude grasslands, where fire and past mismanagement have destroyed even the secondary pine forests.

China's best-known research centers for tropical and subtropical species introductions are located within the Evergreen Broad-Leaved zone—Xishuangbanna in southwestern Yunnan (near the Mekong River) and the Forest Research Institute near Guangzhou. The Xishuangbanna Institute has experimented with more than 2000 exotic species and varieties including economic trees (balsa, rubber, oil species, bamboos, rosewood, mango, coconut species, coffee), oil-bearing indigenous species (120 species from thirty-nine families have been analyzed—see Wang et al., 1982), and medicinal species. Several medicinal extracts are being developed beyond the laboratory, including antitumor agents of *Molinia* spp. Interesting ethnobotanical studies—for which Xishuangbanna is ideally situated in the center of several minorities—have examined traditional timber usage of some seventy-five tree and bamboo species; they reveal a long tradition of continuing fuelwood planting of *Cassia siamea* by the Dai people—perhaps dating back 2000 years (see Yu et al., 1982; Pei, 1985).

The institute at Guangzhou has introduced more than fifty *Eucalyptus* species. *Eucalyptus camaldulensis, E. citriodora, E. exserta,* and the hybrid *E. exserta* × *E. robusta* (=*E. leichow* No. 1) have been extensively used in reforestation of red earth soils. The North American southern pines (notably *Pinus taeda, P. elliottii,* and *P. palustris*) and *Pinus caribaea* are also being planted. Growth rates of *Eucalyptus* species, the introduced *Melia azedarach,* and *Cunninghamia lanceolata* are potentially outstanding—and, in some cases, phenomenal. At the Forest Research Institute, Guangzhou, trial plantations of *Melia azedarach* (clean-cultivated and very carefully nurtured) have reached a mean height of 8 m and a mean diameter of 12 cm in three years from establishment by sowing (and following a thinning at two years); *Eucalyptus citriodora* can average 2.5 to 3.7 m growth in height per year and reach a breast-height diameter of 25 cm in five years. Poplars, too, have shown promise on an experimental scale—*Populus robusta* putting on nearly 2 m annual growth in height and reaching a mean diameter of 20 cm in seven years. At Xishuangbanna, balsa plantations have reached 5 m in one and a half years. Clearly factors limiting the growth of the exotic production forests in southern China are likely to be edaphic and biotic, not climatic.

Tropical Rain Forest

Although rain forest forms part of the evergreen broad-leaved formation, it is distinctive physiognomically and generically. Moreover, the occurrence of laterites within the rain forest belt provides a justification for separating these associations.

The tropical monsoon rain forest type is restricted to South China and forms a narrow belt 250,000 km^2 in area along the southern edge of Fujian and Guangdong provinces, Guangxi Zhuangzu autonomous region, and the autonomous districts of Wen-Shan, Hung-Ho, Ho-Kou, Ping-Pien (all in Yunnan province), and Hai Nan on the island of Hainan (which formerly belonged to the province of Guangdong but is now a province itself). The zone also includes Taiwan, where, as on Hainan, it is restricted to elevations below 500 m. According to Wang (1961) rain forest is also found in northwestern Yunnan, within sight of the permanent snows of southern Quamdo. Hou et al. (1956) and Chang (1962), however, do not extend it north of the Tropic of Cancer in Yunnan province.

Topographically, the area lies below 500 m except for the mountain range forming the main divide on Taiwan, isolated peaks on Hainan, and the southeastern plateau country of Yunnan (800 to 1500 m). The mean annual rainfall ranges from 130 to 350 cm, and no month of the year is without rain. The winters are comparatively dry, however, with generally less than 10 cm of precipitation. It is warmer than the evergreen broad-leaved zone, with a mean temperature of 22 to 26°C, a July average of 27 to 29°C, and a January mean of 13 to 21°C. Frost is virtually unknown, even at the higher altitudes. Soils are variable and include gray-brown podzols in the hills, red earths and noncalcareous alluvial soils in the coastal belt (the latter around the river floodplains), purple-brown forest soils in southern Yunnan, and yellow laterites developed on basalt, granite, gneiss, sandstone, and other acidic materials. Chloride-containing solonchaks occur along the coast.

The tropical monsoon rain forest is almost entirely evergreen and is rich in epiphytes, lianas, climbing palms, and other parasitic plants. Wang (1961:159) describes it as "a community of communities" and points out that epiphytes alone "may far exceed in number of species the total flora of simpler types of forest, such as deciduous or even mixed northern hardwoods, of comparable area." It includes all life forms and is unique in the diversity of its flora. Wang suggests that "the tropical regions in general and the rain forest in particular have been a great reservoir of plant stocks, ancient and modern, primitive and advanced, from which or from whose predecessors all of the other plant communities are derived" (p. 160).

In view of the multiplicity of plant species in the rain forest, to offer a comprehensive description here would be beyond both the ability of the author and the patience of the reader. Certain noteworthy features of the Chinese ecotype, however, should be mentioned. First, in comparison with the rest of tropical Asia, the Chinese forests are conspicuously deficient in members of the Dipterocarpaceae, and their place is taken by genera of the Evergreen Broad-Leaved Forest zone—*Michelia, Cedrela, Chukrassia,* and *Dysoxylum.* It is also remarkable that, despite the diversity of species, the rain forest community shows considerable floristic homogeneity: Wang (1961) lists sixty genera from the canopy trees of rain forest in southern Yunnan; of these, about 85 percent are also found on the island of Hainan, though southern Yunnan is over 1000 km inland from Hainan. Of twenty-eight genera considered by Merrill (1945) to be characteristic of the primary forests occurring between continental Asia and Australia, twenty-six are represented in the Chinese rain forest. Conifers are extremely scarce in the tropical monsoon rain forest, but *Pinus merkusii* is found and *Amentotaxus* and the deciduous *Glyptostrobus pensilis* are featured. The latter, a monotypic genus, is endemic to Guangdong and Jiangxi provinces.

To some extent, Hainan Island forms a distinctive association—many species are either endemic or occur only rarely elsewhere. Among them are *Vatica astrotricha, Alseodaphne rugosa, Pentaphylax euryoides, Tarrieta parvifolia, Hopea hainanensis, Aglaia tetrapetala, Dysoxylum binectariferum, Casearia membranacea, Machilus tsanali, Cryptocarya densiflora, Xanthophyllum hainanensis, Syzygium hainanense, Schefflera octophylla, Canarium album, Quercus bambusaefolia, Q. blakei, Castanopsis echinocarpa, C. hainanensis, Lithocarpus cornea, Adinandra hainanensis,* and *Ficus championii.* The conifers include *Dacrydium pierrei, Podocarpus imbricata, P. neriifolius,* and *Pinus fenzeliana.*

As a measure of the richness of the vegetation, Ren Mei'e et al. (1985) record that the mainland of Guangdong has 1494 genera of seed plants of which 84 percent can also be found in Vietnam, 70 percent in the Philippines, and 64 percent in Malaysia. There are many trees of the following families:

Lauraceae
Moraceae
Anonaceae
Sterculiaceae
Palmae
Myrsinaceae
Dipterocarpaceae
Sapindaceae
Theaceae
Fagaceae
Samydaceae (Flacourtiaceae)
Euphorbiaceae
Myrtaceae
Apocynaceae
Sapotaceae
Rubiaceae
Leguminosae
Meliaceae
Aquifoliaceae
Rutaceae
Proteaceae

Included by Hou et al. (1956) in the tropical monsoon rain forest belt are the littoral forests (mangroves and beach forests) and the vegetation of China's coral islands. The latter vegetation is of no economic importance, but the mangroves of the swamps (primarily Rhizophoraceae) are exploited and the coastal sand dunes are being extensively used for combined protection and production (mainly fuelwood) forestry.

Existing natural forest in the tropical zone is limited to the hills of Fujian, the islands of Taiwan and Hainan, and parts of southern Yunnan. Except for Taiwan, the coastal belt is one of intensive cultivation. The major crops are rice, barley, wheat, sweet potatoes, sugarcane, mulberries (for silk), tea, and soybeans—with significant areas of millet, maize, peanuts, jute, ramie, rapeseed, and tobacco. Among horticultural crops, bananas, pineapple, litchis, grapes, mango, and tung trees are extensively grown. Pigs, water buffalo, and poultry are also raised in large numbers.

The variety of fruit trees provides favorable conditions for a range of wildlife, especially monkeys and gibbons. There is also a big population of snakes (120 to 130 species), which in turn leads to numerous snake-eating birds. Enderton (1985) lists a number of endangered (and protected) animal and bird species, including six primates and several pheasant species.

The agricultural crops can be divided into two types: equatorial crops requiring high temperatures and freedom from frost (rubber, cocoa, pepper, coconut, oil palm, and betel nut) and tropical crops with some resistance to cold (pineapple, mango, banana, papaya, coffee, and aniseed). The region should perhaps be divided into tropical and quasi-tropical zones, but it would be difficult to delineate them on the map. Coconut palms grow normally on the Leizhou peninsula and on Hainan, for example, but in the Guangzhou–Nanning area they will not bear fruit. In the tropical zone, farmers can produce three crops of rice, winter peanuts, and winter sugarcane. But conditions are uncertain—there are occasional cold spells that cause crop losses, there is a dry season, and the South China coast is swept by strong winds and occasional typhoons. The typhoon season runs from May to November; attempts are being made to mitigate damage through the establishment of shelterbelts of *Casuarina* and other species. There has been considerable degradation of natural forests in the south with the exposure of lateritic red earth, so that there is a continuing need for protection forestry.

This region includes China's biggest island, Taiwan. It has a maritime monsoon climate and rainfall varying considerably according to topography. The vegetation is of the quasi-tropical rain forest type. The upper tree layer is mainly *Castanopsis* and *Camphora.* (Taiwan used to be famous for the abundance of its natural camphor.) In the middle and lower layers are *Helicia formosana, Tirpinia formosana,* and the like. There is a shrub layer of species such as *Psidium guajava* and *Rauwolfia verticillata* and ground vegetation of tropical rain forest species. Above 500 m the rain forest runs into evergreen broad-leaved forest and subtropical coniferous forest. Characteristic species are nanmu (*Schima superba*) and *Castanopsis.* There are tree ferns, ligneous vines, and epiphytes up to about 2000 m. Mixed forest comes in from 2000 to 3000 m—deciduous broad-leaved, evergreen broad-leaved, and coniferous species. The deciduous trees include beech and maple; the coniferous species include *Tsuga formosana, Chamaecyparis formosensis, Taiwania,* and *Cryptomeria,* which are of particular value. This forest runs into subalpine coniferous species (including fir, spruce, and junipers) with an understory of rhododendron and honeysuckle. At the highest altitudes there are subalpine shrubs including azaleas and alpine meadows with *Festuca ovina* and *Parnassia foliosa.* The following conifers noted for their high-quality timber are described as "the five woods of the Ali Mountains": *Chamaecyparis formosensis, C. taiwanensis, Taiwania cryptomeria,*

Tsuga formosana, and *Pinus formosana.* Plantation agriculture and horticulture are highly developed on Taiwan.

In the rain forest zone, as in the rest of South China, plantation forestry is assuming prominence (at any rate, in terms of areas planted). The "economic" crops include *Cinnamomum camphora, C. cassia, C. kanahirai, C. micranthum, Morus alba, Citrus* spp., *Casuarina equisetifolia, Bischoffia trifoliata, Sapium sebiferum, Canarium album, C. pimela, Euphoria longana, Mangifera indica, Coffea arabica, Litchi chinensis, Zizyphus spinosa, Ficus lacor, Aleurites cordata,* and bamboos. The principal production forest species are *Pinus massoniana, Cunninghamia lanceolata, Acacia confusa, Eucalyptus exserta, E. citriodora, E. globulus, Liquidambar formosana, Melia azedarach,* and *Schima confertiflora.*

The inland regions of the tropical monsoon rain forest zone are, again, more primitive and more restricted regarding land use. Apart from natural forests, some rice, tobacco, maize, sweet potato, kaoliang, and soybeans are grown, and water buffalo and ponies are raised. Tibetan sheep are grazed in some higher-elevation areas, but, as in most of China, pastoral agriculture is primitive with little scientific base.

NORTHERN STEPPES AND DESERTS

The northern steppes and deserts run in a broad belt around the southern borders of Outer Mongolia. The steppes form a transition between the forest climax in the east and the northwestern deserts, while the latter give way, more or less abruptly, to the high mountains of the far northwest and the Tibetan Plateau to the south. The steppes do not extend south of latitude 33°N, and the western and southwestern limits of the forest climax abut onto the mountain formations along the eastern edges of the Tibetan Plateau. The tree and shrub flora of the formation is of interest in the present context because most of the species used in protection forestry in the arid regions of China have been selected from this association.

Forest Steppes

The forest steppe formations are transitional between the closed forests of eastern China and the true steppe and desert regions of the north and west; they comprise substantial areas of open woodland amid extensive and uniform grasslands. Two subregions may be recognized—that of the northeast, characterized by *Quercus mongolica* and *Betula platyphylla*, and the northwestern forest steppe typified by *Quercus liaotungensis* and *Betula japonica*.

NORTHEASTERN FOREST STEPPE This subregion covers nearly 400,000 km^2 and takes in the Manchurian Plain in the provinces of Heilongjiang, Jilin, and Liaoning (including several Mongol autonomous districts) and the Jehol highlands northeast of Beijing, which include parts of the provinces of Hebei and Liaoning and Nei Monggol. These hills form a southward extension of the Hinggan Mountains, rising to become the eastern edge of the Mongolian Plateau. The subregion lies generally below 500 m, however, except at the southwestern end, where the rolling hills give way to mountains of over 1500 m.

As might be expected, the climate is strictly continental—with very cold, long, and dry winters and short but mild summers. The mean annual temperature varies from 1.0 to 6.4°C, with an annual range of more than 40°C. January temperatures decrease from −13.7°C in the south to −26.9°C in the north, while the July mean ranges from 21.7°C in the north to 25.0°C in the south. The annual precipitation amounts to 65 cm in the east, decreasing to 47 cm farther west, and is very unevenly distributed during the year. The length of the frost-free season is little more than 100 days.

Hou et al. (1956) describe the soils as "unique." While this is not literally true, they are certainly unusual. The soils that have developed under a tree cover are either secondary podzols (gray forest soils) or leached chernozems. The secondary podzols are characteristic of the Hinggan foothills and support, as well as *Betula platyphylla*, various species of willow (*Salix raddeana*, *S. chinganica*), *Populus davidiana*, *Corylus heterophylla*, *Rosa avicularis*, and *Spiraea media*. The ground vegetation includes *Pteridium aquilinum*, *Epilobium angustifolium*, *Polygonum japonicum*, *Bupleurum dahuricum*, and *Adenophora* spp. The leached chernozems are found over basalt and carry dwarf species of the same genera. Thus willows are represented by *Salix cinerascens* and *S. mongolica*, poplars by *Populus maximowiczii*, and the genus *Rosa* by *R. dahurica*. *Quercus mongolica*, *Tilia amurensis*, and *Lespedeza bicolor* are also frequent, while *Betula platyphylla* and *Corylus heterophylla* are common to both soil types.

Soils of the grassland areas are either calcareous chernozems (with a pH as high as 8.0 to 8.5) or solonchaks. In virgin steppe country, *Glycorrhiza uralensis, Astragalus melitotoides, Aneurolepidium chinense, Stipa baicalensis,* and *Artemisia* (many species) are characteristic, with local clumps of trees—*Ulmus pumila, U. propinqua, U. macrocarpa, Populus* spp., and *Salix* spp. The solonchaks support a halophytic vegetation of *Suaeda corniculata, Achnatherum splendens, Cyperus serotinus, Saussurea glomerata, Puccinellia tenuiflora, Triglochin palustre, Nitraria schoberi, Aster tripolium,* and the like.

The woodland areas and the solonchaks have little land-use potential other than for sparse grazing and, in some areas, protection forestry. Cattle, Mongolian sheep, donkeys, yaks, horses (including the famous Hailar horse), and, in the southern part of the zone, camels are raised; but only small areas of these soils are cultivated. The grassland-climax areas, on the other hand, are almost entirely cultivated and undeveloped steppe is difficult to find. Wheat, soybeans, millet, kaoliang, oats, potatoes, and similar northern-latitude crops are common, with smaller areas of maize, sugarbeet, and barley. The Manchurian Plains are suited to large-scale, mechanized farming (if suitable precautions are taken), and the Chinese, following the Russian pattern, have established several state farms in this area. They are economic in terms of labor input—productivity is limited only by the length of the growing season, shortage of fertilizers, and occasional floods.

NORTHWESTERN FOREST STEPPE This subregion forms part of the loess highlands of Shanxi, Shaanxi, and Gansu provinces and the Ningxia autonomous region; it covers some 250,000 km^2. Topography ranges from alluvial plains along the Huang River and its tributaries to the loess plateaus (about 1000 m) and then to the mountains rising over 2000 m above them. The January temperature averages -3.1 to -11.3°C and that for July is 17.9 to 25.4°C, giving a mean annual value of 8.8 to 11.3°C. Precipitation ranges from 30 cm in the west to 60 cm in the east, approximately half of it falling in summer. Thus, in comparison with the northeastern subregion, the northwestern forest steppe has a milder winter and, generally, a wetter summer, resulting in an effective growing season of about 150 to 180 days.

Soils developed from the loess are the so-called calcareous zierokorichnevie types; they support a sparse natural flora that includes *Dicranostigma leptopodum, Rosa hugonis, R. santhina, Andropogon ischaemum, Stipa bungeana, S. grandis, Cleistogenes serotina, C. squarrosa,* and *Artemisia* spp. In the ubiquitous eroded gullies, *Zizyphus sativa* var. *spinosa, Sophora vicifolia, Hippophae rhamnoides, Prinsepia uniflora,* and *Lycium chinense* occur. True forest is found only at higher elevations under fairly humid conditions. Essentially it is deciduous broad-leaved, comprising *Quercus liaotungensis, Betula japonica, B. chinensis, Populus davidiana, Tilia paucicostata, T. mongolica, Acer ginnala, A. mono, Ulmus japonica, U. pumila, Lonicera ferdinandii, Hippophae rhamnoides, Euonymus alata, Corylus heterophylla,* and, locally, many of the species listed earlier as growing in plantations or seminatural stands on the North China Plain (p. 50). *Biota orientalis* is found in the high mountains above the deciduous forest and, even higher, *Picea asperata, P. neoveitchii,* and *Abies nephrolepis* appear.

Formerly rich pastoral country, the flat land of the northwestern forest steppe zone is now extensively cultivated and grows spring wheat, oats, maize, millet, peas, potatoes, flax, kaoliang, and beans. The ravine slopes are terraced and similar crops are cultivated, while the terrace shoulders may be planted with trees such as *Salix matsudana, Populus simonii, P. cathayana,* and *Robinia pseudoacacia.* Away from human settlement, and where soil erosion is not too far advanced, Mongolian cattle, horses, yaks, and Mongolian sheep are grazed. Unlike the northeastern subregion, however, the northwestern forest steppe is highly susceptible to erosion by both wind and water; though fertile, the soils are far from stable, and it is in part from this area that the Yellow River derived its evocative name. Together with the true steppe formation, this zone has a high priority under present-day government policy for water conservation and protection afforestation projects. The species being planted are *Elaeagnus angustifolia, Populus simonii, P. pseudosimonii, P. diversifolia* (*P. euphratica*), *Ulmus pumila, U. laciniata, Sophora japonica* var. *pendula, Robinia pseudoacacia, Haloxylon ammodendron, Tamarix juniperina,* and various species of *Artemisia.*

Within the steppe zones there are two research institutes experimenting with arid-zone species—Lanzhou (Gansu province) is the home of the CAS (formerly Academia Sinica) desert research unit, and Zhangwu (Liaoning) hosts the Liaoning Sand Fixation and Afforestation Research Institute. The Liaoning institute, established in 1952, has reported on trials of twenty-eight species (listed in Table 6). The most successful in terms of survival and growth are *Salix flavida, Lespedeza bicolor, Artemisia halodendron,* and *Caragana microphylla.* Three of these four species appear in a list of species identified for use in sand

TABLE 6
SAND SPECIES TESTED IN LIAONING SINCE 1952

Achnatherum splendens	*Echinops gmelini*
Agriophyllum arenarium	*Eurotia ceratoides*
Amblytropis multiflora	*Hedysarum fruticosum*
Amorpha fruticosa	*Inula salsoloides*
Aneurolepidum chinense	*Ixeris chinensis*
Aristida adscensionis	*Kochia prostata*
Artemisia halodendron	*Lactuca tatarica*
A. wudanica	*Lespedeza bicolor*
Atraphaxis manchurica	*Polygonum divaricatum*
Axyris amantheroides	*P. sibiricum*
Bassia dascyphylla	*Salix flavida*
Calamagrostis pseudophragmites	*Thermopsis lanceolata*
Caragana microphylla	*Trigonella korshinskyi*
Cleistogenes chinensis	*Vicia pseudocracca*

NOTE: These species were tested at the Liaoning Sand Fixation and Afforestation Research Institute, Zhanggutai, Zhangwu county.

stabilization in northwestern China compiled by the Sand Control Team of the Academia Sinica and the Soviet Academy of Sciences in 1958 and annotated in Richardson (1966). Significant industrial development is occurring in the western part of the steppe zone based on coal and oil reserves. The importance of agricultural activity (and hence of erosion control) in support of industry has yet to be fully appreciated.

STEPPES

Much of the steppe region, together with the semidesert and desert, is reclassified in Hou's 1983 scheme as temperate desert. It includes the western parts of Nei Monggol, Gansu, the Qaidam Basin in Qinghai province, and the whole of Xinjiang. For present purposes the old classification is retained.

The steppes cover an area almost a million square kilometers in extent running southwest in a broad belt along the eastern edge of the Mongolian Plateau. They also take in the northern boundaries of Hebei, Shanxi, and Shaanxi provinces, parts of Gansu, and most of the autonomous region of Ningxia. The greater part of the formation is on flat-to-undulating country between 1200 and 1500 m in elevation.

Climatically, the zone is rigorously continental, with a mean annual temperature of −1.8 to 2.5°C in the north and 2.8 to 10.5°C in the south; the July average is 20.9 to 21.6°C in the north and only slightly higher (21.3 to 24.1°C) in the south. January averages, however, are very different, reaching only −29.3 to −25.7°C in the north and −15.1 to −4.9°C in the south. Precipitation ranges from 15 to 40 cm, most of it falling in summer; the frost-free period is from 110 to 140 days.

Edaphically, the region comprises mainly chestnut soils on neutral or slightly alkaline sand dunes and solonchaks bordering parts of the middle reaches of the Huang River and in areas without external drainage. With the exception of atypical sites along watercourses (where *Populus cathayana, P. simonii, P. euphratica, Ulmus pumila, Salix matsudana, S. cheilophylla,* and the like survive) and on the stabilized sand dunes of the northeast (which *Pinus sylvestris* var. *mongolica* colonizes), erect trees are absent from the true steppe. Woody species are confined to dwarf trees and shrubs, but these are fairly abundant—particularly as the steppe approaches the semidesert and desert. They include *Artemisia halodendron, A. frigida, A. adamsii, A. desertorum, A. pubescens, A. sacrorum, A. sibirica, Caragana pygmaea, C. microphylla, Hippophae rhamnoides,* and *Ephedra* and *Atriplex* spp. The prevalence of legumes (*Glycorrhiza uralensis, Astragalus melitotoides, A. adsurgens, Thermopsis lanceolata*) is also noteworthy.

The characteristic species of the formation are grasses—in particular, narrow-leaved species, with *Stipa gobica* and *S. glareosa* replacing *S. capillata* and *S. krylovii* from east to west. *Achnatherum splendens, Stipa* (many species), *Agropyron cristatum, Aneurolepidium pseudoagropyrum, Cleistogenes squarrosa, Elymus dahuricus, Koeleria gracilis, Allium* (many species), *Aster* spp., *Bupleurum* spp., *Medicago ruthenica, Tanacetum sibiricum, Oxytropis grandiflora,* and *Potentilla* spp. are common constituents.

On the chestnut soils of the Huang River terraces, the *Artemisia* and *Allium* species frequently dominate the vegetation type; on the same soil type on the sand dunes of the northeast, however, a greater variety of shrubs occurs. Hou et al. (1956) cite the following: *Oxytropis psammocharis, O. aciphylla, Agriophyllum arenarium, Caragana tibetica, C. microphylla* var. *tomentosa, Atraphaxis mandshurica, Artemisia salsoloides, Pugionium cornutum, Salix mongolica, S. cheilophila,* and *Juniperus chinensis*—to which may be added (from Wang, 1961) *Malus baccata* var. *sibirica, Crataegus dahurica, Ribes diacantha, Papaver nudicaule, Scabiosa isetensis,* and *Veronica incana.* Many of these species are characteristic woodland plants.

The solonchak vegetation is typically halophytic and is strongly affected by relief. Where drainage is good, *Nitraria schoberi, Kalidium gracile, Scorzonera mongolica* var. *putjatae, Lepidium latifolium, Suaeda glauca, Kochia sieversana, Atriplex sibiricum,* and *Salsola collina* are found; where drainage is poor, the common species are *Phragmites communis, Typha davidiana, Scirpus maritimus, S. compactus, Triglochin palustre, Puccinellia tenuiflora,* and *Glaux maritima. Aneurolepidium pseudo-agropyrum* and *Puccinellia distans* are common grasses in the transition between the solonchak vegetation and the *Stipa–Aneurolepidium–Artemisia* steppe.

Traditionally the steppes are the territory of nomadic Mongol and Uighur graziers with some dryland farming along the great loop of the Huang River. In spite of industrial development in parts of the region, the land-use patterns have not greatly changed since the advent of socialism, though attempts have been made to increase cultivation. On the river terraces, precarious crops of millet, potatoes, flax, barley, oats, and wheat are grown and, where irrigation is practicable, soybeans, peas, and cucurbits. Attempts to create "grain bases" have not been successful. Elsewhere horses, cattle, Mongolian sheep, and camels are grazed.

The major problem with pastoral farming on the steppes is the seasonal unreliability of stock feeding due to vagaries of climate. In Outer Mongolia—where nomadic life-styles are similar—a single harsh winter or a failure of the summer rain can so decimate the herds that, without massive assistance by way of fodder and food, widespread starvation would ensue and a period of five to ten years may be needed to restore herd numbers. In Outer Mongolia help is usually forthcoming from the USSR, but China has neither the resources nor the infrastructure to respond to emergencies. Given a balance between pastoral grazing and stall feeding, the steppes could undoubtedly support higher animal numbers than at present. There is significant wildlife—including endangered species such as the snow leopard (*Panthera uncia*), the Asiatic wild ass (*Equus hemionus*), Hillier's gazelle (*Gazella subgutterosa*), possibly Gobi bear (*Ursus arctos bruinosus*), saiga (*Saiga imberbis*), Mongolian argali (*Ovis ammon*), Asiatic ibex (*Capra siberica*), possibly Przewalski's horse (*Equus przewalskii*), and African wildcat (*Felis libyca*). Other more abundant species (the Mongolian gazelle, for example) may have a cropping potential. The marmot (*Marmota sibirica*) has long been harvested for skins and meat. (Buddhism, which forbids the taking of life, made an exception to the marmot since it was believed to bear the reincarnation of evil souls.) There are, however, problems in the arid areas with wolves and foxes, the spread of stock diseases, and rodents.

Based on hunting (and the culture of minorities) there have been recent moves to develop tourism in the steppe region of Nei Monggol. As in the forest steppes, however, there can be little economic development without water conservation and protection forestry. Any kind of cultivation needs shelter from the high year-round winds, and trees and shrubs can also mitigate problems of salinization. Wang (1961:171) points out that the grassland zone has always been of great importance in the national economy and in the minds of ancient Chinese: "During those early periods the favorite national sport was polo and the national beauty . . . was not the willowy weakling but an agile and rather generously proportioned horsewoman on the rolling steppe." If China's current plans in these spheres (water conservation and protection forestry) materialize, the steppe area will be considerably transformed. And the buxom horsewoman could again become the symbol of pastoral productivity.

Semidesert and Desert

Covering more than 1.75 million square kilometers, the semidesert and desert association form the most extensive natural region in China. Administratively, the zone takes in much of the autonomous regions of Nei Monggol and Xinjiang, part of the province of Gansu, the autonomous districts of Akosai, Palikun, Mulei, Ili (all Kazakh regions), Supei, Payyinkuoleng, Poerhtala, Hapukosaierh (Mongol districts), and sundry other minority areas in Xinjiang and Qinghai. Surrounded on nearly all sides by mountains, the desert basin ranges in height from 300 m below sea level (in the Turpan depression) to some 2000 m in the high desert; the average elevation is 1300 m in the north and east, decreasing to 800 m in Tarim and to 280 m in the Junggar.

It is an area of very low rainfall and without external drainage; with a few exceptions, the mean annual precipitation seldom exceeds 10 cm and may be as low as 0.5 cm. The January mean temperature ranges from −5.7 to −19.3°C and July averages from 23.7 to 25.9°C—giving a mean annual value of 4.3 to 12.6°C. Noteworthy exceptions are the Turpan and Hami depressions (both below sea level), where the summers are hotter than in the tropics—indeed, the July mean temperature is over 33°C and the frost-free season extends 240 days. In all seasons, sand and dust storms are frequent.

Not surprisingly, the soils are unstable and characterized by a marked lack of organic matter and by high salt content; they are saline zierozem, desert solonchak, and solonetz soils. The zierozem is most common in the semidesert; the climate does not support forest growth and the sparse vegetation is limited to drought-resisting grasses and shrubs, including some halophytes. *Stipa* (many species), *Cleistogenes* spp., and *Allium* spp. are characteristic, with scrub growth of *Artemisia frigida*, *A. caespitosa*, *A. zerophytica*, *A. incana*, *Tanacetum achillaeoides*, *T. fruticulosum*, *T. trifidum*, *Salsola passerina*, *Caragana pygmaea*, *C. bungei*, *Potaninia mongolica*, and *Sympegma regelii.*

On the true desert soils, the vegetation cover is even more fragmentary. The halophytes include several species from the semidesert (*Tanacetum achillaeoides*, *Anabasis brevifolia*, *Salsola passerina*, *S. ruthenica*, *S. collina*, *Sympegma regelii*, *Eurotia ceratoides*, *Agriophyllum gobicum*, *Echinopilon divaricatum*, and *Kalidium gracile*) together with *Salsola arbuscula*, *Peganum nigellastrum*, *Nanophyton erinaceum*, *Haloxylon ammodendron*, *Calligonum mongolicum*, *Nitraria sphaerocarpa*, *N. sibirica*, *Zygophyllum xanthoxylum*, *Ephedra przewalskii*, *Alhagi camelorum*, and *Caragana leucophylla.* The solonchak soils are essentially similar to those of the steppes and support much the same vegetation.

The unbroken extent of China's desert regions is quite remarkable. The Tarim Basin, for instance, covers more than 300,000 km^2. The Tarim itself is more than 2000 km long and drains a catchment of almost 200,000 km^2 (Tong, 1947). It has no outlet and is entirely absorbed within the basin. Surrounding the central salt lake (the Lop Nor) is the vast Taklamakan Desert; its only vegetation comprises two psammophytes, *Agriophyllum arenarium* and *Corispermum hyssopifolium* (but it may support some of the few remaining wild camels in the world). The sand plain is bordered by concentric belts of fluvio-glacial gravels forming the foothills of the surrounding mountains. Within these gravel belts are found numerous oases that carry remnants of a relatively luxuriant vegetation.

Ren Mei'e et al. (1985) illustrate a typical cross section through one of the desert basins. Starting from the salt lake, the catena comprises salt beach, saline meadow, oasis, inclined desert plain, diluvial fan, running up to the high mountains. The effect of topography on microclimate is severe—there are strong winds bearing snow in winter and violent sandstorms in spring. Owing to the extensive soil erosion, there have been many attempts to establish shelterbelts around the oases.

It is, in fact, only around the oases that remnants of a tree flora are found in the desert, and here land use is so intensive that the original vegetation has been much altered. The common species indicate a deciduous broad-leaved forest type: *Populus euphratica, P. pruinosa, P. alba, P. nigra, Salix* spp., *Ulmus pumila, Tamarix chinensis, T. hispida, T. juniperina, Myricaria germanica, Morus alba, Elaeagnus angustifolia,* and *Lycium turcomannicum. Juglans fallax, Acer turkestanicum, Prunus divaricata,* and *Celtis australis* were also probably once components of the oasis vegetation. The oasis at Lolan was once (in the second century B.C.) on the main route from China to Rome and in the first century A.D. had a population of 17,000. Ruins show that the buildings were constructed from logs of *Populus alba* more than 60 cm in diameter (Wang, 1961).

The oases support virtually the entire indigenous population of the western deserts. Those of the Turpan depression, for example, have ancient and elaborate irrigation systems and even before the Second World War carried a population of over 90,000 (Hwang, 1944). The climate is such that, with adequate moisture, the oases can grow cotton, vegetables, grapes, melons, maize, millet, kaoliang, and even some rice. Maize, cotton, and kaoliang are also raised along the Tarim River at the western end of the basin, and there is a spring wheat zone at the western end of the Junggar Basin. Elsewhere, however, the western deserts support only a sparse, nomadic population of cattle, horse, and sheep graziers. The sheep are mainly Tibetan, while the horses include the once famous Ili race. Camels are raised, as well, and on the desert fringes there is a surprisingly extensive native fauna of gazelles, wild asses, and antelopes. The scattered inhabitants of the region are predominantly Buddhists and hunting used to be rare; as a result, these animals were little disturbed. In view of the extensive industrial development that is taking place along the northern borders of the Tarim Basin, in the foothills of the Tian Mountains, and in the Junggar, this situation is changing. Extensive resources of oil, coal, iron, copper, molybdenum, uranium, and other minerals have been discovered and are being exploited.

The northern deserts contain few oases and—apart from the Ningxia Plain (which, strictly, falls entirely within the steppe region)—are virtually uncultivated. Land use is limited to the pasturage of horses, Mongolian sheep, camels, Mongolian cattle, yak, and cattle/yak hybrids; productivity is not high.

The fringes of the northern deserts and the communication routes to the northwest have a special importance in China's protection afforestation projects. Attempts to border the Gobi Desert with shelterbelt plantings have been widely publicized. Less well known (but equally herculean) is a project designed to protect the northwestern railways from the ravages of wind-blown sand. These endeavors have entailed planting trees on very inhospitable sites, often with species far removed from their natural habitats. Extensive plantings have been made along riverbanks in a corridor arrangement, and the belts are known as "corridor woodlots." The principal species is *Populus diversifolia*—the Euphrates poplar—which can grow to some 10 m in height but with a canopy density below 30 percent.

The Junggar Basin in the extreme northwest (north of Urumqi) has a higher rainfall than most desert areas and supports a scrub vegetation—surprisingly, because of the warm summer temperatures and snowmelt, it is possible to grow cotton and irrigated fruit. Moreover, the combination of high-intensity daylight and low night temperatures produces fruit (such as grapes and melons) with a high sugar content.

There is an unexpectedly rich fauna in the deserts—from rodents to carnivores. It includes various jerboas (*Dipus* spp.) and gerbils (*Rhombomys optimus*), donkeys, gazelles, antelopes, and argalis, the virtually extinct wild camel, wolves (*Canis lupus*), the lynx, the fox (*Vulpes caragera*), steppe cat (*Felis manul*), and tiger weasel. Following developments in Outer Mongolia enabling tourists to hunt desert ungulates, there are proposals for similar moneymaking activities in China.

In the pastoral areas, rodents pose problems. In particular, there is a high population of voles and, where they are present, it is estimated that pastures will support two fewer sheep per hectare. Attempts at mass control of rodents have not been successful. In the Inner Mongolian grassland regions, areas cultivated for cereal cropping have suffered insect damage as well as destruction by rodents.

There is no doubt that with efficient management of the steppe and desert resources (including further development of artesian water supplies and access to emergency stock feed) these vast areas of China could be made more productive, particularly in pastoral use. Overgrazing, especially around oases, has reduced range productivity and transformed good pasture into a wasteland of hardy, poorly palatable, plants. Grazing is, of course, free range and the nutritive value of winter feed is less than half that of spring growth (see, for example, Petrov, 1979). Controlled, rotational grazing, the provision of winter protection, and emergency fodder could undoubtedly increase the region's stock-carrying capacities.

MOUNTAINS AND PLATEAUS OF THE WEST AND SOUTHWEST

The western montane forests of China comprise two subdivisions: the northwest ranges, supporting an extension of the northern coniferous forests, and the forest zones of the Qinghai–Tibetan Plateau, nearly 40 degrees of latitude farther south. They contain several genera of interest to foresters, and it is in the south that substantial timber resources have been discovered recently and are now being exploited.

MOUNTAINS OF NORTHWEST CHINA

The region covers 400,000 km^2 and comprises three unconnected mountain ranges—the Qilian Mountains, marking the boundary between Gansu province and Qinghai autonomous region and forming the northeastern edge of the Tibetan Plateau; the Tian range to the north of the Tarim Basin; and the Altay, north and northwest of the Junggar Basin. In all three, climate and soils show marked zonation with altitude. The mean annual rainfall ranges from 10 to over 50 cm; the January temperature from −48 to −8°C; the July temperature from 10 to 22°C; and the mean annual temperature from −4 to 10°C. From the desert soils and the steppe zierozems, the soils change with altitude through the following sequence: chestnut, gray forest, and, above the timberline, alpine meadow soils. In altitude, the Qilian and the Tian mountains reach over 5000 m (the Tian, indeed, rise to over 7000 m); the Altay range, on the other hand, is nowhere higher than 4000 m.

The forest zone of the Qilian Mountains ranges from 2200 to about 3800 m. On northerly slopes, *Picea asperata* occurs in pure stands with an understory of *Potentilla, Caragana, Spiraea, Cotoneaster, Lonicera,* and *Berberis* spp. Other conifers are found rarely, including *Picea asper-*

ata var. *heterolepis, P. likiangensis* var. *purpurea, Abies faxoniana,* and *A. crassifolia.* On southern slopes, *Juniperus tibetica* occupies the zone between 2700 and 3300 m, with *J. saltuaria* invading the alpine meadow zone for a further 300 to 400 m in altitude. Below about 2800 m, *Pinus tabulaeformis* var. *glacilifolia, P. tabulaeformis* var. *leucosperma,* and *Ulmus pumila* may occur. Wang (1984) notes the destruction of vegetation in the Qilian Mountains in recent years and urges the need for protection.

The dominant species of the north-facing slopes of the Tian range is *Picea schrenkiana,* occupying chernozem soil types between 1500 and 3000 m. The trees reach 75 m in height and 1.5 m in diameter, yielding where accessible an excellent quality timber. The southern slopes at this altitude are grass-covered; occasional clumps of deciduous forest comprise *Juglans fallax, Pyrus malus, Acer turkestanicum, Prunus divaricata,* and *Celtis australis.* Beneath the spruce forest grow *Populus tremula, Salix* spp., *Betula pubescens,* and *Sorbus tianschanica,* ready to form a secondary broad-leaved forest if the spruce is destroyed (as it often is for summer pasture).

Above the timberline, the alpine meadows may be quite moist and even swampy. Wang (1961) lists the following species: *Erigeron pulchellus, Gentiana algida, Lagotis glauca, Papaver nudicaule, Pedicularis cheilanthifolia, Saussurea involucrata, Saxifraga hirsuta,* and *Seseli athamanthoides. Juniperus semi-globosa* is also common on both north- and south-facing slopes. Wang (1984) lists species by altitudinal zones, including Tertiary relict species of walnut, apple, and apricot. Again, he reports serious degradation of these communities and the need for protection.

In contrast to the mountains of the Qilian and Tian ranges, the Altay are richer in tree species (including many more species of *Abies* and, also, *Rhododendron*) but lack *Juniperus* as a subdominant species. The primary components of the coniferous forest zone (2000 to 3000 m) are *Larix sibirica, Picea obovata, Abies sibirica, Pinus cembra* var. *sibirica, Betula verrucosa,* and *Populus tremula.* Below the forest, on chestnut soils and chernozems, the vegetation includes *Artemisia, Potentilla, Spiraea, Caragana, Astragalus,* and other steppe genera; above it lies a transitional zone of *Betula nana, Cotoneaster uniflora, Lonicera hispida, Ribes fragrans* var. *infracanum,* and *Salix* spp., leading to alpine meadows rich in such species as *Aquilegia glandulosa, Viola altaica, Anemone narcissiflora, Callianthemum rutaefolium,* and *Ranunculus altaicus.*

Land use in the northwestern mountains is restricted to some forest production and grazing. As mentioned earlier, where the forest is accessible it is being exploited and sometimes converted into summer pasture for cattle, sheep, and horses—particularly in the Tian Mountains, where the pastures are within reach of the oasis farmers of the Tarim Basin.

The grasslands on the slopes of the high mountains provide good seasonal grazing but, again, pastoral farming has to be supported in isolated regions by access to stock feed. According to Ren Mei'e et al. (1985) there are some 14 million hectares of reclaimable land in northwestern China and another 10 million hectares in Xinjiang. The paramilitary Production and Construction Corps of Xinjiang, with membership reportedly over one million, is attempting to develop northwestern China as one of the "last frontiers" (see Dreyer, 1986). The problems are as much organizational as they are technical.

The Qinghai–Tibetan Region

The Qinghai–Tibetan (Xizang) Plateau is young, huge, and exerts a profound influence on climate. It covers one-fifth of the total area of China and—despite unfavorable natural conditions of high elevation, cold and dryness—it contains China's biggest reserves of natural forest outside of the northeast and includes pastoral areas greater in extent than Nei Monggol and Xinjiang. The forest zone covers some 300,000 km^2 and includes parts of the provinces of Yunnan and Sichuan and several of the Tibetan autonomous districts (Yushu, Kuolo, Apa, Kantzu, Tiching, Muli) and other minority regions (Qamdo, Kungshan, Lichiang, Nuchiang, Tali, and Ninglang). The general level of the plateau is about 4000 m, but it is deeply incised by river gorges of the Chang, Mekong, Salween, Irrawaddy, and Brahmaputra rivers and their tributaries. Furthermore, the high mountain ranges rise to more than 8000 m, and this variation in relief (together with climatic differences) results in a very rich, if complex, flora containing many species endemic to the region.

The mean annual temperature ranges from 4°C in the north to 16°C in the south, with July and January values of 16 to 20°C and 2°C respectively. The annual rainfall varies from less than 20 cm on the Tibetan Plateau to about 120 cm in the southeast of the region. The length of the growing season varies from 90 to 150 days, but rarely is there a period of more than two months that is completely free of frost.

The soils are predominantly podzolic, merging with alpine meadow soils (including lithosols and tundra soils) at high elevations.

The montane coniferous forests of eastern Tibet comprise many species of *Picea* and *Abies,* with occasional admixtures of *Larix* (*L. potaninii* and *L. griffithii*) and *Tsuga* (*T. chinensis, T. yunnanensis,* and *T. dumosa*). The *Picea–Abies* mixture, however, is not a simple one, and many species and varieties are found varying in dominance in different parts of the zone. Their distribution is shown in Table 7. In a recent study of the subalpine forest of western Sichuan—which forms part of the Qinghai–Xizang Plateau—Yang (1986) notes the large number of fir and spruce species, including ten species of *Abies.* He suggests that this edge of the plateau may be the original distribution center for fir and spruce—indeed, he notes 80 formations and 250 forest types, divided on the basis of topography and microclimate. He claims the average growing stock per hectare to be between 300 and 400 m^3; the forests are overmature and the trees of large size. These forests are at risk—in a number of counties, the forest cover has dropped from 20 percent in the 1950s to half that figure at present. Regeneration after felling is difficult because of the high diurnal temperature changes (up to 27.5°C) and high soil surface temperatures (more than 51°C). Natural regeneration is damaged by excessive temperatures, and planted seedlings die from frost heave. Yang suggests that the clear-cut blocks should not exceed 3 ha and that there is a need—in the confusing omnibus phraseology of the ideology—"to take silviculture as a foundation, putting management of the protection forest first" (p. 15).

Although in general the montane coniferous forest is open, the trees grow to a large size: Heights reach 50 m and one-third or more of the trees are greater than 10 cm in diameter in the 51 to 90 cm classes (Kuo and Cheo, 1941). The *Picea–Abies* forest reaches its best development on north-facing slopes above 3500 m. Below it, several species of pine (*Pinus tabulaeformis* var. *densata, P. armandi,* and *P. yunnanensis*) and *Acer–Betula* forest occur sandwiched between the montane coniferous species and the evergreen broad-leaved forest at lower elevations. Above the timberline, *Rhododendron* species often form a transition to alpine scrub and meadow plants. The timberline varies from 3800 to about 4300 m. The south-facing slopes are often characterized by evergreen oaks and junipers, including, in the northeast, the dwarf species *Juniperus lemeana, J. przewalskii, J. zaidamensis, J. formosana,* and *J. tibetica.* Regional segregates of the formation are summarized by Wang (1961).

TABLE 7

DISTRIBUTION OF SPRUCE AND FIR IN THE MONTANE CONIFEROUS FORESTS

Species	East	Northeast	Southeast	Southwest
Picea asperata	x	x		
P. asperata var. *heterolepis*	x	x		
P. asperata var. *retroflexa*	x		x	
P. aurantiaca	x			
P. likiangensis	xxx	x		x
P. likiangensis var. *balfouriana*	xxx			
P. likiangensis var. *hirtella*	x			
P. likiangensis var. *montigena*	x			
P. likiangensis var. *purpurea*		xxx		
P. sikangensis	x			
P. brachytila	x	x	xxx	x
P. complanata	x	x	x	x
P. spinulosa				x
P. neoveitchii		x		
Abies fabri	x		xxx	
A. delavayi	x			xxx
A. chensiensis	x	x		
A. forrestii				x
A. yuana				x
A. faxoniana	x	xxx		
A. georgei	xxx			
A. recurvata	x	x		
A. squamata	xxx			
A. webbiana				x
A. fargesii var. *sutchuenensis*		x		

SOURCE: Wang (1961).

NOTE: xxx = primary forest constituent; x = minor constituent.

The ground vegetation in these forests is sparse, and the forest floor is covered only by mosses and shade-bearing shrubs (*Berberis, Ribes, Cotoneaster, Rhododendron,* and the like). Occasional bamboos are found in the south.

Timber production has recently become a major industry in the forested areas. Logs are extracted by floating on the fast-flowing rivers into Yunnan and Sichuan—but they arrive at utilization centers considerably damaged and, inevitably, logging is limited to species that float. There is some hunting and grazing of yaks, Tibetan goats, and sheep, supported by agricultural crops in the valleys of cold-resistant rape, barley, and roots. Cereal yields are claimed to be exceptionally high (wheat up to 12 tons per hectare and 7.5 tons per hectare for highland barley, according to Ren Mei'e et al., 1985). The forage species are *Kobresia* and annual bluegrass. Certainly the plateaus of Tibet can support a significantly higher population than at present: They are not uniformly harsh; the quality of pasture is good; there is high insolation; and the forage grasses, though sparse, are nutritious. Valley soils are fertile and could be used to grow supplementary stock feed. Moreover, there are possibilities for the development of "*muka*" animal fodder—the product of dried and pulverized coniferous foliage that is used in the USSR and, in pilot projects, in Canada.

Tibetan Plateau

Of little importance in Chinese forestry, the Tibetan Plateau is a lofty and inhospitable habitat covering 1.75 million square kilometers in the autonomous region of Xizang. The average elevation is generally from 4700 to 5300 m, and the plateau is bounded by even higher ranges—the Himalayas to the south and the Kunlun to the north. The climate is cold and semiarid, with severe winters and low summer temperatures. Chang (1962) gives mean annual temperatures of 4 to 8°C, with mean January and July values of −5 to −16°C and −6 to 10°C respectively. The higher figures relate solely to the southeastern edge of the region along the upper reaches of the Brahmaputra. In this area the annual rainfall amounts to over 200 cm, but in the plateau interior it is never more than 30 cm and is often less than 10 cm. The length of the growing season at Lhasa (which has a milder climate than the interior) is 148 days (Wang, 1961).

Forest is developed to some extent in the southeast of the region with evergreen broad-leaved species (*Tilia* spp., *Magnolia* spp., and *Quercus* spp.) in the lower valleys and dense montane coniferous associations higher up. Below the snowline at 4500 m is a narrow belt of alpine vegetation (including *Betula*, *Acer*, and *Rhododendron* spp.) above a coniferous zone made up of *Abies webbiana*, *Picea likiangensis*, *Larix griffithii*, and *Picea spinulosa*. From 3000 m down to 2400 m, *Tsuga dumosa* is dominant. The soils of the coniferous communities are gray-brown podzols and merge, on the plateau itself, with alpine meadow and desert soils. Here the drainage is internal and the vegetation halophytic.

The common species are *Eurotia ceratoides, Tanacetum tibeticum, Artemisia wellbyi, Capsella thomsoni, Cheiranthus himalayensis, Myricaria prostrata, Astragalus* spp., *Thermopsis inflata, Allium senescens, Ephedra gerardiana, Stipa stenophylla,* and *S. purpurea.* In wetter areas, *Ranunculus tricuspis, Triglochin palustre, Juncus thomsoni,* and *Puccinellia distans* occur.

The vegetation of the outer edge of the plateau is rather different. Along watercourses, species of *Juniperus, Salix, Ulmus, Populus,* and *Hippophae* occur, while in the semidesert there is a shrub flora of *Berberis* spp., *Sophora moorcroftiana, Caragana tibetica, C. jubata, Potentilla fruticosa, Salix biondiana,* and *Spiraea, Rosa, Cotoneaster,* and *Lonicera* spp. The meadow soils, on the other hand, support *Kobresia, Carex, Eriophorum, Helictotrichon, Deschampsia,* and *Poa* spp. among a cosmopolitan variety of herbs and shrubs.

The interior of the Tibetan Plateau supports a small nomadic population of graziers and their herds of yaks, goats, cattle/yak hybrids, ponies, and sheep. Herds of wild asses, gazelles, and antelopes are also found, as in the desert fringes farther north, together with Himalayan blue sheep (bharal), donkeys, rodents (voles, marmots, hares), and their predators. The white-lipped deer (*Cervus albirostris*) and the Tibetan antelope (*Pantholaps hodgsonii*)—said to have been the prototype for the unicorn—are endangered. In contrast to the dry interior, the moist-temperate southeastern edge of the region contains numerous permanent pastoral and agricultural settlements such as Lhasa, Gyangtze, and Shigatze. As well as raising yak, yak/cattle hybrids, horses, sheep, goats, and camels, farmers grow crops of maize, millet, barley, tobacco, and tea. Yields are not high, but they are adequate for the low-density population. The forests of this area are not exploited industrially to any extent but, as with the east Tibetan border resources, the development of communications and improved extraction methods may change this picture. One hopes that the recent developments with respect to resettlement, construction, and tourism will not result in exploitative harvesting—a very real danger in view of the demand for firewood that is generated by such changes.

Wang (1984) points out that because of the sparse population in the western uplands of China, it would be easier than in the east to establish (and monitor) protected reserves. Moreover, the harsh climate makes recovery from degradation extremely slow—adding urgency to the need for an overall rational plan.

The Lhasa River Valley is the site of the first reforestation project in Tibet to be assisted by the World Food Program (WFP) of the United Nations. It is part of an irrigation development and aims to establish multiple-purpose belts of trees (for shelter, fuelwood, and canal bank stabilization) across southeast–northwest valleys of the river system. The species are poplars, willow, and nitrogen-fixing species such as *Hippophae rhamnoides*, *Sophora vicifolia*, and *Amorpha fruticosa*. Irrigation and protection from grazing are essential, so costs will be high. The economic benefits, on the other hand, will be substantial—fuelwood is currently trucked into Lhasa from a distance of up to 600 km and sold (on the free market) at a price per cubic meter that exceeds the average annual per capita income in the project area. Moreover, if successful, the shelter will increase both cropping and pastoral areas.

CONCLUSIONS

Vegetation classification in China has grown in realism since the 1960s with the beginnings of land-use planning and the acceptance of "cultural vegetation types." Because so much of China is either marginal for plant growth (deserts and mountains) or greatly modified by human beings, the use of a natural (climax) vegetation system gives a false impression of the extent of forest cover. Although this system affords an insight into history—and the evidence of species survival may provide a valuable basis for making decisions about future land development—alone such a classification is of limited interest. There is a greater awareness of these limitations than twenty-five years ago.

At the same time, there is also greater realism with respect to land-use potentials—as is evidenced by the overall reduction in cropping area, the return of steppe land to pastoral and forestry uses, and the return to traditional areas for food staples. By contrast, there has been some expansion of tree-planting into nontraditional areas and considerable extension in the geographic ranges of some "economic" tree species through agroforestry. What is not clear, however, is the success of marginal land planting and, in particular, the extensive aerial seeding of mountain lands that is reported.

There is now an environmental protection agency in China (see Chapter 7) working to an Environmental Protection Commission of the

State Council. A major focus is the development of an "agroenvironmental" policy. It is concerned about timber cutting in overmature forests (particularly in the southwest where temperature extremes prevent regeneration in rapidly opened areas) and with the increased soil erosion that may result from the expansion of cash cropping under the new incentive system (the downside of decentralization). Elsewhere, national forest destruction through insect damage and fire (which may be related) is extensive (more than 1 million hectares a year) and sufficiently concentrated to prevent regeneration. There may be permanent changes in vegetation cover that, one hopes, will be detected and recorded by the modern survey techniques now available.

There is still an unrevealed wealth of genetic stock in China—including some ancient gene pools for which the world may yet have cause to be grateful. Conservationists need vigorous support if management and land-use practices are to be brought into line with policies. The involvement of the international agencies (notably the World Bank and the World Food Program) in reforestation activities in China is a hopeful sign. By providing alternative wood resources, pressure on the natural forest remnants may be relieved.

3

THE FOREST ECONOMY

LENIN CLAIMED that there could be a transition from feudalism to socialism, bypassing capitalism. This idea was espoused by the Communist parties of the USSR and its first satellite, Mongolia. Although China was never quite so committed to the total abolition of mercantilism, the PRC has adopted many of the trappings of central planning and economic controls. In 1963, the discipline of the regime was apparent. It is less so now and, in its place, a mixture of empiricism and ideology characterizes decision-making. And underlying that decision-making are the complex personal motivations of the players on the stage. Anthony Lawrence (1986:35) writes: "China is a labyrinth of human relationships; factions weaving behind the scenes; an international network of mutual obligations; old men hanging on to perquisites; long-term revenge seekers waiting in the wings. These are the mainsprings of action." His judgment may be extravagant, but the depiction of human frailty is reassuring.

The organization of economic sectors in China is a combination of Soviet-type vertical control systems and "horizontal cleavages" (Wong, 1986) that are administratively demarcated at state (central), provincial, municipality, and county levels. The system results in fragmentation and a pigeonhole mentality that is said to distinguish bureaucrats the world over. Many visitors complain that officials in their sponsoring department are reluctant to arrange visits to closely related organizations in other systems. Even in research, scientists in (for example) the Academy of Forestry have little knowledge of work being done in their own fields in the Academy of Sciences (CAS). China faces no bigger problem than that posed by the restricted vision of its bureaucrats.

The multiple and pervasive roles of forestry in the economy—and its long gestation period—make it more vulnerable than most disciplines to haphazard planning. Forestry contributes to both rural and industrial economic sectors; it has horizontal as well as vertical linkages; it produces raw materials as well as products and goods as well as services. Agricultural productivity depends on water conservation and erosion control—and hence on catchment afforestation and shelter planting. Timber resources are vitally important in the development of communications and industry, which must provide a base for agriculture, and wood is the most widely used source of energy. Hierarchical planning and control systems that derive from Soviet practice are the least appropriate models for forestry in China. Before examining the way the sector works, however, it is necessary to set the scene by outlining what is now widely acknowledged to be a severe supply/demand crisis with respect to industrial wood in China and one that, for the first time in recent history, is being openly discussed (and the solutions argued) in technical journals and, even, the foreign-language press.

Historically China has been short of wood resources—indeed, the fact that Chinese civilization has survived the scarcity of forests is a matter for surprise and even puzzlement. Adshead (1979) refers to the widely held view that, in all traditional economies, wood was a universal raw material and a key energy source: The inability to sustain inputs of wood for construction, shipbuilding, and fuel was a major factor in the medieval decline of Islam. Yet the Chinese depended on wood for construction and shipbuilding much more than the Muslims. He points out that in 1800, some 200 million Europeans used 120 million tons of wood a year while 300 million Chinese consumed only 10 million tons! A partial explanation of the anomaly may lie in the highly productive, intensive tree cultivation that was practiced in parts of China and the thrifty utilization of products which enabled a limited resource to be stretched—in contrast to the more extensive (and wasteful) natural forest management of Europe and West Asia. It is also likely that the difference is exaggerated by incomplete records for China.

THE FOREST RESOURCE

A continuing problem in China lies with statistical definition: Where afforestation is quantified in terms of "numbers of trees planted" rather than "areas established"—and survival is taken for granted—distortions soon multiply. The wild claims during spring and autumn afforestation campaigns in the 1950s and 1960s—with few references to failure—destroyed China's credibility among visiting foresters who looked beyond the showplace stopovers. Statistics of forest areas deriving from "percentage cover of land surface" lack precision. Destruction by fire (which exceeds 1 million hectares a year), insect damage, and so-called "grassland reclamation" must substantially qualify productive area statistics.

Moreover, definitions change. On 28 April 1986, the State Council promulgated regulations for the implementation of the Consolidated Forest Law (*Beijing Review*, 9 June 1986). "Grassland and woodland with 30 percent canopy cover" were in future to be defined as forest resources (formerly 40 percent canopy cover was so defined); this measure, by a stroke, increased the forest land area from 273 to 287 million hectares. (This area probably includes Taiwan.)

My 1965 commodity review (Richardson, 1965) presented a range of published estimates of China's forest area and the volumetric resource. Areas ranged from 47 to 100 million hectares and the standing volume from 4.6 to 7.5 billion cubic meters. Despite regular surveys since then, and the availability of sophisticated remote sensing and mapping technologies, it is doubtful whether much greater precision has been achieved with respect to an estimation of *accessible* and *exploitable* areas and volumes. The most complete—and widely quoted—data until recently were based on a 1976 survey and are summarized in the Japanese "Chugoku Nogyo Chiri Soron" (CNCS); they relate only to "land under forest," not total forest land.

The Ministry of Forestry (MOF, 1985) published *Statistics of China's Forestry*, based on later surveys, which indicated "land for forestry use" at 261.0 million hectares of which 110.1 million was "forested." In addition, there were 2.7 million hectares of "shrub" land, 1.7 million hectares of "open forest," 0.5 million hectares of "unestablished" forest, and 101 million hectares "unforested." Comparable data are presented in Table 8. The principal changes over the period 1976–1981 are an 18

TABLE 8
FOREST RESOURCES

Resource	CNCS (1976) (1000 ha)	% Land Area	MOF (1985)[a] (1000 ha)	% Land Area	Standing Volume (million m^3)	Volume/ Hectare (m^3)
Timber forests	98,000	10.2	80,630	8.4	6,881.9	85
Protection forests	7,850	0.8	10,000	1.0	883.7	88
"Economic" plantations[b]	8,520	0.9	11,280	1.2	n.a.	—
Bamboo	3,150	0.3	3,200	0.3	n.a.	—
Fuel forests	3,670	0.4	3,690	0.4	69.6	19
Special-use forests[c]	670	n.s.[d]	1,300	0.1	143.2	110
Total	121,860	12.7	110,100	11.4	7,978.4	

NOTE: *China Agriculture Yearbook* (1985) gives a figure of 115,250,000 ha for 1984 "forest area."

[a] Surveys were carried out between 1977 and 1981.

[b] Horticultural and multipurpose.

[c] Scientific reserves and the like.

[d] Not significant.

percent reduction in timber forests (3.6 percent annually) and an increase in area of all other categories by 23 percent. The net reduction in area is 11.8 million hectares.

The volumetric resource is also uncertain. The CNCS report distinguished "primary," "secondary," and "man-made" forests (see Table 9)

TABLE 9
TIMBER VOLUMES BY FOREST TYPE

A. STATISTICS OF CNCS (1976)

Forest Type	Standing Volume (million m^3)	Volume/Hectare (m^3)
Primary	7125	189
Secondary	2185	39
Man-made	190	7
Total	9500	

B. STATISTICS OF MOF (1985)

Forest Type	Standing Volume (million m^3)	Volume/Hectare (m^3)
Shelter forest	883.7	88
Timber forest	6881.9	85
Fuel forest	69.6	19
Special-purpose forest	143.2	110
Total	7978.4	

NOTE: *China Agriculture Yearbook* (1985) cites 10.26 billion m^3 standing volume for 1984.

and also presented data by age class. Mature and overmature forests make up 65 percent of total volume, middle-aged and mature represent 28 percent, and young forests constitute 7 percent. Conifers constitute 56 percent of the total. The MOF standing-volume figure (8 billion cubic meters) is 1500 million below the 1976 value and is close to that given by MOF in 1963 (7.5 billion cubic meters; see Richardson, 1966), which at that time exceeded statistics published outside China by some 20 percent.

To put these data into perspective, comparable forest area and volumetric data for a selection of countries are illustrated in Table 10. Only Bangladesh, Pakistan, and Israel have lower areas (per capita) of closed forest. China's forest *land* area (per capita), on the other hand, is relatively high (14th out of 22).

TABLE 10

FOREST RESOURCE STATISTICS OF SELECTED COUNTRIES (IN 1000 HA)

Country	Total Forest Land Area	Coniferous Closed	Non-coniferous Closed	Other Forest Land	% Land Area	Per Capita Closed Forest (ha)	Per Capita Forest Land (ha)	Population (millions)
Australia	106,743	5,151	36,507	65,085	13.9	2.69	6.89	15.5
Bangladesh	2,219	626	1,544	—	47.0	0.02	0.02	99.0
Brazil	553,030	13,520	38,510	157,000	65.0	0.38	4.07	136.0
Canada	438,400	204,255	59,845	174,300	43.7	10.44	17.33	25.3
Chile	8,680	1,230	5,020	2,430	11.5	0.53	0.75	11.9
Finland	23,225	18,294	1,591	3,340	68.9	4.06	4.74	4.9
France	15,075	4,643	9,232	1,200	27.6	0.25	0.27	55.3
Germany (DDR)	2,985	2,080	620	285	27.6	0.16	0.18	16.7
Hungary	1,637	227	1,385	25	17.6	0.15	0.15	10.6
India	77,914	4,937	67,584	5,393	23.7	0.10	0.10	748.0
Indonesia	126,235	4,280	118,995	3,000	66.3	0.71	0.73	173.0
Israel	100	55	25	20	4.8	0.02	0.02	4.2
Japan	25,280	12,000	11,890	1,390	67.9	0.20	0.21	120.8
Mongolia	15,000	8,500	1,500	5,000	9.6	5.55	8.33	1.8
Myanmar	32,101	116	31,985	—	47.4	0.91	0.91	35.3
Nepal	2,308	433	1,695	180	16.4	0.13	0.14	16.5
New Zealand	9,500	1,080	6,120	2,300	35.4	2.18	2.88	3.3
Pakistan	4,080	1,325	2,460	295	5.1	0.04	0.05	88.0
Thailand	16,815	220	10,155	6,440	32.7	0.20	0.33	50.8
U.S.	298,076	83,537	111,719	102,820	31.8	0.83	1.26	236.6
USSR	928,600	593,700	197,900	133,000	41.5	1.19	3.33	278.7
China	261,000	110,100		150,900	12.7	0.11	0.25	1033.0

SOURCES: FAO (1985) for countries other than China; Statesman's Yearbook (1985) for population data.

China's wood consumption statistics distinguish production under the State Plan (about 70 million cubic meters annually), industrial wood used outside the plan (a similar volume), firewood (which may be anything from 50 to 100 million cubic meters), and illegally felled timber. Together with losses from fire, pests, and diseases, this last category may exceed 100 million cubic meters (NFPA, 1986). From Table 9 it is clear that in the near future, the natural forest, supplemented by imports, will have to supply most industrial roundwood needs except perhaps for pulpwood.

In natural and aerially seeded areas of China's forests, accessibility is another unknown: USDA/FAS (1986) postulate that only 33 percent of reserves are accessible; in 1963, I was given a figure of 75 percent by MOF and have not seen or read anything to persuade me that the ministry was wrong. If 75 percent was accessible in 1963, of course, the percentage will be less now—but in any case higher than 33 percent.

Natural Forests

The Chinese perception of what is available in commercial volumes is indicated in Table 11, summarizing the MOF's 1981 inventory by species. Since the forests are all mixed—whether predominantly coniferous or broad-leaved—areas of individual species have little significance; moreover, value depends on the perceptions of the end user. For example, the assumption by NFPA (1986) of declining quality as the softwood (coniferous) percentage drops would not be shared by all foresters—or even timber millers. Walnut, ash, and cork, the "three precious trees" of Manchuria—linked together in Table 11—are all hardwoods, while primary forest oak has a market value higher than most softwoods. Moreover, although in the northeast forests the softwood component of regeneration is reduced (compared with the primary stands) the reverse is true in the south and southwest.

The species balance of Table 11 is different from that observed in 1963. (Indeed, it is sufficiently odd to be questionable—though the data for conifers are identical with figures from another Chinese source cited by NFPA, 1986.) The extent of the Mongolian oak resource was unsuspected in 1963, while Korean pine areas were much more extensive. The relatively high values in Table 11 for spruce and fir may reflect the identification of new primary forest areas in the southwest. It can be expected that the species range will be extended—since 1981, China

TABLE 11
COMMERCIAL SPECIES COMPOSITION OF MAJOR NATURAL FOREST STANDS: 1981

Species	Million ha	Million m^3	% Total
Pinus koraiensis (Korean pine)	0.39	85	1.3
Abies spp. (Fir)	2.69	812	12.6
Picea spp. (Spruce)	4.14	1175	18.2
Cupressus funebris (Chinese cypress)	1.34	45	0.6
Larix spp. (Larch)	9.96	1000	15.5
Pinus sylvestris (Scotch pine)	3.86	35	0.5
P. tabulaeformis (Chinese pine)	11.97	40	0.5
P. armandi (Hua Shan pine)	6.49	33	0.5
P. massoniana (Masson's pine)	14.24	486	7.5
P. yunnanensis (Yunnan pine)	5.78	432	6.7
Cunninghamia lanceolata (Chinese fir)	6.07	237	3.7
Other conifers	14.80	173	2.7
Total conifers	81.73	4553	70.3
Cinnamomum camphora (Camphorwood)	0.32	36	0.6
Phoebe nanmu (Nanmu)	0.01	2	—
Fraxinus mandshurica (Manchurian ash) *Phellodendron amurense* (Cork-tree) *Juglans mandshurica* (Manchurian walnut)	0.21	17	0.3
Quercus mongolica (Mongolian oak)	11.87	812	12.6
Betula spp. (Birch)	4.89	309	4.8
Populus spp. (Poplar, cottonwood)	2.44	107	1.7
Other hardwoods	0.49	25	0.4
Total broad-leaved	20.23	1308	20.4
Mixed deciduous	2.95	102	1.6
Mixed conifer broad-leaved	4.96	470	7.3
Oak coppice	0.24	5	—
Eucalyptus spp. (naturalized)	0.21	6	—
Total	110.32	6444	(99.6)

SOURCE: MOF (1985).

has learned much about the international marketplace. The list of species now being offered for sale by China (Appendix B) indicates thirty-eight species from Yunnan alone; in another publication (TUHSU/TIMEX, 1986), Sichuan offers fifty-one species, though less than half of these can legitimately be described as "commercial."

Table 12 indicates the distribution of forests by province (from a publication probably dating from 1973—see USDA, 1983) together with 1981 areas (MOF, 1985), growing stock data from 1984 ("major reserves"), and 1981 and 1984 timber production. Apart from a major redistribution of forest area from Heilongjiang to Nei Monggol, several features of this table are noteworthy. First, in provinces with significant resources, the forest *area* has been reduced in all but two (Anhui and Hunan) while growing stock has declined substantially in Jilin and Sichuan (though increases in timber reserves are recorded for Fujian and Guangxi); production over the period 1981–1984 increased in all provinces except Henan, Guizhou, and Xinjiang—and substantially in Nei Monggol, Liaoning, Zhejiang, Fujian, Jiangxi, Sichuan, Guangdong, and Yunnan. Second, the provinces recording plantation areas in excess of 1 million hectares are either the border regions (where statistics are less reliable and include significant areas of doubtfully successful aerial sowing) or provinces that have a long tradition of agricultural tree planting—only the latter are likely to yield significant volumes of usable timber within the foreseeable future. Finally, the unchanging growing-stock figures for Yunnan and Xizang—both provinces known to have received substantial in-migration during the period—cast doubt on the reliability of inventory data in the remoter areas.

Man-Made Forests

China's achievements in reforestation are impressive—a claim that could not be sustained in 1963. Then, through the campaigns to "mobilize the masses," prodigious numbers of seedlings were raised and planted, but sites were not well chosen, seed sources were unselected, and tending after planting was virtually ignored. Many visitors between the 1960s and the present (see Westoby, 1975; Grainger, 1980; Dickerman, 1980; FAO, 1982) attest to the greater attention now being given to these shortcomings. Nonetheless, senior MOF officials—following the lead of former Deputy Minister Yong Wentao—have claimed that of the published area of plantations, only 30 percent can be regarded as successful. Again, establishment has been more successful in traditional forest provinces; in the northeast, however, growth rates are slow.

Table 13 gives an MOF cumulative time series for plantations together with the proportions of varying types. There was a sharp drop in afforested area between 1957 and 1964—due in part to felling of young

TABLE 12

PROVINCIAL FOREST STATISTICS

Province	Forest Area (million ha)		Growing Stock (million m^3)	Major Reserves (million m^3)	Annual Production (1000 m^3)		
	1973	1981	1973	1984	1981	1984	%Total
Hebei	2.02	1.68	73	—	104.5	122.0	—
Shanxi	n.a.	0.81	n.a.	—	124.0	128.5	—
Nei Monggol	0.34	13.74	9	850	4,271.5	4,784.7	7.5
Liaoning	4.15	3.65	98	—	443.4	688.9	1.1
Jilin	7.56	6.08	730	660	6,144.3	6,334.2	9.9
Heilongjiang	25.20	15.29	2324	1577	15,399.9	16,682.7	26.2
Jiangsu	0.34	0.32	13	—	n.s.[a]	n.s.[a]	—
Zhejiang	3.96	3.43	82	—	641.0	2,037.2	3.2
Anhui	1.75	1.79	47	—	360.4	464.0	0.7
Fujian	5.90	4.50	243	300	3,668.9	7,280.1	11.4
Jiangxi	6.11	5.46	263	240	2,695.7	3,744.0	5.9
Shandong	1.32	0.90	23	—	44.5	106.3	—
Henan	1.79	1.42	79	—	114.4	99.1	—
Hubei	4.36	3.78	96	—	645.1	731.0	1.1
Hunan	6.58	6.87	189	—	2,078.4	3,759.8	5.9
Guangdong	7.49	5.88	177	—	3,270.3	4,515.9	7.0
Guangxi	5.51	5.23	193	220	1,597.0	2,040.0	3.2
Sichuan	7.46	6.81	1347	1050	3,440.7	4,569.7	7.2
Guizhou	2.56	2.31	159	—	826.5	830.8	1.3
Yunnan	9.56	9.20	989	989	1,997.5	3,066.2	4.8
Xizang	6.32	6.32	1436	1436	187.6	218.7	0.3
Shaanxi	4.59	4.47	244	250	332.2	458.3	0.7
Gansu	1.88	1.77	98	—	449.9	547.6	0.8
Qinghai	0.19	0.19	31	—	50.0	63.0	—
Ningxia	0.08	0.09	5	—	7.5	42.6	—
Xinjiang	1.44	1.12	237	—	519.2	447.0	0.7
Totals	118.46	113.11	9185	7572	49,418.7	63,788.0	98.9

SOURCE: MOF (1985) and USDA (1983).

NOTE: *China Agriculture Yearbook* (1985) cites 10.26 billion m^3 for 1984 growing stock.

TABLE 13
AREAS OF MAN-MADE FORESTS (IN 1000 HA)

Year	Timber Forests	Economic Crops	Shelter	Other	Total
1957	1733	1350	993	279	4355
1964	1392	824	437	258	2911
1975	3651	532	428	363	4974
1978	3130	881	418	67	4496
1980	2927	824	514	287	4552
1981	2531	629	637	313	4110
1982	2631	652	863	350	4496
1983	3805	1100	822	597	6324
1984[a]	3500	600	1000	500	5600
1985[a]	4000	500	1000	500	6000

SOURCE: MOF (1985).
[a] Provisional from Zhong Maogong.

shelter plantations to fuel the backyard industries called for by the Great Leap Forward—and there was another drop in 1984. What is remarkable about the figures, however, is their realism in relation to past claims: In the 1960s, forested areas of several million hectares annually were reported (though not by the central government). There are discrepancies between Tables 13 and 8 with respect to "economic" crops—which may in part derive from differences in definition (since the source of both tables is the same MOF publication). Table 13 is only concerned with man-made *plantations;* there is no reason why some natural stands (coppice, for example, and nut and fruit trees) should not also be classified as "economic."

Another feature of Table 13 is the apparent readjustment of areas between the different plantation types. This is not surprising. Often shelter trees and economic species are interchangeable and there is evidence in the field of reclassification of timber forests as shelter. Indeed, this has been done extensively in the provinces coming within the "Three Norths" shelter project. The policy of allocating forest land to households and farmers for forest management has also led to boundary readjustments.

Substantial changes in planted areas can be expected in the future. According to *China Daily* reports, by the middle of 1986 more than 30 million hectares of mountain land had been allocated to households for afforestation and management—under thirty- to fifty-year renewable and heritable leases (see Chapter 5). This is generally poor land, however, and the increment in growth will not be commensurate with additions to the forest area.

Annual Growth

It is difficult to estimate the overall annual increment of China's forests—the requirement that provincial production limits must not exceed their annual growth is impossible either to enforce or to monitor. Various statistics have been published by FAO, the World Bank, and the MOF, but they generally relate to plantations, not natural stands, and highlight *potential* rather than actual growth rates. In most countries of the world, research sample plots tend to be biased toward fully stocked areas that have been exceptionally well tended; they overestimate increment. Figures from natural, unexploited forests are only approximations since future growth will be determined by treatment. Table 14 gives a token selection of mean annual increment data for three provinces in which a World Bank forestry development project has been prepared. These figures are optimistic and cannot be applied to existing plantations—especially in the northeast, where rotations are more realistically sixty to eighty years rather than forty to fifty and silvicultural practices are conservative. Data for other species (grown in plantations) and for other provinces in China are presented in Table 15.

In "Four Around" and line plantings, growth rates are more spectacular—but since many such trees are privately owned, or belong to highway bureaus, they are likely to be cut well before maturity for private use. The timber will not contribute much to planned production of industrial roundwood, unless the state is prepared to buy it on the free market and unless there are drastic changes in the utilization development model (see Chapter 4).

Various statistics are cited for total annual growth in China's forests; they are on the order of 200 million cubic meters with a range from 150 to 300 million. They have little value in view of the unknown (and unknowable) premises on which they rest. They could be used to argue that the annual harvest is in excess of annual growth, but that fact is known (China is importing logs on a big scale) and agreed by the Chinese. They cannot tell us (or the Chinese) when demand and supply

TABLE 14
CHINA FORESTRY PROJECT:
TREE SPECIES, MEAN ANNUAL INCREMENTS,
AND ROTATIONS BY PROVINCE

Species	Mean Annual Increment[a] (m^3/ha)	Rotation (years)
Heilongjiang		
Larix gmelini (Larch)	6.0	40
Pinus koraiensis (Korean pine)	6.0	40
P. sylvestris var. *mongolica* (Scotch pine)	4.0	50
Fraxinus mandshurica (Manchurian ash)	4.0	50
Juglans mandshurica (Manchurian walnut)	4.0	50
Sichuan		
Cunninghamia lanceolata (Chinese fir)	7.5	25–30
Pinus armandi (Armand's pine)	6.0	30–35
Cryptomeria japonica (Japanese cedar)	11.0	30–35
Guangdong (North)		
C. lanceolata (Chinese fir)	5.0	25–30
Pinus massoniana (Masson's pine)	6.0	30–35
Guangdong (South)		
Eucalyptus exserta	12.0	15
E. citriodora	15.0	20
E. leizhou No. 1	15.0	20

SOURCE: World Bank (1985d).
[a] With fertilizers and more intensive management, growth rates should increase over the rates shown.

are likely to balance. That crucial determinant will be the success of new plantations. In 1981 the minister of forestry (Yong Wentao), discussing the need for improved tending of young plantations, claimed that the "average stocking" was only 1.4 m^3/*mu* (about 21 m^3/ha). If this is true, even the 2.5 million hectares of man-made timber forest (in 1981) would represent little more than the annual official harvest.

The successful integration of trees with both agricultural and urban land use is the most striking feature of forestry in present-day China. If similar success could be assured with areas of barren hillside leased to households and farmers, China could become self-sufficient in timber by the year 2000. The 1986 revision of this target date to 2040 is recognition, though, that the objective is unrealistic.

TABLE 15
GROWTH RATES OF COMMONLY PLANTED TREE SPECIES

Species	Spacing (m)	Rotation (years)	Mean Annual Increment (m^3/ha/year)	Type of Forestry
Henan				
Paulownia fortunei	4 × 3	20	10–25	2-row, 4-around
Robinia pseudoacacia	3 × 2	20	5–10	2-row, 4-around
Populus tomentosa	3 × 3	20	10–15	2-row, 4-around
R. pseudoacacia	3 × 2	15–25	3–6	Sand dune
Jilin				
Pinus sylvestris	2 × 1	120	1	Secondary
Larix dahurica	2 × 1	60–70	1–2	Secondary
Liaoning				
Populus simonii	1 × 1.5	40–50	5–9	Research plot
Larix olgensis	1.5 × 1.5	50	3	Research plot
Pinus sylvestris	1.5 × 1.5	80–100	2–3	Research plot
Larix spp.	1 × 1.5	50	3	Research plot
Populus canadensis	3	20	17	Single row
Hunan				
Cunninghamia lanceolata	n.a.	20	7.5–15	Coppice
Guangdong				
Eucalyptus spp.	3 × 3	25	15–25	Fuelwood
Pinus massoniana	1 × 1	50	2	Hill site
P. massoniana	2 × 2	40	5–12	Research plot
Yunnan				
Cassia siamea	n.a.	25	16	Coppice

SOURCES: Field visits (1986).

Log Size

Changes are evident in the species mix of the forest harvest and, with pressure to increase production, in log size. Changing log size classes cannot be quantified, but the increasing scarcity of top-grade softwood logs is widely attested in China. Indeed, a reason given for changing the grading system was to accommodate smaller logs. Only a fraction of industrial production derives from plantation-grown logs, but, increasingly, secondary forest—first logged in the 1940s and 1950s—contributes to the northeast harvest. In some nine logyards first visited in 1963 and again in 1986, I noted evidence of reduced log size. Domestic logs compare poorly with North American softwood logs and with Southeast Asian hardwoods. There is little difference between domestic and Russian (Siberian larch) imports, and easily the best-quality softwood logs seen in China were imports from Outer Mongolia. Hardwood logs are very variable and there is no segregation for domestic use; veneer-grade logs are mixed indiscriminately with sawlogs. Certainly one hopes the development of markets overseas for hardwood products will promote more rational utilization. As we shall see, changing log quality and origin have implications for the restructuring of the forest industries.

Special-Purpose Planting

A notable change in China since 1963 is the increase in planting by nonforestry organizations for their own use. The Ministry for Railways (as in other countries) has for a long time established tree plantations along its tracks. Originally trees were planted for fuel (when trains were pulled by wood-burning steam engines) and for sleepers; they also served to lay claim to land for eventual double-tracking. (Similarly, road reserves have been planted with trees.) Other organizations (including the Ministry for Railways and highway bureaus) undertake significant housing programs for their workers and are establishing their own plantations to yield timber for construction. Mining operations are encouraged to grow trees for props (as well as housing), and the Ministry for Light Industry sponsors pulpwood plantations. Even before 1979, coal mine enterprises had legal title to more than 1 million hectares of reforested land. The Ministry of Water Resources and Electric Power operates sawmills and is involved in replanting forests. The PLA not only plants trees in the border regions and is responsible for ambitious aerial seeding programs but has its own logging and sawmilling operations. It is a major user of lumber for construction; it manufactures furniture; and, again in the remote regions, it processes a multitude of "minor forest products" (foodstuffs, medicines, and the like) for sale. The PLA even farms deer and operates a safari holiday venture.

The new reform policies have encouraged self-sufficiency among enterprises, and there are increasing references in the Chinese press to organizations—some of them unlikely—engaging in tree planting. It is impossible to estimate their contribution to industrial wood production.

TIMBER PRODUCTION AND CONSUMPTION

Despite increasing publication of production statistics, data relating to timber production and consumption in China are inadequate; projections must be hedged around with caveats—partly because of poor definition of categories. The major divisions are industrial wood (which serves as a raw material) and fuelwood (which is difficult to measure). Both categories are produced within and outside the State Plan, but only industrial wood is properly monitored.

INDUSTRIAL ROUNDWOOD

As with forest resources, my 1965 commodity report (Richardson, 1965) offered a selection of estimates for 1960 industrial roundwood production in China—ranging from 35 to 48 million cubic meters. There was variation of even greater magnitude in estimates available in 1985—depending on what is included within the definition. The FAO statistics are usually "secretariat estimates" and include data from Taiwan. This is misleading, however, particularly with respect to panel products (plywood, particleboard, and the like) and trade. A time series based on FAO data *omitting* Taiwan is given in Table 16, together with two MOF series. The first gives targets for the State Plan production, including pulpwood and industrial fuelwood; the second is the reported plan achievement but excluding pulpwood and industrial fuelwood allocations. Both relate only to roundwood reported to the Ministries of Forestry and Light Industry.

There is reasonable consistency between Chinese statistics for roundwood production under the State Plan. The first MOF series in Table 16 is from Zhong Maogong (pers. comm., 1986) and is identical with the time series published in *Statistics of China's Forestry* (MOF, 1985); the figure for 1985 was provided by Zhong himself. This series plus imports is used in the NFPA (1986) report to relate roundwood consumption to GNP (GVIAO—gross value of industrial and agricultural output).

Data for production outside the State Plan and for illegal harvesting are more problematic. Zhong gives an ex-plan figure for industrial wood of 66 million cubic meters, which is similar to a value cited by the former minister of forests (Yong Wentao, 1982), who postulated total

TABLE 16

INDUSTRIAL ROUNDWOOD PRODUCTION: 1975–1985 ALLOCATIONS UNDER THE STATE PLAN (IN MILLION M^3)

Source	1975	1976	1977	1978	1979	1980	1981	1982	1983	1984	1985
FAO	48.40	50.82	53.78	56.03	58.82	61.71	64.14	67.30	70.67	71.65	n.a.
MOF (1)	47.03	45.73	49.67	51.62	54.39	53.59	49.42	50.41	52.32	63.85	63.23[a]
MOF (2)	41.29	n.a.	n.a.	44.16	46.85	46.69	42.49	43.35	44.81	52.31	53.14[a]

NOTE: MOF (2) excludes pulpwood and industrial fuelwood.

[a] Personal comm., Zhong Maogong, 1986.

wood consumption in China at 200 million cubic meters—of which one-third was "outside the plan" and one-third fuelwood. Zhong also estimates an additional 50 million cubic meters cut for fuelwood but argues that total "forestry resource consumption" might be as much as 290 million cubic meters—though he provides no evidence why recorded consumption should be increased by a further 50 percent. The NFPA report accepts 290 million cubic meters for total domestic consumption and postulates the following breakdown:

State Plan	67
Outside the plan	70
Fuelwood	50
Illegal felling, fire, and other losses	103
	290 million m^3
Imports	9.3 million m^3
Total consumption	299 million m^3

To derive future demand, NFPA took a base statistic of 187 million cubic meters (290 million cubic meters less the 103 million cubic meters from alleged illegal felling, fire, and disease loss). Then, accepting planned growth of GVIAO and a postulated wood input of 0.056 million cubic meters per 1 billion renminbi (RMB) of GVIAO (reduced from a supposed minimum input of 0.08 million cubic meters per 1 billion RMB of GVIAO), consumption in the year 2000 was projected to reach 266 million cubic meters (156 million State Plan plus 70 million ex-plan plus 40 million fuelwood). The NFPA accepts that these estimates involve some quite unjustified—and overly optimistic—assumptions with respect to the likely reduction of illegal felling. Moreover, the fuelwood figure is probably an underestimate.

The earlier (1965 and 1972) commodity projections could not distinguish between State Plan and ex-plan production. They were based on per capita apparent consumption of all industrial wood (0.042 m^3 in 1963 and 0.0647 m^3 in 1972) and varying assumptions with respect to population growth rate and demand levels (annual demand growth rates of 2, 3, and 4 percent and population growth rates of 1.5, 2.0, and 2.5 percent—each decreasing by 2 percent of the initial level annually—in all possible combinations). The projected demand figures for 1985 on these bases ranged from 80 million to 120 million cubic meters and for the year 2000 from 122 million to 264 million cubic meters. With hindsight it is evident that the quality of the basic data did not justify the sophistication of the forecasts attempted; nonetheless, the higher figures in each set are remarkably close to USDA estimates of 130 million cubic meters for 1985 (excluding imports) and to Zhong's projection of 267 million cubic meters for the year 2000. It is not suggested that the relation is other than coincidental.

In 1981, MOF announced (*People's Daily,* 7 November 1981) that from 1982 to 1985 annual allocations of roundwood would be reduced by 4.5–5.5 million cubic meters. To mitigate the effects of this reduction, effort was to be directed to closer utilization (especially of thinnings and forest residues) and the opening up of overmature forests. The reduction appears to have operated during 1981 (though it was partly compensated by a rise in imports). Since 1981, the domestic cut has gradually risen and imports have grown even more. Official production statistics for 1981 and 1984 by province are presented in Table 12. There are inevitably some discrepancies, however, since they derive from two sources. For 1984, the national total from MOF is some 10 million cubic meters higher than in Table 16 and is assumed to include industrial fuelwood.

Apart from discrepancies in statistics of various origins, there are other reasons for exercising caution in assessments. Cadres, used to reporting achievement or overachievement of targets, have a natural bias to upward rounding. In the current situation, with official policies designed to *reduce* production running counter to incentives to overproduce for sale at above quota prices, there could be significant variation in the accuracy of official returns. It seems prudent to assume that published figures may vary by 25 percent. For practical purposes, though, the level of accuracy is acceptable.

FUELWOOD

Fuelwood consumption in the economy cannot be ignored. The need for firewood throughout China (not only in the north) is growing despite increased production of coal and oil. (As noted earlier, China's may be an "energy-constrained development"—Field, 1982—and both transport and the opportunity costs of using coal and oil are high.) Moreover, the need for fuel in remote areas is a reason for the continuing (illegal) destruction of timber forests. There have been many exhortations to substitute coal for fuelwood (see, for example, FE, 21 March 1980).

The FAO fuelwood data are based on surveys carried out in other countries of household consumption and related to conditions in China. The endeavor is heroic and the estimate (154 million cubic meters in 1984) is as good as any. The USDA estimate (very tentative) is 70 to 90 million cubic meters. Shiraishi (1983), in a five-part review of timber supply and demand, estimates a "shortage" of firewood of 134 million cubic meters—based on statistics published in the Chinese *Renmin Baozhi* of 13 November 1982 and an assumed per capita demand of 500 kg annually in rural areas. High priority has been given to creating fuelwood forests (several fuelwood projects are being carried out with foreign aid); but the forestry policy over thirty years with respect to fuelwood has been described by Yong Wentao as "an extremely large failure." The *Beijing Review* of 10 August 1981 claimed that *rural* fuel consumption for household use was equivalent to 290 million tons of coal (one-third of China's coal production in 1986; FEER, 19 March 1987).

The Zhong report uses a figure of 50 million cubic meters for 1985 firewood production, declining to 40 million cubic meters by the year 2000. Neither level is supported by evidence, and it is simply not possible for a country as big as China to make realistic estimates of domestic firewood use. In a book entitled *Uncertainty on a Himalayan Scale*, Thompson et al. (1986) responded to agency pressure to quantify fuelwood consumption in Nepal by publishing the factor by which previously published estimates had varied—it was 67! Even if the upper estimates were omitted as typographical errors, the per capita annual consumption ranged from 0.1 to 2.6 m^3 (with the majority around 1.0 m^3). An estimate of 1.0 m^3 would be of the same order of magnitude as 500 kg and, even if applied to China's *rural* population only, would exceed 800 million cubic meters annually. Despite the frequent appeals in the Chinese press to substitute coal for wood fuel, it is difficult to imagine the conservation controls that would be needed to reach the fuelwood levels postulated by Zhong or the NFPA report. In Tibet, where there is no coal, the price of firewood in the Lhasa free market in 1988 per cubic meter exceeded the average annual per capita income.

General

Also of interest is the volume of industrial timber cut and utilized outside the State Plan. Zhong uses Yong Wentao's estimate of 66 million cubic meters, but he offers no firm evidence in support. Hardly a model of precision, a party directive on afforestation (FE, 21 March 1980) points out that official timber harvest statistics "do not count the timber consumed by various localities, communes, and production brigades." In most provinces now, significant volumes of timber are cut from roadside trees under the control of the highway bureaus. Harvesting is not reported to the MOF even though, in some cases, MOF sawmills purchase logs from this source.

Nor, of course, do statistics cover illegal fellings—which certainly increased during the Cultural Revolution and perhaps even more so as a result of the post-Mao decentralization policies (in particular, the freeing of the market). One component of decentralization enables manufacturing enterprises to make supplementary production plans (see, for example, Lee, 1986); in many cases—including a number of provincial sawmills, joineries, and furniture factories—it became difficult to ensure material supplies without having recourse to illegal timber procurement. During my visits to autonomous areas in 1986 there was widespread evidence of illegal felling—even in showplace reserves such as the Wo-long Giant Panda sanctuary in Sichuan. As noted earlier, the PLA has been involved in illegal logging and, during log floating operations in the southwest, losses of 15 to 18 percent were reported. As noted in Chapter 5, Ross (1988) reports numerous references to newly privatized forest holdings being cut by disputing neighbors of the new owners, as well as premature felling in such holdings to generate cash flow.

Given the uncertainties of data relating to nonplan timber and fuelwood production—and the fact that a significant discrepancy between timber supply and demand is now openly acknowledged by China's economists (in contrast to the situation obtaining in the 1960s and 1970s)—it is questionable whether any purpose is served by attempting to quantify other than State Plan roundwood production.

Table 17 illustrates per capita consumption of forest products in a selection of countries. The caveats outlined earlier must be borne in mind in the interpretation of these data. Newsprint consumption, of course, is an important indication of quality of life and, because of imports, the statistics may be more reliable than for other wood products. China's consumption per capita exceeds that of Pakistan, Nepal, Indonesia, Myanmar, and Bangladesh but is less than that of every other country in the list. Sawnwood consumption is higher than in the countries of the Indian subcontinent (the figure from Myanmar is probably a gross underestimate), while panel consumption is too small to draw

TABLE 17
APPARENT PER CAPITA CONSUMPTION OF FOREST PRODUCTS IN SELECTED COUNTRIES: 1984

Country	Sawnwood (m^3)	Panels (m^3)	Paper (kg)	Newsprint (kg)	Printing and Writing (kg)	Other (kg)	Industrial Roundwood (m^3)	Population (millions)	Annual GDP Growth (%) 1970–1982	Annual GDP Growth (%) 2000[a]
Australia	0.275	0.060	135	36.64	26.31	72.35	0.523	15.5	2.7	2.8
Bangladesh	0.002	n.s.[b]	1	0.46	0.70	0.24	0.004	99.0	3.5	4.2
Brazil	0.115	0.016	24	1.99	5.79	16.35	0.387	136.0	7.3	5.3
Canada	0.543	0.147	185	34.86	54.23	95.69	5.976	25.3	3.5	2.8
Chile	0.094	0.011	25	5.04	6.22	13.78	0.632	11.9	1.8	3.7
Finland	0.708	0.133	271	41.42	87.96	141.22	8.772	4.9	3.1	2.5
France	0.189	0.050	123	10.45	43.80	68.92	0.462	55.3	2.9	2.6
Germany (DDR)	0.230	0.100	85	8.32	12.99	63.23	0.490	16.7	4.6	2.0
Hungary	0.190	0.042	66	6.51	14.24	45.28	0.245	10.6	4.5	2.0
India	0.015	n.s.[b]	3	0.73	1.15	1.10	0.021	748.0	3.2	4.9
Indonesia	0.024	0.004	4	0.58	1.10	1.89	0.142	173.0	7.3	5.5
Israel	0.071	0.037	77	11.90	22.86	42.38	0.053	4.2	4.8	4.5
Japan	0.275	0.079	161	23.04	36.11	101.62	0.592	120.8	4.0	3.6
Mongolia	0.215	0.002	8	2.22	1.67	4.44	0.542	1.8	2.5	3.5
Myanmar	0.009	n.s.[b]	1	0.28	0.42	0.37	0.051	35.3	4.4	4.4
Nepal	0.013	—	—	—	—	0.12	0.026	16.5	2.3	3.1
New Zealand	0.525	0.076	145	44.50	15.80	84.80	2.113	3.3	2.4	2.7
Pakistan	0.002	0.001	2	0.39	0.96	1.09	0.004	88.0	5.1	5.7
Thailand	0.028	0.005	9	2.15	0.91	6.38	0.044	50.8	7.7	5.7
U.S.	0.483	0.134	290	50.41	76.43	163.60	1.299	236.6	2.4	2.4
USSR	0.320	0.043	34	4.21	5.41	24.60	0.619	278.7	4.7	2.0
China	0.015	0.002	9	0.64	1.85	6.80	0.061	1033.0	5.6	6.0

SOURCE: FAO (1986a), except China; Cotchell Pacific (1987) for China.

[a] Projected.

[b] N

any conclusion other than to note the gap between industrialized and less developed countries. All product values emphasize the magnitude of the task facing China if it is to "catch up" with the modern world.

CONSUMPTION BY END USE

The MOF has provided "three-year average" volumes of wood distributed to various end uses under the State Plan. These figures do not include imports, though all imports are centrally allocated. Table 18 (column 1) prorates the MOF volumes over the 1985 roundwood production from Table 16. Imports are used in furniture, construction, packaging, railways, and "other" categories. Of total log imports in 1985 of 9.33 million cubic meters, USSR larch (nearly 2 million cubic

TABLE 18

1985 ROUNDWOOD CONSUMPTION BY END USE: STATE PLAN DISTRIBUTION OF DOMESTIC PRODUCTION AND IMPORTS

Subsector	Roundwood (1000 m³)	Roundwood Including Imports (1000 m³)	% of Total
Construction	12,940	17,560	28
Mining	5,670	6,110	9
Railways	560	620	1
Pulp and paper	3,600	3,800	6
Transport	2,240	2,240	4
Transmission poles	450	450	n.s.[b]
Packaging	4,310	5,310	8
Furniture	1,980	2,490	4
Plywood	690	2,130	3
Miscellaneous[a]	4,310	4,310	7
"Other" uses	16,390	17,450	28
Total	53,140	62,470	98

NOTES: The "other" category covers a wide range of wood manufacturers, of which (based on time series to 1978) some 13 million cubic meters can be described as "agricultural" but excluding housing (Shiraishi, 1983); it also includes shipbuilding. As noted in the text, less than 60 percent of planned production is allocated directly by the state; the rest is under the control of provincial governments.

[a] Includes textiles, shoes, musical instruments, charcoal, and particleboard.

[b] Not significant.

meters) is used for railway sleepers, mine props, and urban construction; Chilean Radiata pine (800,000 m^3) is predominantly for packaging and pulp—though a small quantity is used in furniture; in 1985, the United States supplied 4 million cubic meters of softwood logs (Douglas fir, hemlock, balsam) and Canada supplied 500,000 m^3 (roughly half logs, half lumber), including both cottonwood and basswood as well as conifer logs; the balance came from dipterocarp logs from Southeast Asia and trial shipments of tropical logs from Brazil. In column 2 of Table 18, note that 1 million cubic meters is added to plywood and the remainder (5.63 million cubic meters) is prorated to the construction, furniture, plywood, and 25 percent of "other" categories.

The breakdown is not directly comparable with the earlier analyses, since it only details centrally allocated wood (excluding major uses such as rural housing and small-scale industries). The allocation levels—though not the percentages—are reasonably consistent with data published from time to time in China (see *China Weekly*, 22 October 1982) but, compared with USDA estimates, are low with respect to mining usage, pulp and paper, and transmission poles. Compared with estimates for 1962 (Richardson, 1966) the breakdown presented in Table 18 shows a massive increase in "miscellaneous" and "other" categories—indicating differences in definition—while percentage allocations to construction, mining, and railways have dropped. There are significant increases in pulp and paper (3.7 to 6 percent), packaging (1.7 to 8 percent), and furniture (0.7 to 4 percent).

CHINA'S SUPPLY AND DEMAND CRISIS

It is now recognized in China that the annual timber harvest is significantly in excess of increment and that the shortfall is unlikely to be made up by new plantation production in the present century. Until recently it was assumed that the mass afforestation schemes launched with such fervor in the 1950s and 1960s would provide all the raw materials needed for China's forest industries by the year 2000 and that per capita consumption of forest products would rise from 0.056 m^3 in 1958 to 0.34 m^3 by 1990. (See Richardson, 1966, citing ministerial forecasts.) On the basis of Chinese data, advisers from the international agencies recommended increased felling in the dwindling, but overmature, primary forests and an increase in imports to cover the (strictly temporary) deficit in supply and demand.

At first view the reasoning seems sound: There is little economic value in remote and overmature forest, in which the increment is balanced by volume loss from death and decay.[1] But the timing and planning of utilization require accurate inventories and realistic assessments of plantation survival and growth rates. Until 1981, these data were not available to Chinese researchers and, even now, predictions lack authority.

An article in the China *Economic Daily* (cited in *China Daily*, 6 September 1986) projected a continuing imbalance of supply and demand to the year 2040—and well beyond that date unless strict measures are adopted to curb "wanton felling" and to protect newly planted areas. It indicated a decline in forest cover from 12.7 percent in 1978 to 12 percent in 1981 and a drop in reserves (state forest only) of 850 million cubic meters. Of 131 state forestry bureaus, 61 have overcut and 25 have virtually no more reserves; within "ten years or so," two-thirds of all forestry bureaus will have no more production capacity. In the major forestry provinces of Heilongjiang and Jilin, net volume reserves are declining by some 10 million cubic meters annually and production exceeds annual growth by 38 percent. Ross (1988) cites the example of Ychun, where primary forest reserves have fallen by nearly 40 percent and there is a net decline in forest area by 80,000 ha annually. As noted earlier, even where the cut-over forest is regenerated, secondary growth in the Mixed Coniferous and Deciduous Broad-Leaved formation contains more of the pioneer hardwoods (birch, poplar, and others) than the more highly valued oak, ash, and walnut. Ross also refers to concern in the northeast over unemployment and increasing reliance on sideline production.

In provinces visited in 1986, the following statistics were provided by Chinese foresters:

Sichuan: Annual felling, 30 million cubic meters; annual growth, 15 million cubic meters; resource life, ten years (see also FE, 7 January 1981).

Henan: Already a net importer to the tune of 1.3 million cubic meters annually.

Guangdong: Accessible primary forest exhausted (0.47 million hectares in scientific reserves); a net importer of industrial wood.

Yunnan: Annual harvest more than three times that prescribed under the State Plan (6.7 million cubic meters as against 1.9 million), not including fuelwood gathering or forest clearance for agriculture.

Heilongjiang: Annual felling, some 2 million cubic meters in excess of increment; by 2000, only two or three forest bureaus out of forty will have a primary resource—in the province that in 1984 contained 16.2 percent of China's total stock.

Jilin: Harvest exceeds growth by 2 million cubic meters annually; natural forest has a life of twenty years (from 1981), but the oldest plantations are only twenty-eight years, with a rotation age of sixty.

Liaoning: Demand exceeds 2 million cubic meters annually, but the official supply is only 0.5 million cubic meters—one-third from plantations; mature plantations make up only 2 percent of the total.

Keep in mind that these data relate only to legally felled, commercially traded timber. There are frequent references in China's press to illegally felled timber (4 million cubic meters seized in thirteen provinces, 49,000 personnel dismissed, and free markets closed in 1980; FE, 28 February 1981); violence to forest guards by timber thieves (FE, 4 October 1983); illegal logging operations by the PLA (Dreyer, 1986); and 3.1 million roadside trees illegally felled in Hunan province between 1979 and 1981, with a value of 26 million yuan (FE, 23 March 1982). Moreover, environmental problems resulting from forest destruction are causing alarm inside and outside China (Lampton, 1986; Leung, 1986). A consultant to the World Bank, Vaclav Smil, presenting a bibliography of 249 publications on environmental aspects of economic development in China, says that "deforestation appears to be the country's most critical environmental problem" (Smil, 1982). Commercial forestry does not, of course, involve deforestation, but significant areas of commercial forest are depleted by uncontrolled felling—and much of it is not accounted in official statistics.

An estimated 1 million hectares of commercial forest are destroyed annually by fire (Cotchell Pacific, 1987)—and, as any visitor quickly observes, damage by insects, especially to plantations, is widespread. It is alleged that damage was even more serious during the Cultural Revolution, when protection virtually ceased.

In May/June 1987, Heilongjiang province lost 3 percent of its forest reserves in the most disastrous fire ever recorded in China (Richardson and Salem, 1987). The fires raged for three weeks and more than 200 lives were lost. In the recriminations that followed this tragedy, the minister of forests (Yang Zhong) and several senior officials were dismissed.

The causes of active forest depletion are several. First, as noted elsewhere, there is a continuing shortage of industrial timber—and fuelwood—throughout China. Historically, China's economy has long suffered from timber supply constraints. Adshead (1979:4) writes: "Shortage of wood would appear to have been a serious constraint on the Chinese economy in the period of modernization." He was referring not to post-Mao China, but to the seventeenth century! Shortages of shipbuilding timber are documented from 1533, and since the year 1600 difficulties had been experienced in obtaining wood for palace construction, mining, and the salt industry. Adshead poses the question, "How did a civilization which made so much of wood manage to survive with so little of it?" The query is still appropriate. (See, for example, Delfs, 1986a.)

Second, the dominance of the state in commercial timber production (state bureaus account for 53.4 percent of total forest cover and 70 percent of production forest reserves) has led to gross inefficiencies in the wood-using bureaucracies, appalling waste, and low profits. Prices for "plan" production are fixed and—because of a complex and unrealistic grading system—bear little relation to market values. To maintain cash flows, forests are being overcut perhaps by as much as 30 percent—the amount by which quotas may be overfulfilled in northeastern China without penalty. The primary forests are clear-cut and the secondary areas selectively felled ("creamed"); there is no incentive to minimize deterioration of logs or sawn wood, and log and lumberyard hygiene is poor.

Finally, much of China's remaining primary forest is in autonomous regions or prefectures, where the writ of Beijing does not extend in full force. Thus 70 percent of Yunnan's forest area is in minority areas (40 percent of primary production reserves are in the three-county autonomous prefecture of Xishuangbanna); in Sichuan, the three autonomous prefectures of Aba, Ganze, and Liangshan contain a majority of the remaining primary forests; and Nei Monggol has recently regained control of what may be the most important timber production bureau in China (contributing 11 percent of the primary forest harvest)—Yakashih. For political reasons, conservation regulations are enforced less vigorously in these areas and, in the southwest, shifting cultivation (swidden) is still practiced. In-migration of Han Chinese has not been discouraged and there has been increasing pressure on forests for non-productive uses. In Yunnan, valuable decorative hardwoods (including rosewoods) are being cut and burned for firewood. (Since they do not float, they cannot be extracted where there are no roads.) The Forest

Law (see Appendix C) contains clauses—Articles 7 and 41—that excuse the autonomous areas from its provisions.

There is ample evidence, too, that reforestation is being undertaken with somewhat less zeal than was the case in the 1960s. In the central and southern provinces—on good sites—there are no problems in persuading households to plant trees: The rotations are short and the financial yields promising. There is less enthusiasm, however, for tree-planting in poor soils, remote areas, and sites where long rotations preclude early returns. Even though households are being given security of land tenure under "contract responsibility" schemes, there is reluctance to divert effort from more immediately profitable pursuits to tree-planting. Moreover, the instrument of organizing mass participation in campaigns—the commune, which was such a prominent feature of Mao's China—is no more. There are in effect no more masses. (They have become successful and motivated individuals or households.) An alternative force for public works has yet to emerge.[2]

Growing unemployment in isolated, single-industry communities is a depressing feature of present-day China—nowhere in evidence in 1963. The earlier visitor to the northeast, for example, was struck by the feverish activity that characterized even rural communities: Mobilization of mass labor for tree-planting was undertaken with vigor and enthusiasm—if somewhat erratically (and, often, misguidedly). But there were jobs for everyone. Now, despite rural prosperity, there is unemployment among middle-school leavers which, because there is no labor mobility, has to be accommodated locally. In the forest communities, special training facilities have been established to cater to this modern problem. Their success, however, is variable (see Chapter 4).

China's response to the impending timber supply crisis has been realistic. First, it has been acknowledged: There is professional debate about how to meet it (reports of seminars in various issues of the *Journal of Forest Economics*, for example), and steps have been taken to secure supply bases overseas (in the Pacific Northwest of North America, in Brazil, and in the South Pacific). Next, for the first time, an "obligatory"[3] tree-planting program was given legislative enactment by the NPC (13 December 1981) with the prospect of legal sanctions in the event of noncompliance (reportedly on the personal initiative of Deng Xiaoping—see Ross, 1983). The Eight Obligations in China—to which has been added tree-planting—are: work; education; family planning; child care; support for the integrity and unity of the state;

support for China's security, honor, and interests; military service; and payment of taxes. Third, as detailed in Chapter 5, regulations issued in April 1986 for the implementation of a new Forest Law (which came into force on 1 January 1985)—again for the first time—prescribed penalties of reparation and restitution for damage to forests, illegal cutting, burning, or theft (*People's Daily,* 15 May 1985).

Finally, the search for substitute materials for wood is being stepped up: The Western visitor may be surprised to discover that the major objective of forest products research organizations in China is to identify wood substitutes. Moreover, regulations promulgated in 1983 by the National Conference on Wood (comprising representatives of the State Economic Commission, the State Planning Commission, the Materials Supply Bureau, and other bodies) prohibit the use of wood in the following applications: floors, stairs, and windowsills; power substations and utility poles; railway sleepers (except crossing ties); bulkheads, deckheads, and decorative applications on ships; mine props and piles; bridges; industrial fuel; and coffins. Exemptions are allowed (with government permission) and strict regulations are impossible to enforce; but the lengths to which they go indicate the concern.

Apart from outright prohibition, targets have been set for partial substitution of wood in construction outlets and other subsectors. Thus, in 1985, the use of timber in concrete shuttering (formwork) was to be reduced to 50 percent, in scaffolding to 25 percent, and in joinery from 20 to 60 percent. Allocations for pit props were set at 5 m^3/1000 tons of coal (the norm was 23 m^3 in 1961); textile usage (looms and bobbins) was to be reduced by 30 percent and matches by 18 percent; and plastics were to be increasingly used in manufacturing—the glass industry, defense, and bicycles were specifically targeted.

In construction, the principal substitutes are steel and concrete, and guidelines have been issued by the Ministry of Urban and Rural Construction and Environment Protection (MOURCEP) with respect to material inputs. Depending on the type of structure, the timber norms range from 0.04 m^3 per square meter of building area in multistory residential construction to 0.12 m^3 for "small-size embassies." Steel inputs in high-rise residential buildings are 39 to 65 kg and range up to 95 kg in hotels.

For several reasons, there is resistance to the increased use of steel and concrete in residential construction and, outside the state sector, regulations are widely disregarded. First, in buildings below six stories, materials must be hoisted manually and steel formwork, for example, is heavy and cumbersome. Second, in a continental climate, steel doors and windows tend to jam and are poor insulators. In low-rise dwellings, moreover, structural timber performs better than steel in fire and is preferable to concrete in earthquakes. Third, wooden construction is easier (and cheaper) for the tenant to extend than steel or masonry. In urban centers throughout China, closed-in balconies and tiny rooms cantilevered out from windows—to create more living space—are a common sight. Though these innovations are strictly illegal, the authorities turn a blind eye (even in Beijing) and work units may distribute small quantities of timber to favored workers for use in such do-it-yourself applications. (In August 1986, my interpreter in Beijing learned that her husband had secured a few planks of rough-sawn larch—she contemplated the redesign of a tiny balcony with the exuberance of Corbusier.) The Chinese like wood; to a foreign visitor, a wooden floor in an urban dwelling is displayed with pride in affluence, valued somewhere between a video recorder and a piano.

It is impossible to envisage success for China's campaign "to design wood out of the economy." Nor can an overpopulated country (with most of its land resource incapable of growing food) ignore the importance of renewable energy resources. There was a time in the 1960s when scientists were able to contemplate a world without wood—a world in which, it was claimed, 98 percent of existing timber use could be substituted by derivatives of the "AlFeSiCa" minerals in the earth's crust, combined with atmospheric gases. Technically, it was (and still is) a feasible option. But it requires an unlimited supply of cheap energy. The 1970s oil crisis removed fossil fuel as a possibility and 1980s global environmentalism has effectively eliminated nuclear alternatives. China has a continuing need for wood: Regeneration of harvested forests and the establishment of fast-growing plantations and multiple-purpose crops are as imperative now as in the 1950s and 1960s.

MINOR FOREST PRODUCTS

Traditionally the forests of Asia have produced a wide range of nonwood products. In the early years of this century massive tomes were published on the by-products of the forests of the Indian subcontinent, the Philippines, and Malaysia. There was significant international as well as domestic trade in furs and skins, fruits, resins (naval stores, copal), fungi, wild honey, medicines, aphrodisiacs, oils (*Cassia, Citronella*), sandalwood, bamboo, and rattan ware. That trade has never ceased—but it has been taken to new heights by China. Probably no country in the world processes as many wild products, and there is growing interest worldwide in China's natural foodstuffs and medicines. China's pharmacopoeia is little understood outside its frontiers but is certainly the most extensive in the world.

The Guangdong branch of the China National Native Products Import and Export Corporation (TUHSU) advertises 159 crude medicinal herbs (many of them without English or pharmacopoeial Latin names), 80 essential oils, 7 medicinal minerals, and 195 natural product (patent) medicines—for the treatment of every conceivable condition (and some beyond imagination). The major buyers are Japan (which imports some US$50 million worth annually) and Hong Kong. An exhibition in Beijing in 1986 of export products manufactured or processed by corporations of the PLA comprised over five thousand products including 150 natural food and medicinal products and a dozen different fungi. Since PLA units are concentrated in the border provinces of China, some of these products are commercially unique. Many, if not most, of the sources are woody plants.

The Kunming Botanical Institute is only one of many institutes in China attracting overseas finance for medicinal products research. Following the isolation of cynanchine—an epilepsy treatment—the Yamanuchi Co. of Japan is investing US$1 million in projects there. In 1963, most botanical gardens were undertaking screening programs for medicines (especially contraceptives); judging from the products now available, that research was successful, but much of it has yet to escape from China.

Also of interest are various forest fruits (which are dried, tinned, or candied for sale), oils, and edible tree seeds. Walnuts are traditional and production is recorded by the MOF. Output reached 128,000 tons (in-shell weight) in 1985. Resin (naval stores) production was nearly 400,000 tons; *Thea* oil seed, 536,000 tons; and tung (*Aleurites*) oil seed, 362,000 tons. There is a growing and high-value market for pine nuts both in Japan and in countries with a health food awareness. China is the world's biggest producer of *Pinus koraiensis* seeds—one of the bigger-seeded species—and seeds of *Pinus cembra*—the Siberian equivalent to the edible European nut species (*P. pinea*)—are also exported. China is also a major producer of shellac (4000 tons in 1985) and bamboo (more than 90 million cubic meters in wood volume, in addition to edible shoots). There is a shellac research institute under the Academy of Sciences, and virtually all of the thirty national botanical gardens (as well as countless provincial ones) are working on minor products.

Finally, a market with enormous growth potential is that for flower and shrub gene stocks. During the past two decades, for example, more than forty new *Camellia* species have been named in southwestern China as well as new orchids, azaleas, rhododendrons, begonias, lilies, and other highly valued plant species.

Table 19 lists trees commonly *planted* in China that yield products other than timber. Even in 1963, some two or three thousand plant species were thought to have an economic potential, and more than 500 were regularly used. A list of more than 100 desert species with dual sand-fixation and economic functions has been published (Richardson, 1966). The advocacy of multiple-purpose tree-planting in North America is a modern phenomenon, hailed as a "new direction"; in China it has an ancient history.

TABLE 19
COMMONLY PLANTED ECONOMIC WOODY PLANTS AND THEIR PRODUCTS

Species	Deciduous	Evergreen	Forest Product
Acacia catechu	+		tannin, medicinal
• *A. mearnsii*		+	tannin, medicinal
Aleurites fordii	+		industrial oil
A. montana	+		industrial oil
• *Amorpha fruticosa*	+		forage, osiers
• *Anacardium occidentalis*		+	nuts, timber
Areca catechu		+	nuts
Artocarpus heterophyllus		+	timber, fruit
Camellia oleifera		+	cooking oil
Canarium pimela		+	cooking oil, nuts, timber
• *Carya illinoensis*	+		nuts, timber
Castanea dandoensis	+		nuts, timber
C. henryi	+		nuts, timber
C. mollissima	+		nuts, timber
Cinnamomum cassia		+	medicinal
Cleidrocarpon cavalerei		+	cooking oil
• *Cocos nucifera*		+	cooking oil

TABLE 19 (*Continued*)

Species	Deciduous	Evergreen	Forest Product
Cornus walteri	+		cooking oil
Dalbergia balansae	+		wax
D. obtusifolia	+		wax
D. szemaoensis	+		wax
Elaeagnus mohls	+		cooking oil
• *Elaeis guineensis*		+	cooking oil
Eriolaena malvacea	+		wax
Eucommia ulmoides	+		medicinal
Fraxinus chinensis	+		wax, timber
Ginkgo biloba	+		nuts, timber
Hevea brasiliensis	+		rubber
Illicium vernu		+	spice, timber
Juglans regia	+		nuts, oil, timber
J. silillato	+		nuts, oil, timber
Listera chinensis		+	fiber, leaf
Litchi chinensis		+	fruit, timber
Lycium barbarum	+		medicinal
Magnolia officinalis	+		medicinal, timber
Mangifera indica		+	fruit, timber
Morus alba	+		fruit, silk
• *Olea europaea*		+	cooking oil
Pistacia vera	+		nuts, cooking oil
Prunus amygdalus	+		nuts, cooking oil
Rhus vernicifera	+		latex
Salix purpurea	+		osiers
Sapium sebiferum	+		wax
Thea sinensis		+	tea
Torreya grandis		+	nuts, timber
Trachycarpus fortunei		+	fiber, leaf
Xanthoceras sorbifolia		+	cooking oil
Zanthoxylum bungeanum	+		spice

SOURCE: FAO (1982).
• Exotic species.

The collection and utilization of various forest products are closely linked with "sideline" production—which, in Mao's China, was controversial, whether done by hunters and gatherers in the natural forests or by farmers on private plots. In 1963, there was a good deal of commune activity processing oil-bearing and aromatic tree species, and there were hydrolysis plants (producing ethyl alcohol), units preparing animal fodder from dried foliage, bamboo utilization operations, and a wide range of woodcrafts (cooperage staves, agricultural implements, tool handles, cabinetwork, furniture, ornaments, even fiberboard). At that time, the collection and processing of minor products were organized by the communes and the proceeds from their sale went—with minor exceptions—to the communal coffers. Controversy related to the distribution of the proceeds from the sale of products collected not from communal property but from plant and animal material deliberately grown on "private" plots. And because of conflicting and changing policies over these "tails of capitalism," peasant farmers were reluctant to grow anything more long-lasting than annual plants. When they were allocated land already supporting trees, they moved quickly to cut them in case the policy changed.

The agricultural development policy ratified by the party leadership in December 1978 reasserted the legitimacy of private plots, "domestic side occupations," and rural free markets. According to Travers (1986) the response was rapid—net income to peasant farmers from the sidelines jumped from 10 yuan per capita in 1978 to 51 in 1983 (FBIS, 1 November 1984). Moreover, the displacement of the inflexible "supply and marketing cooperative system" enabled free markets to grow from 33,000 in 1978 to 48,000 five years later. They were responsible for over one-third of total sales in the predominantly rural sector. According to Travers (1986:386): "The role of the peasant as entrepreneur, on or off the farm, is still tenuous but certainly more secure than at any other time in the last thirty years."

In some forest nurseries and experiment stations, sideline production is almost out of hand, with more time and effort devoted to marketable goods (the proceeds of which are shared between the organization and the individuals) than to the formal purposes of the institution. For example, the Liaoning Sand Fixation Afforestation Research Institute at Zhanggutai serves visitors with a banquet of dishes grown entirely on the station; the Beijing Zoo offers sideline production of meat and poultry from exotic zoo animals and birds; the "three treasures of the northeast" are ginseng, deer antler (for velvet), and the fur of the marten, all grown or gathered as sidelines.

The An Qian Secondary Forest Management Unit (Heilongjiang) established a ginseng garden in 1980—allegedly to promote year-round employment for forest workers. Nursery workers contract to grow a ginseng quota (2.5 *jin*/m^2) on wages, and any surplus is shared 40/60 between grower and the unit. The management provides seed, insecticide, and fertilizer (seeds of *Perilla,* also grown as a sideline). The project was so successful that tree seedling production suffered and contracts had to be amended to ensure sufficient stock to meet the planting program.

In Gaoyao county of Guangdong, a Hill Forest Unit contracts with families to raise cinnamon and *Pinus massoniana.* The cinnamon leaves are distilled for oil (using wood fuel) and the stems are dried for the spice. Each individual in the village is entitled to 5 *mu* (one-third hectare) of hill land for private use, and each family may obtain a further allocation (20–150 *mu*) of land already planted with trees (or up to 250 *mu* if the land is bare). A land rental is paid and the family retains the income generated. Cinnamon is a five-year crop and commands a much higher price than the fuelwood—the result is an evident reluctance to grow pines and the likely consequences of soil erosion.

Other "success stories" of the contract responsibility system applied to sideline production relate to bamboo, pine seeds (for eating), osier weaving, and the cultivation of medicinal *Cannabis sativa.* At the trade exhibition of PLA products organized in Beijing noted earlier, the Wuhan Edible Fungus Group Co., comprising several university departments and academic research units, offered the following fungi: *Auricularia auricula* (Jew's ear), *Agaricus bisporus* (common agaric), *Volvariella volvacea* (paddy straw mushroom), *Tremella fusiformis* (white wood), *Hericium erinaceus* (monkey head), *Flammulina velutipes* (silver coin), and *Pleurotus* spp. (oyster mushrooms). Sideline activities (not highlighted by enterprise management) include the cultivation of shiitaki and Jew's ear fungi on degrading logs in state woodyards. There is an annual migration of beekeepers to the northeastern forests of Jilin and Heilongjiang. Their hives are railed from the south and for a period of two months are established in yards to gather nectar, primarily from flowers of *Tilia amurensis.*

Finally, an activity that is fast becoming a major production operation—and one highlighting the growing importance of "specialized" households—is deer farming. Deer have long been well regarded medicinally in China, and the classic early materia medica cataloged the beneficial uses of pilose antlers—the invigoration of reproduction, nourishment of marrow and blood, strengthening of muscles and bones, and improvement of debility, hearing, vision, dizziness, and dysentery. In ancient China, most parts of the deer had medicinal value—tail, genitals, sinews, blood, meat, and bones; even embryos and slinks (aborted fetuses) had their own merit. Although deer have a high feed/meat conversion ratio, venison is not much eaten in China. Rather, the meat is dried and stewed with rice wine to fortify it; the meat may then be dried again and powdered to make tonic pills and to flavor herbs. One reason for venison's lack of appeal to the Chinese is the very property that accounts for its success as a farmed product in Europe and New Zealand—its failure to lay down fat. The burgeoning demand for lean meat in the industrialized countries is creating demands that China hopes to service, however, along with traditional markets for antler products. There may now be more than a million farmed deer in China.

China has eighteen varieties of deer, but the red (*Cervus elaphus*) and spotted sika (*C. nippon*) are the commonest domesticated species. There are over 150 deer farms in Heilongjiang and Jilin provinces and more than 30 around Beijing. Most of the northeastern farms are run by the MOF (which is also responsible for controlling production from wild deer and their capture), but there are many farms owned by counties, villages, and, indeed, other central departments (the PLA, provincial and municipal zoos, pharmaceutical enterprises). The Forestry Department and the Pharmaceutical Department of TUHSU have research farms. One of Beijing's biggest is at Longshan—an enclosed farm of about a thousand deer that has its own antler-processing plant. Other farms comprise herds domesticated from wild deer—initially as a sideline but now in a more highly organized way by specialized households. Live deer and some by-products (meat, sinews, pizzles, and embryos) are sold on the free market, but velvet and tails have (in theory) to be disposed of through the pharmaceutical administration.

CONCLUSIONS

China's forest economy is more open than in 1963—a characteristic matched by the awareness among foresters of the impending supply/demand crisis. National statistics were not available earlier, and there could be little meaningful discussion of "incomplete data." Foresters are more talkative now, and more concerned, but perhaps less confident than in the euphoric days of "afforestation campaigns" and the regular reporting of the "one millionth tree" planting by counties and, even, communes.

Ross (1983) in his analysis of the "obligatory tree planting" program points out that professional foresters were excluded from policymaking and that this may have seemed sensible because "ministry bureaucrats were unlikely to look favorably on a new program outside their orbit that might absorb scarce resources." When foresters are not consulted about forest policy, there is cause for professional concern. In 1963, forest policy development was firmly in the hands of professionally trained men. The older men have now gone, however, and their successors lack the vision that comes from experience. That they are hardworking is undoubted. But their life's experience may be limited to a single field and a single region of China. As is happening worldwide, China must begin to involve nonforesters in forestry and introduce foresters to the broader horizons of policy.

There is little doubt that the crisis is real, despite the apparently enormous extent of China's forest resources in absolute terms. In Table 10, we see that the forest land area is more or less equal to that of the United States and is exceeded only by the USSR, Brazil, and Canada. Moreover, the natural forests are diverse—ranging from tropical rain forest to the great Northern Coniferous Forest formation. But when the forest area (and standing volumes) are expressed per capita, the crisis comes into focus.

Forestry's several roles remain as important as ever in the economy and—given the rising unemployment among school-leavers in single-industry communities (exacerbated by a virtual ban on labor mobility)—may grow in social as well as economic value. A notable feature of China revisited is the extent and variety of sideline production of "minor" forest products. In part, this new direction reflects underemployment. The increasing complexities of policies and planning indicate a need for the enhancement of forestry's professional status in China. This may come—as it has in some countries—from the involvement of nonforesters in forestry decision-making. It would be assisted by greater exposure to international forestry through increased intergovernmental cooperation in forest science and technology—in a recent review of China–U.S. cooperation involving 500 programs and over 5000 participating scientists (Siddiqi et al., 1987), forestry does not receive a single mention.

Efforts to resolve the supply/demand crisis may also be hampered by China's limited experience of international forestry and the wood trade—and, above all, by the stovepipe vision of the bureaucrats, motivated by a complex of factors. If the enthusiasm of some specialized households (and individuals) involved in sideline production could be diffused through the entire sector, the outlook for forestry in China would appear much brighter than it does. At the same time, the energy generated by the new mercantilism, unless it is controlled, prompts fears of overexploitation of resources and environmental degradation. There is, perhaps, particular cause for concern about the commercial activities of the PLA in the border provinces.

Urging the substitution of a renewable resource (fuelwood) by a nonrenewable one (coal) is anomalous, but economic development is not always characterized by common sense. In Sri Lanka, imported kerosene is used to cure tea (and its cost subsidized) instead of wood. And imported canned fish is a staple of many islands possessing fishery resources.

Finally, it is difficult to see any easy accommodation with the increasing global concern for the establishment of protected natural ecosystems where human activities are minimized. China has a loose system of preserves (see Chapter 7), many of them including forest land, but the degree of protection currently provided is inadequate to satisfy the international conservation movement.

NOTES

1. This argument ignores values other than those of marketable products; it is not a concept that appeals to ecologists, environmentalists, and many foresters.
2. As noted elsewhere (Lardy, 1984; Richardson, 1987a), the absence of an alternative to the commune affects other matters besides tree-planting. There is no effective litter disposal outside the cities; China is not as clean as it once was; and there are doubts about the future of cooperative water conservation and education projects.
3. The Chinese adjective carries the connotation of "obligation" to the community rather than "compulsion."

Drying stacks of bamboo in Sichuan province.

An unusual plantation of the living fossil Metasequoia glyptostroboides *in Jiangsu province. This deciduous conifer was known from fossil records, and found to be still living in 1947 by Cheng Wanchun.*

Overcut and overgrazed natural forest in Nei Monggol (Zizhiqu). Spruce and true fir are the species.

Primary forest in the mixed coniferous and deciduous broad-leaved forest type in Jilin province in northeastern China. The conifer in the foreground is Pinus koraiensis, *in front of* Betula ermanii.

Log transport in a Shenyang logyard (Liaoning province).

Chinese-built 5-ton tractor, with the deck apron partly raised. The back plate can be lowered to take the butts of tree-length logs when it is used as a skidder.

A new house in the northeast built of traditional materials but with metal joinery. In the background are plantations mutilated by the collection of firewood.

A village sawmill operated by a "specialized household." The bandmill is Chinese-built, simple, but lacking any safety features. The bench is manually operated and, in terms of conversion yield and cost-effectiveness, it is more efficient than the so-called integrated wood complexes.

A fifteen-year-old plantation of Robinia pseudoacacia *in Henan province. The leaf litter is being gathered to feed animals. Foresters have tried unsuccessfully to end this traditional practice, since it results in soil impoverishment.*

Living accommodations in a suburb of Shenyang (Liaoning province). Because space is in short supply, the balconies are often filled in to provide additional room.

A log depot at Changchun (Jilin province). Yard hygiene leaves much to be desired and, characteristic of state enterprises, waste is high.

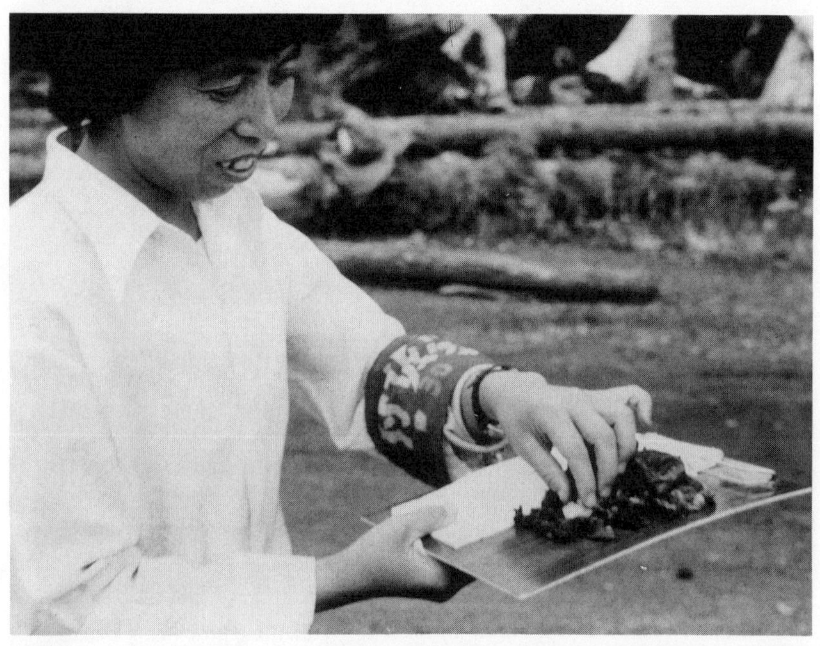

"Sideline production" in one of the state logyards includes the harvesting of edible fungi from rotting logs.

Traditional coffins made from Chinese fir (Cunninghamia lanceolata) *in Henan province. The prodigal use of wood is a demonstration of wealth and a practice the Chinese have attempted, in vain, to abolish.*

Spruce and fir logs on arrival at Chengdu (Sichuan province) after shooting the rapids on their journey from Qinghai. The logs are often split and the gravel embedded in them does considerable damage to the saws in the mills.

The forest nursery at Dailing (Heilongjiang province). In 1963, at this nursery, Russian-made knapsack sprayers were proudly demonstrated in the pouring rain. The nursery now has irrigation facilities and ornamental plants are grown for sale on some 25 percent of the area.

Multistory construction using solid wood concrete formwork and metal joinery. Safety is not a feature of Chinese building sites.

4

THE FOREST INDUSTRIES AND TRADE

CHINA'S ACKNOWLEDGMENT of a timber supply/demand crisis, together with the country's open-door policy, has had a massive effect on forest products trade. Imports of logs (mainly from North America) increased eighteenfold between 1979 and 1985. China has purchased cutting rights—and pulp mills—in foreign countries and has negotiated timber harvesting investment in Russia. Moreover, China is seeking to identify items it may be able to sell overseas and to organize continuing supply channels. An adjunct to this trade is the interest shown by foreign suppliers in China's future needs for woodworking machinery and equipment.

As in other developing countries, the wood-processing industry in China varies enormously in sophistication. Pit sawyers operate alongside computer-controlled bandmills; plymills using air-dried veneers coexist with modern particleboard and medium-density fiberboard (MDF) plants (and may, indeed, be part of the same complex). Since it is the likely development of the primary processing sector (notably sawmilling) that will determine the extent and mix of China's trade, the forest industries must be assessed in the Chinese context. China's policy of importing logs, rather than lumber, is economically and financially sound; the PRC has—or can develop—adequate processing capacity; and it is coming to appreciate the inappropriateness of forest industry development based on either the Russian or North American models.

The state sector of the forest industry exemplifies the problems—posed by China's bumbling bureaucracy—of "vertical controls and horizontal cleavages." The industry is characterized by appalling waste, despite the scarcity of raw materials. In my 1963 assessment (Richardson, 1966:97–101), I noted it was "unlikely that the plants visited are representative of the whole country. Like all hosts, the Chinese try to show their visitors only the best, so it is safe to assume that the operations . . . represent above-average conditions for the industries. . . . In spite of the *caveat,* the forest industries in China do not give the impression of thriving . . . machinery is old and in poor repair, and what remains in working order is often not used efficiently. . . . Factory hygiene and safety precautions leave much to be desired."

In 1986, opportunities to visit the wood-using industries were again limited, but the impression of viewing only the best was less strong. My earlier conclusions, outlined above, were still appropriate to the large-scale plants I saw in 1986 (and must be supplemented by the observation of prodigal yard practices). Village and suburban sawmills, by contrast, are models of thrifty utilization. In villages, visits to small mills and workshops were made on the spur of the moment and unheralded. This could not have happened in 1963.

WOOD PROCESSING

For present purposes, wood processing covers logging, sawmilling (including the so-called integrated wood factories), and pulp and paper production. The following paragraphs also consider working conditions in the state sector and the likely future demand for wood.

HARVESTING

The northeast since 1963 has seen the following changes:

Forest roads have replaced tramways.
Tractors (Chinese-built) are gasoline-fueled rather than wood-burning.
Skidders are replacing haulers.
Cable logging systems are being introduced.

Bucking and transportation scaled for longer logs are replacing short-log preparation as trucks and trailers become heavier.
Tree lengths may be recut into logs at railheads.
Loggers are less conscious of waste than they used to be.

The last observation is the result of contrast: Logging in 1963 was an object lesson in thrifty utilization. After clear-felling in 3–4-ha coupes and saw-log extraction, branchwood down to a 3-cm diameter was harvested for mining timbers, pulpwood, charcoal, and—in the case of hardwoods—tractor fuel. Even the twigs and hardwood foliage were collected for goat fodder. There is little evidence of such clean harvesting nowadays—though it is of interest that in plantation areas farther south, tree litter is garnered for fodder.

Felling coupes now vary in size up to 30 ha and are separated by buffer strips about 20 m wide (from which the biggest trees are removed). Clear-felling is the rule. Formerly, a few broad-leaved trees were left for soil improvement and to provide tractor fuel—but there do not appear to be any wood-burning tractors still operating, and the amount of debris left behind perhaps partly obviates the need for soil improvement.

Some 75 percent of state (MOF) felling is reportedly mechanized (compared with 60 percent in 1963)—the standard saw is Chinese-built but modeled on the 6-hp Russian Druzhba. It has a 60-cm cut, a "stand-up" frame, and weighs nearly 15 kg. The Forest Research Institute at Harbin has developed a new chain that promises improved efficiency, but there is scope for a new design of machine. (In view of the development of lightweight chainsaws during the past thirty years in North America, it is doubtful whether the stand-up frame—a marked advance on the earlier 50-kg two-man machines—now serves any purpose.) Most nonstate felling is still done by handsaws, and in the southwest and northwest there is evidence of ax-felling and crosscutting. In inaccessible areas, high-value logs may be felled and pit-sawn in place, and the sawnwood packed out by workers or animals. There is interest in China in the use of portable chainsaw mills in these situations—an interest that could quicken once the values of decorative and special-purpose species approach international levels.

In the northeast, logging is carried out in winter (November–March) because the fire danger season imposes restraints on summer logging. Tree-length skidding is the norm, using tracked logging tractors

(Chinese-built but modeled on the Russian TDT-40) with a tilting deck apron, a power winch behind the cab, and a fairlead on the apron. The tracks are 50 cm wide, with the front idler and driving sprockets raised from the ground, to enable the machine to traverse rough country. The deck apron, or back plate, can be used in the lowered position to lift the butts of tree-length logs off the ground for skidding, or it can be fully raised for transporting loads of short logs. China also now manufactures articulated wheeled skidders. (There were none made in 1963.)

Logging trams are still important in the northeast for log haulage but, increasingly, roads are being built and heavy-duty Japanese and Scandinavian logging trucks are being employed. In the south and southwest, log floating and rafting are practiced but losses (and damage) are high. For the first time in 1986, the contract responsibility system was introduced for rafting. If the log floating corporations were made at least partly responsible for *deliveries* (rather than being paid for inputs) and a value/grade payment system operated (both were reportedly being considered), standards might improve. Rafting of bamboo is common throughout central and southern China.

Some of the northeast forest areas selectively felled by the Japanese in the 1940s as well as the earliest post-1950 Chinese plantations are now being thinned (though most of the stands visited in 1963 have since been clear-felled, somewhat prematurely, for pulpwood). Cable logging systems are being used on a trial basis and small logs are sometimes floated in the spring and summer months. Elsewhere logging is very labor intensive.

In 1984 (MOF, 1985) the MOF controlled 10,500 km of rail-tramway and operated more than 700 locomotives. By contrast, the ministry had 6500 skidders but only 16,000 chainsaws. These statistics reflect a slow rate of change in the logging industry.

SAWMILLING

Statistics on sawmills in China have been published by FAO (1979), and MOF has provided updated figures for state operations. There are reportedly 298 sawmills operating with an output capacity ranging from 30,000 to 250,000 m^3. MOF estimated a further 21,000 sawmills with an average annual capacity of 10,000 to 20,000 m^3; this estimate does not include many of the village mills increasingly being operated by specialized households or as sidelines. The self-reliance policies of Mao

Zedong during the 1970s—and a growing sawmill engineering capacity—encouraged the installation of commune (now village) benches. They are adept at converting small, misshapen logs to service rural manufacturing industries.

The bigger mills are often part of the so-called integrated wood industries that formed the focus of an FAO/UNDP study tour in 1976 and a report published in 1979. In the northeast and the coastal cities of Shanghai and Guangdong, there is a long history of sawmilling under foreign control. The older mills were rehabilitated and expanded during the 1950s, and others were established with Russian assistance. By the end of the First FYP (1953–1957), sawnwood production had more than doubled (plywood production increased sevenfold). Since 1958, sawnwood production has increased at a much slower rate—to nearly 16 million cubic meters in 1985. (Wood-based panel production, however, has increased from a mere 140,000 m^3 of plywood to 1.7 million cubic meters of plywood, fiberboard, and particleboard.)

In general, the reconstructed and enlarged mills are located in the larger coastal cities (Shanghai, Tianjin, Guangzhou, Beijing) and in the northeast provincial capitals (Harbin, Shenyang, Changchun). New sawmill-based plants were built close to the forest areas being logged in the late 1950s and early 1960s. None has been designed with a view to ease of handling residue.

Many of the complexes are integrated in name only. They may comprise one or several sawmills and a host of small-scale ancillary plants—each unit representing an ad hoc addition to an original sawmill. Where there is more than one sawmill, it is because the ancillary plants (designed to use the sawmill waste) have grown too big to be supplied by the original mill and there is no space to enlarge the existing mill. An additional sawmill then creates additional waste—which calls for another recovery plant, and so on. The complexes resemble a Chinese banquet: a host of bite-size operations going on in the same arena, not obviously related, and leaving a mass of debris and waste.

Commonly throughout China in the state mills, and particularly where there are no free-market outlets, product quality is poor and wastage appalling. Not only do logs rot in the yards (as noted in Chapter 3, "sideline" production at some logyards includes the gathering for sale of edible wood-rotting fungi) but sawn timber—heaped rather than stacked—shows excessive deterioration. Few mills have drying kilns (even fewer operate them), and timber treatment is found only in mills operated by specialist units (such as the Ministry of Railways for sleepers). Panel products are exposed to sun and rain, corrugating from inadequate stacking. The big sawmills are invariably bandmills, though steam-driven frame saws and circular headrigs are in use in many smaller operations. They still cut all species (hardwoods and softwoods) with the same saws, speeds, and settings. Hardwoods and softwoods are often not segregated until they emerge from the mill and sometimes not even then. The sawmills have one or more lines comprising a headrig, two- to six-band resaws, and docking saws. They are very simple and, except in a few modern mills, material flows are entirely manual. A major problem is the imbalance brought about between headrig and resaws because of manual cross-transfers. Often there are four resaws to every headrig. Although the Chinese possess the capability to make modern machines and control systems, they lack flexibility because of electric power shortages. (Even in hydroelectric areas, factories are rostered to close up to three days a week to avoid overconsumption.)

Moreover, the economic inefficiency of the state sector does not encourage investment in the forest industry. The timber industry was the only one of eleven production sectors to show a consistent decline in GVIO per worker since 1952. The *China Daily* for 6 September 1986 recorded the after-tax profit (per employee) of state forestry enterprises at 194 yuan for 1984 and 219 yuan for 1985. ("Profit" calculations, of course, disregard costs of capital.) Despite the existence of national timber standards, most mills do not (or cannot) cut to them—though China Timber Corporation (CTC) officials claim that 60 percent of State Plan sawnwood is cut to the standards (10 million cubic meters in 1985).

In contrast to the big complexes, the stand-alone sawmills are generally cost effective and much less wasteful (because they are located in areas where waste is readily disposable). They are, moreover, forced to cut poor-quality logs and maximize conversion. Despite claims of 70 percent lumber yields reported by the NFPA (1986), no state mills that I visited in 1986 would have exceeded 55 percent volume recovery. Many of the smaller suburban mills, on the other hand, approach 70 to 75 percent.

The village benches are 10–20-cm bandsaws (with tables or manually operated carriages) where there is an electric power supply, or circular where there is not. They cut low-grade logs to service local furniture shops, joinery, coffin makers (still a major wood use in rural China), implement manufacturers, and others.

There are reportedly 260 "integrated" wood factories under the MOF in China—including 87 plymills, 223 fiberboard plants, and 22 particleboard lines. In addition, 102 plymills, 141 fiberboard plants, and 89 particleboard lines are run by other ministries or as stand-alone operations by the MOF. Panel production was 1,659,000 m^3 in 1985 and is planned to increase to 2,500,000 m^3 annually during the seventh FYP. Though there is strong demand for plywood, the bulk of the increase in domestic production of panels is likely to be of particleboard and other reconstituted wood panels. There are over thirty new lines under construction or planned—including oriented-strand board (OSB), medium-density fiberboard (MDF), waferboard, wood/wool cement, and flakeboard. The two MDF plants under construction in 1986 were of 50,000 and 30,000 m^3 capacity respectively, but the plymills and particleboard plants are generally smaller. Because of the poor quality and high cost of fiberboard made in China, particularly in the very small plants (some produced less than 300 m^3 annually), it is likely that other products will displace it. Many of the fiberboard lines seen in 1963, despite their acclaim then as outlets for waste, are no longer operating.

There have been problems of design and planning with some of the newer plants, for which the Forest Industry Design Unit of the MOF in Beijing—as well, perhaps, as some equipment suppliers—must be held partly responsible. The following paragraphs cite examples of poor design among the factories I visited.

The *Kunming Timber Processing Factory* (Yunnan) expanded from a sawmill built in 1960 into furniture and then plywood and particleboard. Slice veneering equipment (Italian and West German) was being installed in 1986. Steaming and drying capacity is 9000 m^3, but the slicing capacity was only 3000 m^3 and there was no sander; another slicer would be needed as well as a flitching mill. The vice-director of the complex complained of having seven "mothers-in-law"—seven separate committees involved in decision-making, none of which would forgo its right to involvement: the MOF; Kunming City Council; Yunnan Province Economic Committee; Yunnan Provincial Planning Committee; Kunming City Economic Committee; Kunming City Planning Committee; and Kunming Furniture Manufacturing Corporation. It was expected that at least two years would be needed to commission the new installation. The factory had been designed without any long-term assurance of sufficient veneer logs.

The *Chan Jing Integrated Wood Factory* (Yunnan) was a sawmill producing 12,000 m^3 annually with a particleboard line (3000–5000 m^3

annually) and a three-shift plymill making 1000 m^3 plus 500 m^3 veneer. The plymill used sliced veneers only, air-dried and laid up in a single-daylight, manually operated press. The line was to be supplemented by a twelve-daylight press and full-scale dryer, which would have six times the capacity of the slicer. Again, no sander had been ordered (or budgeted) but a veneer lathe had been proposed. By the time it is installed, the high-cost dryer and press will have been idle for over two years. As in all state operations, costings made no allowance for cost of capital.

The *Chengdu Comprehensive Timber Factory* (Sichuan) incorporates a 100,000-m^3 output capacity sawmill (two shifts), a 7000-m^3 plymill (two shifts), a 4000-ton fiberboard mill (three shifts), and a 3500-ton pulp and paper mill. The 1986 log allocation, however, was only 80,000 m^3. The factory purchased 3000 m^3 in the free market and had a deal with Hubei province to exchange Chinese fir logs for Indonesian lauan; nonetheless, it operated at well below capacity. There were no veneer splicers, and the high-speed sander could only operate for two days a week. Since the wood and power supplies were insufficient, the complex could meet its production targets with single-shift operations—but because the work force is fixed, the plant operated for the equivalent of two half-shifts. As in Kunming, the director was aware of deficiencies but unable to take action to overcome them.

The *Langxian Particleboard Mill* (Heilongjiang) was a new installation alongside an original logyard and a sawmill (in 1986 operated with Canadian aid). The particleboard plant was built in 1984 with predominantly West German equipment. Even at full stretch it could reportedly reach only 75 percent capacity, because a dryer was omitted in the design. A sander had only just arrived (in September 1986) and had not been installed. There were also power shortages at Langxian.

Most of these complexes feed joinery, box, and furniture plants of varying standards. There is little to change in the comments I made in 1963:

> The furniture and joinery factories also show little evidence of efficient mass production, though the quality of individual workmanship is high. Factory layout is haphazard with much wasted space and even converging assembly lines carrying different products. The Peking wood factory, for example, produces cupboard units, chairs, tables, and a variety of miscellaneous items; any or all of these may arrive at the finishing shop at one time, giving an impression of utter confusion. Lack of "good housekeeping" on the shop floors creates a

> high fire hazard. Mechanization is primitive and is usually limited to devices designed and built by the factory employees. One cannot help but admire the ingenuity of some of these precarious pieces of equipment—and the enthusiasm of their designers—but they scarcely represent realistic examples of modern industrialization and mass-production techniques, which is what the Chinese often claim for them. In general, apart from the initial cutting and planing, all joinery operations are manual, including assembly, painting, and polishing. Again, the Shanghai factory (which produces only sewing machine cases) appears to be more efficient though here, also, most of the operations are manual and factory hygiene leaves a lot to be desired. The same criticisms can be leveled at the box factories. [Richardson, 1966:101]

Some malpractices stem from the "cost-plus" system of pricing output. A state mill in Guangzhou was cutting battens (purlins) from 35-cm-diameter imported Douglas fir logs—reportedly, the price set for purlins was 10 yuan higher per cubic meter than for wide boards. In Changchun, drying costs were 80 yuan/m^3 (equal to more than twice the basic monthly wage). Thus "there was no demand." And where kilns were operated, moisture content often could not be measured and no assurance of quality standard given. (Yet, in Harbin, the Forestry Research Institute is working on microwave drying of wood—a world away from industrial practice.)

A frightening disregard for safety is a feature of many mills—straw hats and open-toe rubber shoes are often seen on shop floors. The practice (on a rural bench) of touching up the bandsaw with a sharpening stone while it was still in operation was, fortunately, unique.

I do not claim that the examples cited here are representative of China's forest products industry. There are well-run plants that stand comparison with the best in the developing world. Particularly in Shanghai, Shandong, and Fujian, there are traditions of high-quality wood manufactures, including panel products. And, in all provinces, there are specialist woodworkers whose craftsmanship would not have disgraced the master carvers of ancient China. The points to be made, however, are these:

1. The hierarchical structure of the ministerial systems in China (and the cleavage between hierarchies) does not facilitate rational and integrated planning.

2. The multiple and pervasive roles of the "integrated complexes" in the communities dominated by them make these communities more vulnerable than most to haphazard planning.
3. Management is constrained by cumbersome decision-making structures (the "mothers-in-law"). In these circumstances, smaller and simpler plants function more effectively than large, complex models. As a result, medium-sized, stand-alone operations are more cost-effective than the large, supposedly integrated, operations.
4. Optimum operational scales for sawmilling and reconstituted wood products do not coincide.
5. The new reform policies call for greater awareness of the economic costs of capital investment—and the opportunity cost of failing to utilize fully both expensive machinery and scarce entrepreneurial manpower. Such awareness has yet to touch the bureaucrats in low-status industries such as forestry.

Another constraint to which many Western visitors to China refer is the unclear managerial role of the party cadres (and committees). They are present in every enterprise, and relations between them and management are not clear-cut. Some of the cadres, however, are good managers as well (or ill) trained in management as the technicians. (In the forest industries there is no formal managerial training. In seven provinces, only Guangdong provides training for forest industry management per se—and that is restricted to a maximum of three-month courses.) Again, problems stem from the division between two hierarchies: party and ministry. As noted in Chapter 1, despite the clear intention of the State Council Regulation of May 1984 that the operational authority of management should override that of the party, practical issues have yet to be resolved at provincial and enterprise level.

Pulp and Paper

I lay no claims to expertise in pulp and paper technology or economics. It is a specialist area and particularly so in China: No country in the world uses so wide a range of raw materials or boasts a variation in mill size (and technology) from less than 1 ton per day (TPD) to the latest expansion of Guangzhou mill to over 500 TPD. Virtually every known pulping process is used in China.

Chinese sources usually distinguish only machine-made and handmade paper. There are, however, big differences within the machine-made sector; the large "modern" mills (100 TPD plus) use Scandinavian technology of the 1950s (before there was much global concern about pollution and environmental protection); small plants (1–20 TPD) use mainly nonwood raw materials to produce a great variety of low-grade papers. Traditional handmade papers use no wood (except recycled fiber) but a wide range of agricultural residues and bamboo. There are three kinds of handmade paper—art and calligraphy grades; run-of-the-mill printing papers; and sanitary and packaging papers. Thus handmade papers can substitute for machine papers at all levels.

Statistics of mills and capacity vary according to source. The American Paper Institute estimates "approximately 4000" pulp and paper mills (CBR, 1981). Chinese sources claim "more than 1500 paper mills above the county level" (Anon., 1982). A brochure issued with an announcement of "the first international pulp and paper technology exhibition and conference" held in Beijing in June 1987 refers to "some 1600 paper and pulp enterprises under the Ministry of Light Industry and the Ministry of Forestry." Qu Geping, chief of the Environmental Protection Agency, has referred to "over 8000" pulp mills, lamenting the "hectic problem" they create for his agency (Qu, pers. comm., 1987).

Some thirty mills have capacity in excess of 100 TPD and account for 25 percent of the national output: They include mills at Qingzhou (Fujian), Liuzhou (Guangxi), Yueyang (Hunan), Jiamusi (Heilongjiang), Jilin, Shixian, and Kaishantun (Jilin), Guangzhou (Guangdong), and newsprint mills at Qiqihar (Heilongjiang), Nanping (Guangxi), and Beijing. The first four of these mills have recently been modernized using U.S. and Japanese technology. There are also major mills in Anhui, Fujian, Gansu, Hebei, Hubei, Jiangsu, Liaoning, Shanxi, Shandong, Shanghai, Sichuan, Tianjin, Yunnan, and Zhejiang—mostly in the capacity range 50 to 100 TPD. The initial focus of the "modern" industry was in the northeast and the coastal cities of Shanghai, Tianjin, Hangzhou, and Guangzhou—reflecting the use of wood fiber and imported carrier pulp. Outline descriptions of mills in the main centers (Beijing, Guangzhou, and Shanghai) are contained in a series of "Reports from China" in *Pulp and Paper International* (Kalish, 1982).

FAO estimates of production and consumption of wood-based mechanical and chemical pulp, together with newsprint and other papers, are summarized in Table 20. The 1984 data compare reasonably well with figures provided by MOF except for printing and writing papers (for which the MOF production figure is 1,480,000 metric tons). Provisional total paper production for 1985 (MOF) was 9,110,000 metric tons, and an increase in capacity for machine-made paper of 21 percent was projected. It is likely that official statistics underestimate actual paper production—perhaps by up to 2 million metric tons (according to a Canadian estimate).

Less than 30 percent of the raw fiber used in papermaking is wood. The early northeast mills (wood-based) belong to the MOF, but the proliferation of small, nonwood mills led to the involvement of the Ministry of Light Industry. Inevitably there have been problems of bureaucratic cooperation—creating anomalies in materials allocation and waste. A Canadian pulp and paper mission to China in 1977 expressed astonishment at good-quality sawlogs (up to 30 inches in diameter) being chipped for pulp, reporting that "the experts with whom we had contact, being responsible for pulp and paper . . . were not involved in forestry." Other visitors (see *World Wood,* August 1984) remark on the fact that very little by way of sawmill residue goes to the pulp mills. There are reports, too, of individual plants vying with each other for resources. In Hunan, there are six state mills of 30–40 TPD, fifty-nine small "county" mills, and ninety-five collective mills. The county and collective mills have no recovery systems and pollution problems are acknowledged (FE, 4 March 1981; Zhong, 1985); they compete successfully with the state mills for wood and agricultural residues.

Residues comprise reeds (which are farmed), bagasse, rice and wheat straw, flax, kaoliang, soybean fiber, ramie, and cotton stalks. Except for reeds, the availability of residue fluctuates from year to year according to the harvest; for this reason, technicians would like to build up the supply of "economic crop" raw materials—including *Miscanthus* reed, Kenaf (*Hisbicus*), and plantation-grown pine and eucalyptus. In Guangzhou, special crushing and shredding techniques have been developed to prepare *Pinus massoniana* bolts using cement and brick rollers. The use of wood, however, is more energy-demanding, and wood is more costly to transport than agricultural wastes. On the other hand, wood pulp poses fewer pollution problems (both air and water) than do crop wastes. In general, fiber lengths of nonwood materials are shorter than wood raw materials and contain many more nonfiber cells (some twenty times more). Higher silica content in black liquor and high pectin

TABLE 20
PULP (WOOD-BASED) AND PAPER PRODUCTION AND CONSUMPTION: 1976–1985 (IN 1000 METRIC TONS)

Product	1976	1981	1982	1983	1984	1985
Wood Pulp						
Mechanical						
Production	224	330	330	340	373	403
Net imports	2	3	8	3	31	16
Consumption	226	333	338	343	394	419
Chemical						
Production	526	778	778	780	832	938
Net imports	181	439	300	500	544	454
Consumption	707	1217	1078	1280	1376	1392
Total production	750	1108	1108	1120	1205	1341
Total consumption	933	1550	1416	1623	1770	1811
Nonwood Pulp				3608	4147	5024
Total Pulp				5231	5917	6835
Paper						
Newsprint						
Production	328	323	352	396	413	425
Net imports	20	165	100	100	50	236
Consumption	348	488	452	496	463	661
Printing and writing						
Production	1143	1557	1698	1906	1574	1889
Net imports	n.s.[a]	n.s.[a]	n.s.[a]	n.s.[a]	35	27
Consumption	1143	1557	1698	1906	1539	1916
Other paper and paperboard						
Production	1872	3522	3840	4311	5573	6790
Net imports	77	371	256	356	435	263
Consumption	1949	3893	4096	4667	6008	7053
Total production	3343	5402	5890	6613	7560	9104
Total net imports	97	536	356	456	520	526
Total consumption	3440	5938	6246	7069	8080	9630

SOURCE: FAO for 1976–1983; MOF for 1984–1985 (pers. comm., 1986).
[a] Not significant.

(especially in cotton stalks) lead to high alkali consumption and create difficulties in recovery that have not yet been solved (see, for example, Wang et al., 1986). Nonetheless reed, grass, and straw pulps are suitable for low-strength papers. (See also Yu, 1983.) In view of China's big resources of bamboo—and the country's requirements for long-fiber pulp—surprisingly little use is made of it. The recent installation of a chemimechanical pulping (CMP) pilot plant in Beijing may enable the use of a wider range of species.

Administrative squabbles contribute to the low level of development in China's pulp and paper industry. Table 21 illustrates industrial growth of selected light-industry products from 1970 to 1981 according to a World Bank assessment. The switch from heavy to light industry is evident. Of light-industry products, however, paper and paperboard gave the poorest performance. The paper industry is one of ten light industries singled out for priority importation of machinery under the Seventh FYP (EIU, 1986). The output of machine-made paper and paperboard is targeted at 10 million tons by 1990 and 17 million tons by the year 2000.

The major technical problems facing China's pulp and paper makers relate to energy and pollution. Electric power supplies are unreliable in most of China, and there is temptation to use lower than optimum temperatures in paper-machine head boxes and for pulp washing and other operations. As a result, the consumption of fresh water is excessive and recycling inadequate. Where there is concern with recovery and

TABLE 21
INDUSTRIAL GROWTH:
1970–1981

Category	Annual Change in Gross Output (%) 1970–1977	1978	1979	1980	1981
Type of Industry					
Heavy industry	9.3	15.6	7.7	1.4	−4.4
Light industry	7.8	10.8	9.6	18.4	14.1
Total	8.7	13.5	8.5	8.7	4.1
Selected Products					
Cotton cloth	1.5	8.7	10.2	10.9	5.9
Chemical fibers	9.4	49.9	14.4	38.0	17.1
TV sets	60.2	81.8	157.1	87.5	116.5
Cameras	29.5	−27.5	33.0	56.7	67.0
Bicycles	10.5	15.0	18.2	29.0	34.7
Sewing machines	8.8	14.7	20.6	30.9	35.3
Machine-made paper	6.6	16.4	12.4	8.5	0.9

SOURCE: World Bank (1985c).
NOTE: 1970 prices to 1980; then 1980 prices.

recycling, it is for the sake of saving materials rather than for environmental protection. Thus calcium-based wood-pulp mills process waste liquor for ethyl alcohol, fodder yeast, and vanillin whereas magnesium-based residue mills concentrate waste liquor for dispersants and road binding materials. Most mills, however, sewer their cooling liquor untreated. Given the generally small mill size, levels of water consumption are acceptable. The expansion envisaged, however, will pose major problems, and China looks to modern pulping technology to help resolve them. There is a thermomechanical pulp mill (TMP) operating in Jilin and a chemi-TMP pilot plant (funded by UNIDO) processing agricultural residues and the like in Beijing. China intends to develop chemimechanical pulping for bamboo.

The way in which China's pulp and paper industry has developed has (inevitably) involved a mixture of technologies and equipment. The 1977 Canadian mission noted a mix of Czech, Finnish, Japanese, British, French, Swedish, Russian, and Chinese equipment, posing obvious problems of replacement parts. From Zhong (1985) it appears that current modernization may involve more standardization.

Apart from technical problems, there are administrative constraints to modernization. Most small pulp and paper mills belong to collectives, not the state sector, and are subject to less central direction. They generate no foreign exchange and are at the end of the line for modernization. They are inefficient energy consumers and major environmental polluters; yet they produce more than 60 percent of China's printing and writing papers. Zhong (1985:V63) observes that "their continued existence will certainly bring about serious problems in the modernization schemes of the industry as a whole . . . most will be able to survive for an extended period of time, mainly to meet the demand of rural areas . . . nothing seemingly can be done about them until perhaps one day they become conscious of their noncompetitivity and decide to shut down." A possible—and sensible—compromise would be to close down their pulp mills (for environmental reasons) but continue to produce paper from commercial pulp.

Labor Organization and Working Conditions

My earlier survey (Richardson, 1966) gave examples of wage rates and living standards and discussed the role of trade unions in the administration of social amenities in the woodworking complexes. There have been few changes between 1963 and 1986. At the Beijing No. 1 Wood

Factory, for example, the minimum wage in 1963 was 33 yuan and the average was 62 yuan. Throughout the industry, eight grades of skill were recognized and paid accordingly and there were supplements for experience and qualifications. At the same factory in 1986, there were still eight categories of skill and the wages ranged from 38 to 125 yuan per month with bonuses for "quality" reportedly amounting to 10 to 20 yuan per month. (The minimum wage was 30 yuan with a night allowance of 0.5 yuan per night.) Thus there have been no changes in wage rates—improvements, rather, have come through bonuses and in the value of social amenities. (Housing is free for single men, for example, and families are charged at a rate of 2 yuan per room or 4 yuan per room if in a high-rise apartment block and exceeding 15 m^2; medical treatment is free; and food is provided at 0.4 yuan per meal.)

Walder (1987) makes the point that there was a virtual wage freeze in China from 1963 to 1977. In state industry, the real wage index of 1956 = 100 actually dropped during the 1960s and 1970s, only regaining the 1956 level by 1984. Bonus payment systems and piecework were revived in 1978, and by 1984 over 40 percent of the total wage bill was in bonus form. In any comparison of living standards (temporal or spatial), official wage data can be misleading because it is virtually impossible to compare the roles and value of welfare and amenities. Lardy (1984) expresses astonishment that discussions of income levels in China usually ignore significant differences between state employees and peasants with respect to subsidies—which, even by socialist standards, are substantial. Apart from the customary education and welfare services, they cover a range of consumer goods (including cereal staples, edible oil, and coal) sold at highly subsidized prices through work units or the trade union system for state employees. In 1978, such subsidies amounted to 82 percent of the official average wage.

Moreover, the trade unions (which operate only in the state system) administer retirement, long-term disability, maternity, and death and survival benefits, all underwritten by the state. Work units provide health care, recreation, childcare facilities, and housing; state employees may also qualify for travel allowances (if required to live more than three bus stops from the place of employment) and the costs of traveling home for a funeral or visiting a spouse. As noted earlier, work units can obtain occasional rations of timber and plywood for allocation to their members. Despite extensive discussion in China's press about the need to reduce the level of subsidies, according to Lardy (1984) their value (together with fringe benefits) to "nonagricultural workers" at the end of 1983 approached 1000 yuan—double that in 1978. There would obviously be resistance to change and, especially, to any attempt to replace benefits by cash—since tax is levied on incomes more than one and a half times the average.

Eberstadt (1986) draws attention to certain paradoxes with respect to China's performance in alleviating material poverty. He points out that while the Chinese have reached a life expectancy enjoyed by only a few low-income countries (65 years compared with an unweighted average of 64 for thirteen other developing countries), this achievement has involved considerable loss of life from famine and malnutrition. China has achieved it, moreover, at a low level of gross domestic product (GDP) per capita (US$310 compared with an unweighted average of US$1520 for the other developing countries). Second, while education tends to rise with national income, in China enrollment at the primary level has apparently fallen since the surge in GDP began in 1978—perhaps because of a declining birthrate and a reluctance on the part of peasant families to forgo the earning potential of their children. Finally, whereas development normally reduces the mortality difference between rural and urban populations and between male and female fractions, China does not follow this pattern. Eberstadt suggests that these unusual features reflect the power of the state in imposing its own social preferences and development priorities on its population. The statistics show, too, that there is considerable scope for increasing the efficiency of the manpower resource—at least through improved technology and the allocation of other inputs.

As well as quantity, the quality of amenities provided by enterprises can vary substantially. When the emphasis is on production, services like childcare tend to decline in quality and, particularly during the Cultural Revolution, the enterprise kindergartens became little more than babysitting services. Improvements have been selective and, as noted in Appendix A, some amenities are said to be allocated on the basis of seniority and political influence rather than merit. Differences are most noticeable between, say, factories and universities, but there are also variable standards within similar types of work unit. As in all institutions, quality depends on the conscientiousness and morale of the management.

Most recent studies of "quality of life" in China concentrate on urban standards. In the rural areas economic development is undoubted, but it is less clear that the standards of social welfare and amenity which the communes used to provide are being maintained.

Future Demand in China

In the 1965 and 1972 commodity reports, much attention was given to demonstrating that China would emerge as a major market for wood. It is no longer necessary to argue this point; nor is there any need to quantify with great precision the volume of likely total demand. The NFPA report (NFPA, 1986) relates wood use to GVIAO (the Chinese equivalent of GNP)—and, even assuming a *reduction* in wood consumption in relation to economic growth, there will be a dramatic increase in total wood demand as long as GVIAO continues to follow the trend of the last decade. Restricting consideration to roundwood consumption equivalent to that currently allocated within the State Plan, the NFPA accepts MOF projections of a demand for 156 million cubic meters by the year 2000 and imports equal to 34 percent of the projected demand/supply deficit of 56 million cubic meters (that is, some 20 million cubic meters annually). The NFPA projection of wood consumption related to planned growth of GVIAO is illustrated in Figure 1.

The assumption of a reduction in wood consumption per unit of GVIAO (from 0.08 million to 0.056 million cubic meters per billion RMB) is ultraconservative at this stage of China's development. It would be reasonable perhaps in societies with a zero population growth rate and an advanced infrastructure in place, but China is still growing and has increasing per capita demands for the products of major industries that use wood. Despite pressures to use steel and concrete instead of wood in construction—to design wood out of society—substitution equivalent to nearly 70 percent of the current plan harvest is unreasonable. Nor are certain other postulates of the MOF paper (on which the NFPA forecasts are based) very convincing. For example, it is estimated that by 2000 the opening of new logging areas will increase the harvest by 10 million cubic meters a year, that thinnings and fast-growing plantations can yield 27 million cubic meters a year, and that wood (and fuelwood) harvested ex-plan, but turned over to the plan, will reach 13 million cubic meters. As noted earlier, the assumption that illegal logging can be eliminated is also without foundation.

Given the more reliable population and economic data base now

FIGURE 1.
FORCAST OF WOOD CONSUMPTION BASED ON PROJECTIONS OF GVIAO

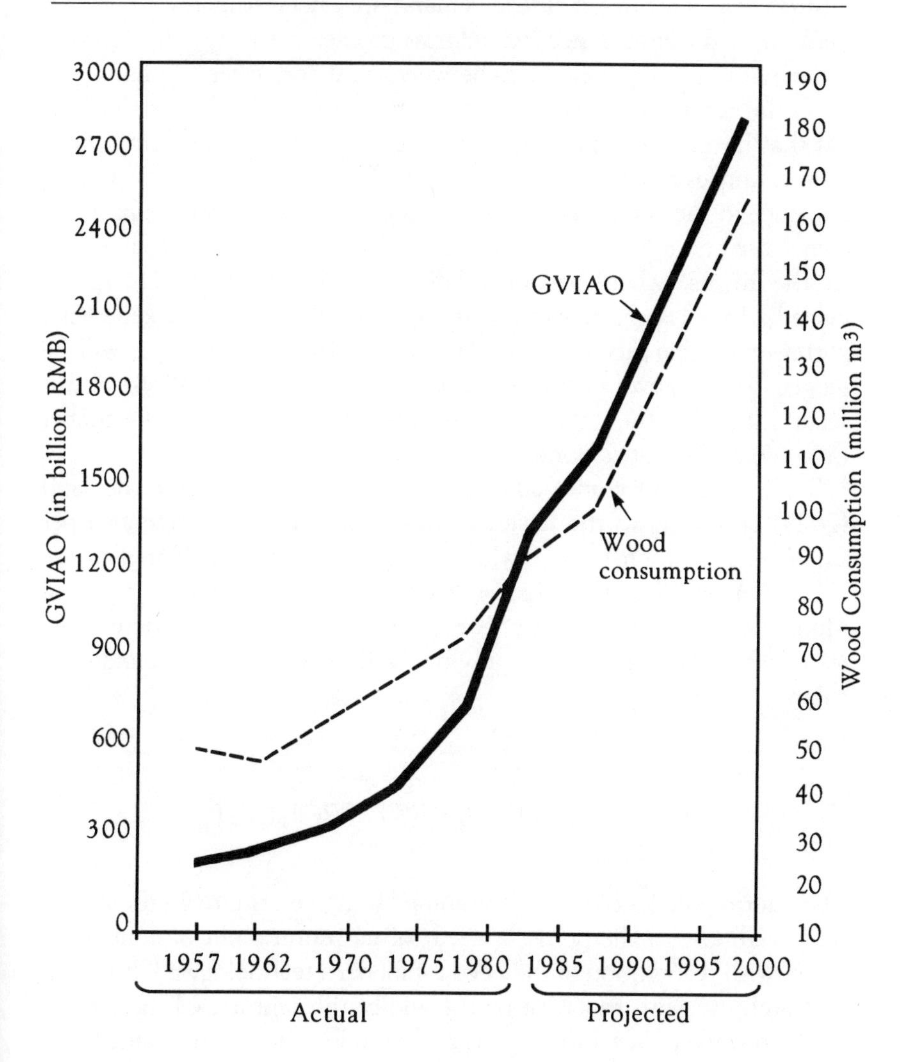

Source: NFPA (1986)

available, it is perhaps sufficient to project future *total* industrial demand on the basis of population and per capita consumption projections. Appendix A discusses population projections and analyzes the major wood-using sectors (construction, furniture, transport, mining, plywood, and "other sectors") in some detail. One must conclude that China's own forecasts of future demand are underestimates—not least because of the urgent need to address problems arising from growing differences in living conditions between rural and urban China.

To the returning visitor there is no more dramatic change than the rift that has developed between town and country—a rift that in most poor countries favors the urban dwellers. But in China there is growing affluence in the countryside and—no doubt because of private property ownership and inheritance—savings are being channeled into housing. In the towns, where accommodation is state-owned or enterprise-owned, there is overcrowding and dilapidation (despite the superficial glitter of the big city centers). The success of reform policies will be judged by the government's response to this unusual duality—and it seems likely that the state may have underestimated the need for timber in residential construction.

Table 22 shows roundwood production (1984) and projected demand (2025) according to the lowest of three scenarios. It represents a per capita value of 0.2 m^3 a year, which is on the 1984 level of Hungary (see Table 17) and less than half the current apparent consumption of Mongolia or the USSR. By 2025, given existing production forest resources, only Heilongjiang province will have a positive roundwood balance.

TRADE AND FOREST PRODUCTS

The reform policies of post-Mao China have given rise to an upsurge in trade—almost to the point, indeed, of its running out of control. It began with recognition of the need for modern technology if China was to "catch up" with the West by the end of the century. Chinese references to trade policy still emphasize technology but, at the same time, there is a quiet acknowledgment of China's need for certain raw material imports and a readiness to export scarce materials if cheaper substitutes can be developed locally or imported. This reasoning underlay China's decision in the 1970s to export petroleum to high-priced overseas markets and to develop resources of coal and other fuels for domestic power. It is also the rationale behind China's present eagerness to export high-value forest products and to replace them with cheaper imports.

TABLE 22
ROUNDWOOD PRODUCTION AND DEMAND PROJECTIONS BY PROVINCE

Province	1984 Production (million m^3)	Demand Projections for 2025 (Lowest Scenario) (million m^3)	Balance
Hebei	0.122	15.3	−15.2
Shanxi	0.129	7.3	−7.2
Nei Monggol	4.784	5.6	−0.8
Liaoning	0.689	10.5	−9.8
Jilin	6.334	6.6	−0.3
Heilongjiang	16.683	9.6	+7.1
Jiangsu	0.350	17.5	−17.1
Zhejiang	2.037	11.4	−9.0
Anhui	0.464	14.4	−13.9
Fujian	7.280	7.5	−0.3
Jiangxi	3.744	9.6	−5.9
Shangdong	0.106	21.5	−21.4
Henan	0.099	21.5	−21.4
Hubei	0.731	13.8	−13.1
Hunan	3.760	15.6	−11.8
Guangdong	4.516	17.2	−12.7
Guangxi	2.040	10.5	−8.5
Sichuan	4.570	28.8	−24.2
Guizhou	0.827	8.3	−8.2
Yunnan	3.066	9.4	−6.4
Xizang	0.219	0.5	−0.3
Shaanxi	0.458	8.4	−8.0
Gansu	0.548	5.7	−5.2
Qinghai	0.063	1.1	−1.0
Ningxia	0.043	1.1	−1.1
Xinjiang	0.477	3.8	−3.3

SOURCE: Cotchell Pacific (1987).

China's Trade

Table 23 illustrates changes in China's trade since 1979; Table 24 summarizes trade values with China's major trading partners in 1984–1985. Two-way trade has more than doubled but was fairly evenly balanced between imports and exports until the runaway import spree of 1985. China's imports are predominantly from the industrialized countries, but most of its exports are to the less developed countries, primarily in Asia.

Two-way trade with the socialist countries is declining. In fact, the

TABLE 23

TRADE GROWTH: 1979–1985

(IN MILLIONS OF U.S.$)

	Exports							Imports						
Category	**1979**	**1980**	**1981**	**1982**	**1983**	**1984**	**1985**	**1979**	**1980**	**1981**	**1982**	**1983**	**1984**	**1985**
Trading Partner														
USSR, Eastern Europe, etc.	1,211	1,191	782	894	1,032	1,251	2,031	1,338	1,310	811	1,251	1,360	1,535	2,144
Albania	—	—	—	—	4	3	10	—	—	—	4	7	2	6
Bulgaria	—	35	29	22	24	13	20	—	30	22	41	56	36	39
Cuba	93	77	117	97	95	117	118	127	189	268	201	133	108	—
Czechoslovakia	113	147	68	72	90	113	225	162	131	56	187	113	140	251
Eastern Germany	198	169	100	69	59	93	112	198	260	92	136	187	138	233
Mongolia	—	4	3	2	2	2	5	—	4	2	2	2	2	2
North Korea	317	374	300	281	273	226	239	330	303	232	304	254	272	245
Poland	143	141	83	187	164	122	267	167	190	65	66	100	142	245
USSR	242	228	123	143	319	585	1,037	250	264	154	243	441	670	1,017
USSR, etc., not specified	105	—	—	—	—	—	—	114	—	—	—	—	—	—
Country or area not specified	1,143	—	3	4	12	7	16	632	—	535	374	297	233	69
Memorandum Items														
EEC	2,363	2,502	2,168	2,508	2,232	2,283	3,354	2,814	2,714	2,178	3,390	3,323	6,151	—
Oil-exporting countries	437	661	1,116	999	848	736	714	135	265	122	359	319	409	614
Non-oil developing countries	5,227	8,178	10,136	10,598	10,881	12,423	13,163	2,489	3,575	4,075	3,907	4,831	5,866	9,888

Percentage Distribution														
Industrial countries	41.3	44.7	44.0	42.9	42.2	41.9	41.7	70.7	73.6	74.4	68.9	68.1	69.0	70.1
Developing countries	41.5	48.7	52.4	53.0	53.1	53.0	50.8	16.7	19.7	19.4	22.5	24.2	24.2	24.7
Africa	0.7	2.7	3.2	3.5	2.4	2.2	1.5	1.0	1.5	1.1	1.4	1.5	1.2	0.7
Asia	36.0	34.6	33.0	34.6	37.7	39.2	4.7	8.7	10.9	13.4	12.7	16.0	16.7	—
Europe	4.4	3.5	2.3	1.7	1.7	1.6	1.8	4.7	4.1	3.1	3.1	2.6	2.5	2.6
Middle East	3.5	4.4	9.9	12.5	12.4	9.8	6.5	1.2	1.8	1.0	1.4	1.4	1.1	0.5
Western Hemisphere	1.0	2.2	2.5	2.4	2.0	1.8	1.8	5.2	3.7	3.3	3.2	6.1	3.4	4.3
USSR, Eastern Europe, etc.	8.9	6.6	3.6	4.1	4.7	5.0	7.4	8.5	6.7	3.7	6.6	6.4	5.9	5.0
Annual Change (%)														
World	40.1	32.8	18.4	1.8	1.1	12.3	10.1	43.6	24.4	10.9	−12.5	12.6	21.8	63.9
Industrial countries	59.0	43.8	16.4	−0.7	−0.5	11.6	9.6	47.6	29.5	12.1	−19.0	11.3	23.5	66.5
Developing countries	33.2	56.0	27.3	3.1	1.1	12.2	5.5	51.6	46.4	9.3	1.6	20.7	21.8	67.4
Africa	−21.5	433.0	40.4	12.4	−30.9	3.4	−22.6	44.7	79.8	−20.2	14.8	19.1	−0.4	−8.0
Asia	49.8	13.8	−2.8	6.0	22.1	14.5	44.1	132.5	39.0	7.5	6.7	52.8	71.4	—
Europe	17.2	4.7	−21.9	−26.5	4.0	2.6	26.7	64.4	8.3	−16.0	−10.9	−8.3	20.4	67.7
Middle East	41.7	66.8	163.9	28.6	0.6	−11.2	−27.6	−11.0	86.9	−35.2	22.5	7.9	−4.9	−30.7
Western Hemisphere	417.6	199.2	35.5	−1.4	−18.1	4.0	11.6	77.1	−12.3	—	−16.6	116.1	−31.2	105.7
USSR, Eastern Europe, etc.	18.6	−1.6	−34.4	14.4	15.5	21.2	62.3	39.7	−2.1	−38.1	54.3	8.7	12.8	39.7

SOURCE: DTS (1986).

TABLE 24

CHINA'S MAJOR TRADING PARTNERS: 1984–1985
(IN MILLIONS OF U.S.$)

Country and Region	Import and Export Value (1985)	Export Value (1985)	Import Value (1985)	Change (%) Over 1984		
				Import and Export	Export	Import
Japan	20,112	5,962	14,150	44.6	13.2	63.7
Hong Kong and Macao	11,882	7,196	4,686	17.6	−0.7	63.8
U.S.	7,169	2,254	4,915	16.0	−4.0	28.2
EEC	7,959	2,171	5,788	42.1	−4.4	73.8
Belgium	415	158	256	9.8	12.1	8.0
Denmark	172	70	102	13.9	25.0	7.4
U.K.	1,048	343	705	0.1	−1.2	25.2
West Germany	2,932	707	2,225	36.8	−12.8	67.0
France	907	214	693	45.6	−11.9	82.4
Ireland	13	6	7	−13.3	−14.3	−12.5
Italy	1,154	282	873	47.2	−12.4	89.0
Luxembourg	44	3	41	12.8	—	5.1
Netherlands	570	312	258	13.5	−4.3	46.6
Greece	86	20	65	53.6	11.1	71.1
Canada	1,346	227	1,119	−4.3	−18.6	−0.8
Australia	1,260	182	1,077	6.5	−20.5	12.9
New Zealand	185	32	154	2.2	—	3.4
ASEAN	3,841	2,764	1,077	33.5	36.4	26.7
Indonesia	431	119	312	44.1	72.5	35.7
Malaysia	368	182	186	−9.6	−11.2	−7.9
Philippines	406	311	94	23.0	26.9	10.6

Singapore	2,265	2,039	226	62.5	63.8	51.7
Thailand	368	111	258	−17.1	−57.3	40.2
Brunei	3	3	—	—	—	—
Middle East	718	551	168	−20.0	−19.6	−20.8
Saudi Arabia	149	128	21	−11.8	−7.9	−3.0
Iraq	121	115	7	86.2	101.8	−12.5
Kuwait	89	71	18	−23.9	−19.3	−37.9
Qatar	39	4	36	−44.7	−33.3	−43.8
United Arab Emirates	82	60	21	3.8	−7.7	50.0
Iran	89	84	5	−44.7	−47.8	—
Algeria	62	51	11	−36.1	−42.0	22.2
Libya	52	22	30	−46.9	−64.5	−16.7
Oman	9	9	—	−30.8	−18.2	—
Bahrain	26	7	19	−7.1	−12.5	−5.0
Socialist Bloc	5,097	2,199	2,898	32.0	30.2	33.5
USSR	1,917	968	949	45.0	52.4	38.1
Romania	788	244	543	4.4	−18.7	19.3
Yugoslavia	218	8	211	257.4	−11.1	305.8
Poland	466	244	223	64.7	84.8	47.7
Hungary	176	75	101	69.2	87.5	57.8
Czechoslovakia	426	197	229	56.0	66.9	47.7
Bulgaria	50	14	36	−2.0	7.7	−5.3
East Germany	377	109	268	57.1	11.2	88.7

SOURCE: Cotchell Pacific (1987).

NOTE: Totals may not add due to rounding off.

shift from trade heavult dependent on the socialist countries began in the 1960s when the balance of exports moved from 70:30 (socialist:other) to 25:75; imports into China moved from the same ratio to 15:85 for socialist versus nonsocialist partners. In 1970, however, China's trade (US$4225 million) represented only 0.7 percent of total international trade—and was substantially below that of the tiny colony of Hong Kong. Because of the small role played by international trade in the Chinese economy, it did not reflect the stop-go nature of development in the earlier years. The value of trade has since grown tenfold and it is unlikely that trade would fail to respond to changes in political direction. Nor would such changes go unnoticed by the outside world for as long as did, for example, the 1960s break with Russia or the Cultural Revolution. The global publicity afforded the events of June 1989 in Tiananmen Square was an astonishing contrast to these earler happenings.

Changes in composition of trade have been analyzed in two CIA papers prepared for the 1986 Joint Economic Committee's review (Davie, 1986; Noyes, 1986). The change in composition of imports from the United States is notable—from predominantly raw materials in the 1970s (grain, chemicals, synthetic fibers, agricultural products) to industrial machinery, technologies (notably oil-related), and services. (The increase in log imports to China from the United States runs counter to this trend and is discussed later.) China's biggest trading partner—and rival of the United States—is Japan.

China is undoubtedly attempting to diversify its international trade—including trade with its socialist trading partners—and is determined never again to put all its eggs in one trade basket, as it did with Russia in the 1950s. This policy may to some extent constrain China's trade in the future—especially with trading partners with whom China is unable to countertrade.

Trade in Forest Products

Trade in forest products until the 1970s was dominated by imports of pulp and paper—pulp from Scandinavia (mainly unbleached sulfate and bleached sulfite), corrugated paper products from Japan, and some kraft from Scandinavia and New Zealand. Canada reentered the China market in the 1970s—following the establishment of diplomatic relations in 1970—but its expectations of supplying lumber were not realized. China imported logs sporadically in the 1970s—softwoods from Siberia and hardwoods from a range of tropical countries in Asia, Africa, and Latin America. But even at the end of the decade these imports amounted to less than 1 million cubic meters. Exports of forest products were nugatory—token quantities of plywood to Europe and wood manufactures, minor products, and waste paper to Hong Kong.

In 1980 there began an explosion in growth of wood imports into China that reached nearly 10 million cubic meters in 1985 (see Table 25). Softwoods comprised 85 percent and logs became the United States' number two export to the PRC (after wheat), worth US$287 million. The USSR supplies some 20 percent of the softwood market; Chile supplies 7 percent and Canada 3 percent. Hardwood log and timber imports are from Malaysia, the Philippines, Indonesia, Brazil, Peru, Ghana, and Canada. Of softwood lumber imports totaling 300,000 m^3, Canada supplies 250,000 m^3. With such small quantities, market shares mean little. It is questionable whether China needs to import any sawnwood (see later), but it perceives a need for an alternative supplier to the United States for softwood logs. The export of logs is a sensitive issue in British Columbia with both the International Woodworkers Union and the British Columbian government. (Restrictions date from 1901 and are governed by statutes of the Forest Act.) By lowering prices to some 80 percent of U.S. export lumber prices and by the industry's determination to link log and lumber exports, Canada has

TABLE 25
WOOD IMPORTS TO CHINA:
1979–1985

Year	Volume (million m^3)
1979	0.58
1980	1.81
1981	2.02
1982	4.88
1983	6.65
1984	8.24
1985	9.62

SOURCE: NFPA (1986).

been able to maintain sawnwood sales. Imported sleepers (about 100,000 m^3 annually) have been supplied by the United States.

Plywood imports have increased, as well, and may approach 1 million cubic meters in 1986. (NFPA, 1986, estimates 800,000 m^3 for 1985.) This is virtually all hardwood from Indonesia and, until 1985, had to be transshipped through Singapore and Hong Kong. Some plywood (in small quantities only) is imported from Malaysia, the Philippines, Taiwan, and New Zealand.

Exports from China have scarcely featured in international trade in forest products. Beginning in 1985, however, forestry bureaus were urged to identify export products—and during 1986 promotion (aimed primarily at Japan and the United States) featured everything from ginseng to naval stores. In September 1986, in Tokyo, TUHSU held a "timber products export symposium." Its purpose, according to a vice-chairman of TIMEX (Su, 1986), was not only to generate foreign currency but also to improve processing technology. From 1985 exports worth US$22 million (80 percent to Japan), China set a 1986 export quota of US$45 million and a target of US$100 to US$200 million during the Seventh FYP. Markets targeted are Europe, America, and the Middle East. Items on show were mainly of hardwood (oak, walnut, ash, elm, *Paulownia*, birch, and camphor) but included pine and eucalyptus chips, short clears cut from imported logs of Douglas fir, hemlock, and Radiata pine, as well as panel products and moldings. The quality was not high, but neither were the prices. Another exhibition—advertised as China's First International Wood and Forestry Industry Conference and Exhibition—was held in June 1987 in Beijing (concurrently with the First International Pulp and Paper Technology Exhibition).

China's economic openness has effectively removed political considerations from trade. Quite simply, China wants the best deal it can get. Increasingly, however, this may involve barter and trade/aid "packages." Barter deals involving logs have been concluded with the USSR, Outer Mongolia, Brazil, and Papua New Guinea, and in 1987 such arrangements were being considered with Chile, South Africa, New Zealand, Ghana, and Indonesia. Trade/aid deals are in place involving Sweden, France, West Germany, Romania, Hungary, Canada, and Japan. Apart from these countertrade deals, there is much discussion among Chinese economists and technicians of the relative merits of hardwood and softwood imports, the role of tariffs, industrial scale, and appropriate technology.

Hardwoods and Softwoods

In the temperate zone, the distinction between hardwoods and softwoods tends to be rigid. In the realm of solid wood, producers and users have their own trade organizations and markets. (The NFPA report, for example, carefully avoids comment on hardwood markets in China or the possible substitution of one for the other.) In tropical countries, the distinction is much less firm; there is interchange between hardwoods and softwoods and both may be used in the same product. In countries where there are only rudimentary standards—which may not be mandatory or enforced—availability and cost are more important determinants of wood use than technical properties.

China has a range of National Wood Standards, promulgated by a State Bureau of Standards. They cover logs, lumber, plywood, packaging, pallets, furniture, railway sleepers, truck beds, and poles, as well as design and construction standards. They are little known (and little used) among woodworkers in China, however. Even in architecturally designed and engineered building construction, hardwoods and softwoods are sometimes interchanged—and in joinery and furniture such substitution is general. (Joint promotion of both hardwoods and softwoods may help to explain Canada's success vis-à-vis the United States in selling lumber to China; and combined shipments of hardwood and pine have helped Chile to increase sales.)

Chinese sawmills and plymills process hardwoods and softwoods together. Reconstituted wood products (fiberboard and particleboard) use both, as do many pulp mills. China imports mixtures of many species (up to sixty-five species have been brought in from West Africa and almost as many from Papua New Guinea); the range in species mix is a factor in determining price, to which China is very sensitive; and its growing interest in Pacific Island forest resources (Papua New Guinea, Western Samoa, Fiji, and the Solomon Islands) is due as much to their low log prices (lower than the more uniform Malaysian mixtures) as to the extension of supply sources.

The intense competition that developed in 1985 between American and Soviet log producers in the Japanese market squeezed the tropical hardwood log trade, and at least three plywood manufacturers switched—at any rate temporarily—from using hardwoods to softwoods (FEER, 15 May 1986). Under U.S. pressure, Japan has agreed to reduce tariffs on plywood in a way that may increase pressures to reduce hardwood prices. Thus 12-mm plywood (mainly softwood) declined from 17 percent to 13.5 percent on 1 April 1987 and went to 10 percent in 1988. On the same schedule, 3-mm plywood (entirely hardwood) fell from 20 percent to 17.5 percent and then to 15 percent. The capacity to use species mixtures and to interchange hardwoods and softwoods is also a feature of the China market.

Tariffs and Foreign Exchange

In China, tariffs are not viewed primarily as performing an industrial protection function—rather, their purpose is to conserve foreign exchange. Since the excesses of 1985, the Chinese have become determined to generate or conserve foreign currency.

Import duties on forest products (as of 1986) are outlined in Table 26. The *Official Chinese Customs Guide 1985–1986* is published by Longmans, and extracts are presented in the NFPA report (1986). In addition to import duties, however, Chinese importers must pay a products tax of 5 to 10 percent (see Table 26). The contrast between low levels of duty on logs and penalty rates on finished lumber underlines the impor-

TABLE 26
CHINA'S IMPORT DUTIES ON FOREST PRODUCTS

Description of Goods	Tariff No.	Minimum Import Duty Rate (U.S. Rate)	Product Tax Rate	Total Tariff
Softwood logs	44.03–2	3%	10%	13%
Hardwood logs	44.03–3	3%[a]	10%	13 or 35%[a]
Roughly squared	44.04			
Softwood	44.04–1	3%	10%	13%
Hardwood	44.04–2	3%[a]	10%	13 or 35%[a]
Lumber rough sawn	44.05–1	9%	10%	19%
over 5 mm	44.05–2[a]	9%	10%	19%
Lumber planed and grooved	44.13	40%	10%	50%
All plywood	44.15	12%	3%	15%
Veneers	44.14	30%	10% (?)	40%
Rail ties	44.06/06	9%	5%	14%

SOURCE: NFPA (1986).

[a] Certain hardwood species are taxed at a higher rate: teak, camphor, sandalwood, garoowood, "redwood," rosewood, and a few others. A minimum duty of 30 percent is imposed on rough-sawn lumber of these species; logs and cants are 25 percent.

tance China attaches to unprocessed imports. The NFPA report argues for tariff reductions on rough-sawn squares, balks, and "heavy lumber"—on the ground that there would still be scope for adding value in remanufacture and maintaining employment. This contention, as we shall see, ignores many of the arguments in favor of log imports.

The concern with foreign exchange has had certain unhappy consequences. The use of a forest nursery to raise ginseng (at the expense of tree seedlings) has been mentioned. More serious, perhaps, is the readiness of some forestry officials to turn a blind eye to wildlife conservation regulations in order to obtain foreign exchange from safaris, shooting fees, and the sale of trophy heads—including those of endangered and supposedly protected species. A similar attitude is displayed in the offer for sale by TUHSU of 400-year-old *Torreya* logs (Appendix B).

A problem that could arise in future relates to quarantine. In theory, timber imports must pass a "combination" check by three separate ministries—forestry, agriculture and fisheries, and customs. Moreover, a phytosanitary certificate mandated by the central ministries is not accepted by many port authorities. In theory, the port authority can require logs to be debarked *before* unloading—and prohibits dumping. As far as is known, such requirements have not been enforced; but it is not unknown elsewhere for such provisions to be used in restraint of trade.

Scale and Appropriate Technology

China's wood-processing industry was established along guidelines provided by the USSR (see FAO, 1979). The guidelines were threefold: Wherever possible, processing should be done at log source to reduce transport costs; existing mills should be enlarged where feasible; and new integrated units should spearhead the industry, with smaller, older plants in a backup role.

Some results of these policies were not anticipated. The notion of shedding waste close to the point of harvest to reduce transportation costs makes good sense in countries rich in forest resources (as is the USSR), but it is inappropriate in wood-starved China. Moreover, as the plants have expanded, the primary forest resource has retreated—requiring ever-increasing transport distances and leaving single-industry centers isolated from employment outlets. Dailing, for example, in Heilongjiang, was in 1963 within five hours by steam-driven tram of the logging operation that then sustained it; in 1986, the primary resource was a two-day haul by truck (at six times the speed of the tram) and the industrial processing plant has increasingly to use logs from secondary forest and plantations. The population has doubled and the resource is shrinking. Throughout the northeast, special workshops—"recovery" sawmills, joinery shops using undersized logs, and even pulp and paper mills—have been set up to absorb the growing numbers of middle-school leavers who have no work. In the cities (Harbin, for example) where they serve as stepping-stones to the more profitable use of acquired skills, these workshops are dynamic and successful. In communities like Dailing they lead nowhere—and, since there is no labor mobility in China, they serve little purpose.

Another shortcoming of the complexes has been the failure to realize the limits of labor/capital substitution. In sawmilling and its associated yard operations, labor can readily substitute for capital and economies of scale play a minor role. Residue utilization, on the other hand, calls for capital-intensive technology and demonstrates obvious scale economies. Attempts to substitute labor for capital have resulted in a rash of small plants producing wood-based panels that the free market is now showing to be of unacceptable quality.

Labor productivity in China is low. The NFPA reports a range of 0.5 to 3.0 m^3 per man/day in sawmilling—compared with "international" levels of 7 to 10 m^3 per man/day. The comparison is hardly valid (except that MOF planners claim a commitment to reach international levels by the year 2000), and the figures warrant no criticism. Economically and politically in China, the creation of employment—even at low productivity—is a valid alternative to the provision of the massive welfare subsidies for which some other countries opt.

Of the inputs to production, China has abundant labor but is very short of raw materials and capital. The choice of labor-intensive technology in sawmilling is entirely appropriate. This strategy does not apply to panel product manufacture (other than perhaps blockboard); nor does it mean that there is no scope for the introduction of modern technology in sawmilling. In fact it offers more scope for certain technology (such as monitor and control systems involving miniaturization) than the alternatives. But high technology is needed to improve product quality and to ensure consistency, rather than to increase labor productivity.

A too-familiar sight in China's department stores is a customer carefully examining every vacuum flask on the shelf before purchase to ensure that the vacuum is intact. Similarly, both shoes of a pair must be measured for size. And every piece of timber—or sheet of fiberboard—in a parcel may vary in size and grade. In the small town of Simao (Xishuangbanna, Yunnan) imported plywoods commanded high local premiums because of their quality and consistency.

CONCLUSIONS

There is clearly a need for a revision of the guidelines used for China's forest industry development policy. There is evidence that—in the provinces at any rate—deficiencies in the earlier philosophy (favoring large, "integrated," forest-based complexes) are coming to be appreciated. But quick remedial action is needed if huge waste and pollution problems are to be solved and China is to cope with the rapidly approaching timber supply/demand crisis. It seems probable that the MOF forecasts of wood consumption are conservative. Alternatives, based on population growth and increasing per capita consumption, are no more than indicative, but it can be said with fair certainty that China faces a growing deficit in its domestic wood supply that is underestimated by its economic planners. China cannot significantly increase its harvest—nor, if it is to reach its economic and industrial goals, can it realistically "design wood out of the economy."

The massive growth of log imports—to reach the second biggest U.S. supply item to China (approaching $300 million in value in 1986)—and a deficit situation forecast to remain until at least 2040 underline the urgent need for improved efficiency in the forest industries. At present, the state sector exemplifies the worst problems of bureaucratic self-containment and absence of managerial or entrepreneurial incentive. This would not matter, perhaps, if China had more abundant forest resources or if wood were not so pervasive a raw material in the economy. But given the supply/demand realities and the growing cost of imports, China cannot afford *any* waste—let alone that of its highest-grade wood raw material. The fact that in Tibet the free-market price of one cubic meter of firewood (RMB 350) exceeds the annual per capita income for the WFP project area (Richardson, 1989) suggests the desperate extent of the crisis. Evidence that improved utilization of forest and mill residues can be achieved is apparent in the smaller, stand-alone mills of suburbs and villages. There the greatest need is for the technology to improve product quality and—above all—consistency.

Pulp and paper production poses particular problems. High transport and energy costs support the case for a large number of small mills, rather than a small number of large mills. But state-of-the-art technologies for pollution abatement (and chemical recovery) favor ultra-large, high-speed operations. Small mills—especially those using agricultural residues rather than wood—are major water and air polluters. To control them, China will have to develop (and apply) its own technology. Until it has done so, the optimal solution to the problem might be to concentrate on paper production—using imported pulp or, at least, a much higher proportion of carrier pulp. China could learn much from countries that have successfully developed and applied technologies to the pressing problems of industrial scale and quality control; Israel comes to mind as a sectoral model.

Finally the likely continuation of log imports—and China's preference for countertrade—offer possibilities for enhanced trade with the forest-rich countries of the Third World. In this context, Latin America could certainly supply candidates.

5

ADMINISTRATION, POLICY, AND LAW

IT IS NEVER EASY for an outsider to understand the governmental and administrative mechanisms of a foreign state. Many of us, indeed, have only a sketchy idea about how our own system works—particularly those components of it with which we have no direct experience. And in a country with a political structure in which national executive controls reflect a mixture of empiricism and ideology, in which a multitude of hierarchies abound (and in which everyone belongs to several organizations), it is easy to draw false conclusions. Rapid and sweeping changes can occur from time to time in China, too; often they happen without apparent warning.

Chinese society has always been rigidly organized. Religious traditions (including the secular state religion of Confucianism) produced an inflexible society with clan-based divisions of scholar, farmer, artisan, and merchant: The reforms of socialism were perhaps more readily absorbed than would have been the case in less organized societies. It has to be remembered, too, that before the revolution life for millions of peasants and urban slum dwellers was precarious and harsh. Freeberne (1971:348) observes: "Many peasants were tenant farmers or landless laborers, working tiny, fragmented plots with poor equipment, without draught animals, and unable to invest in fertilizers. Soils were exhausted, slopes stripped of vegetation through fuel scavenging, and soil erosion was menacing. . . . The peasants were heavily taxed many years in advance and were forced to pay extortionate interest rates to money lenders." Any change in the system would have been welcome.

FORESTRY ADMINISTRATION

This section outlines the overall planning structure and the particular problems posed by forestry; it also examines the impact of recent bureaucratic changes.

PLANNING AGENCIES

The National People's Congress (NPC) is the highest organ of state power. Its permanent body is the Standing Committee, composed of members elected by the provinces, autonomous regions, and municipalities (directly under the central government) and by the People's Liberation Army (PLA). The NPC appoints the State Council, which in theory plans and promulgates national policies. It is assisted by a number of planning commissions—illustrated in Figure 2. The most important are the State Planning Commission (SPC) and the State Economic Commission (SEC). The commissions are primarily staff departments, translating policy decisions into short-term and long-term plans; they are not executive bodies but report back to the State Council, which then controls the implementation of the plans through the economic ministries (of which the Ministry of Forestry is one) and various regional and local organizations. In the provinces, the autonomous regions, and the autonomous municipalities (Beijing and Shanghai, for instance), local government organizations parallel those at the national level. Local governments are responsible and accountable to the local congresses at their level; again, they are the executive bodies of local organs of state power as well as the local organs of administration.

Planning takes place at several levels and is complicated by the fact that various national and regional plans may cover different periods of time. For example, while overall economic development follows a series of relatively detailed five-year plans (begun in 1952), agriculture works to a twelve-year plan and forestry to a ten-year plan. In some areas—including timber supply—indicative plans have been prepared for up to fifty years. The five-year plan is published, and annual national economic plans are prepared for production, material allocations, wages, labor, and so forth. The five-year plan is the responsibility of the SPC, but it incorporates the work of several levels of planners. At the local level, the planning bureau in each county prepares a plan that is guided by provincial planners and incorporated into their provincial plans. In the case of state enterprises directly under economic ministries or commissions, a plan is first prepared at the enterprise level and forwarded to planning bureaus of the industry for incorporation into national plans. The SPC is responsible for integrating provincial plans and submitting the final version for approval at the NPC. When approved, plans are issued by the State Council and the implementation is then in the hands of the economic ministries and the Provincial People's Councils (PPCs). The ministries draw up regulations designed to achieve the objectives of the plan and assign production targets and the like that are promulgated by the State Council to the PPCs. The regulations are often exhortatory rather than prescriptive, introduced for trial periods, and issued with a good deal of latitude in their interpretation.

Economic planning follows the general system of Soviet planning but is rather less detailed in China and, certainly, less closely monitored. No attempt will be made here to describe the process in any comprehensive way. Perhaps the best English-language description of its application in forestry and the forest industry is contained in Blandon (1983). Particular problems for forestry posed by the economic conventions of socialism relate to the valuation of natural resources and to pricing. As Blandon points out, the omission of considerations of relative scarcity in pricing prevents prices from serving as signals for the allocation of resources. Moreover, the pricing of timber from natural stands at levels equal to the costs of exploitation[1] does not generate cash flow sufficient to enable reforestation. Stumpage taxes (or silvicultural fees) have been introduced to enable reforestation but, in both the Soviet Union and China, they are totally inadequate for the purpose. Other problems cited by Blandon (problems that are also being discussed in the Chinese technical journals) relate to the effectiveness of new technology, the calculation of rotation length, and deficiencies in socialist methods of investment appraisal—which reflect the fear that acknowledgment of risk and uncertainty may imply planning failures. Problems relating to discounting (in the absence of market-determined interest rates) also pose special problems for forest economists in the Soviet Union, but

FIGURE 2.
CHINA'S GOVERNMENT STRUCTURE

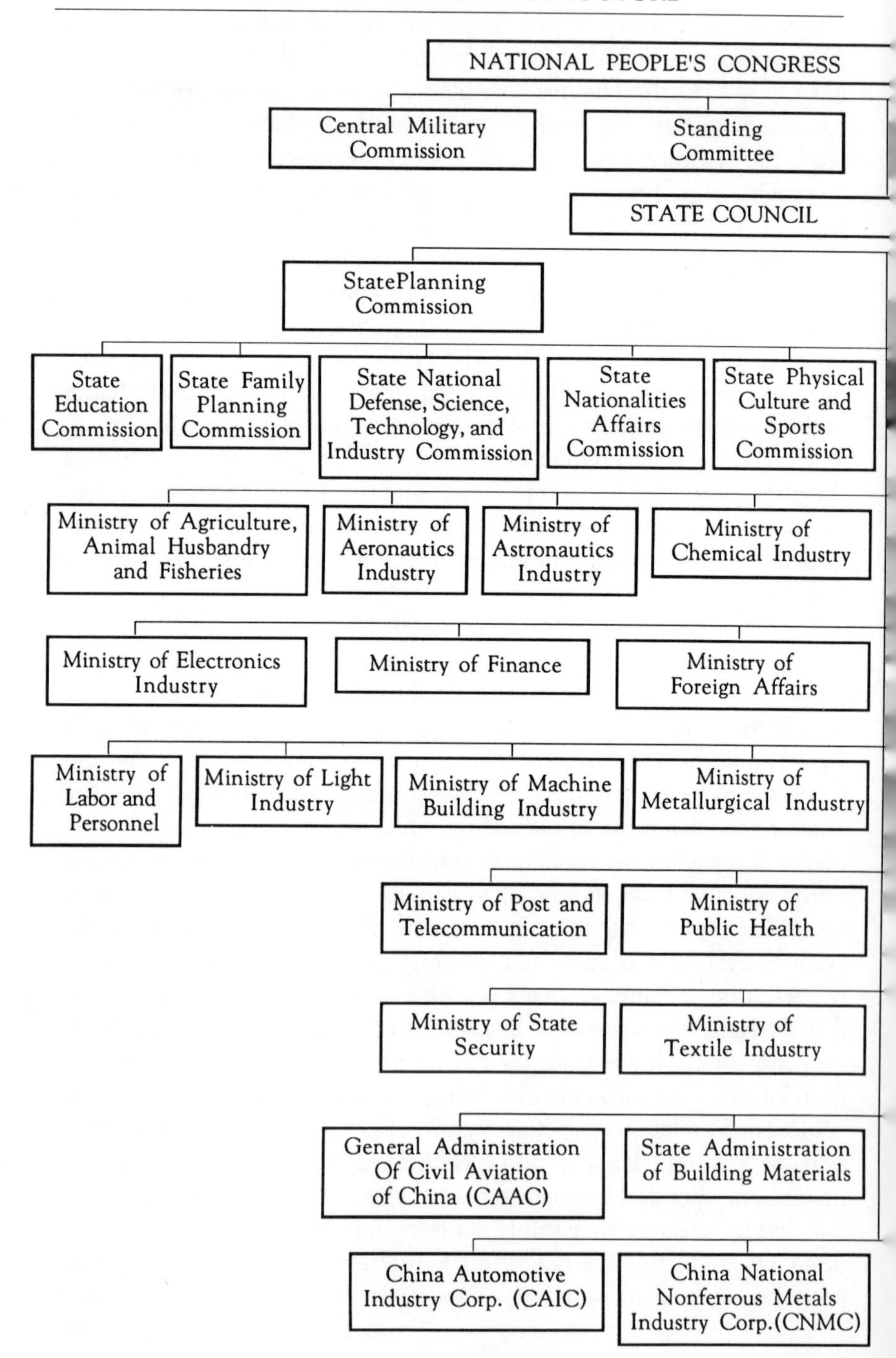

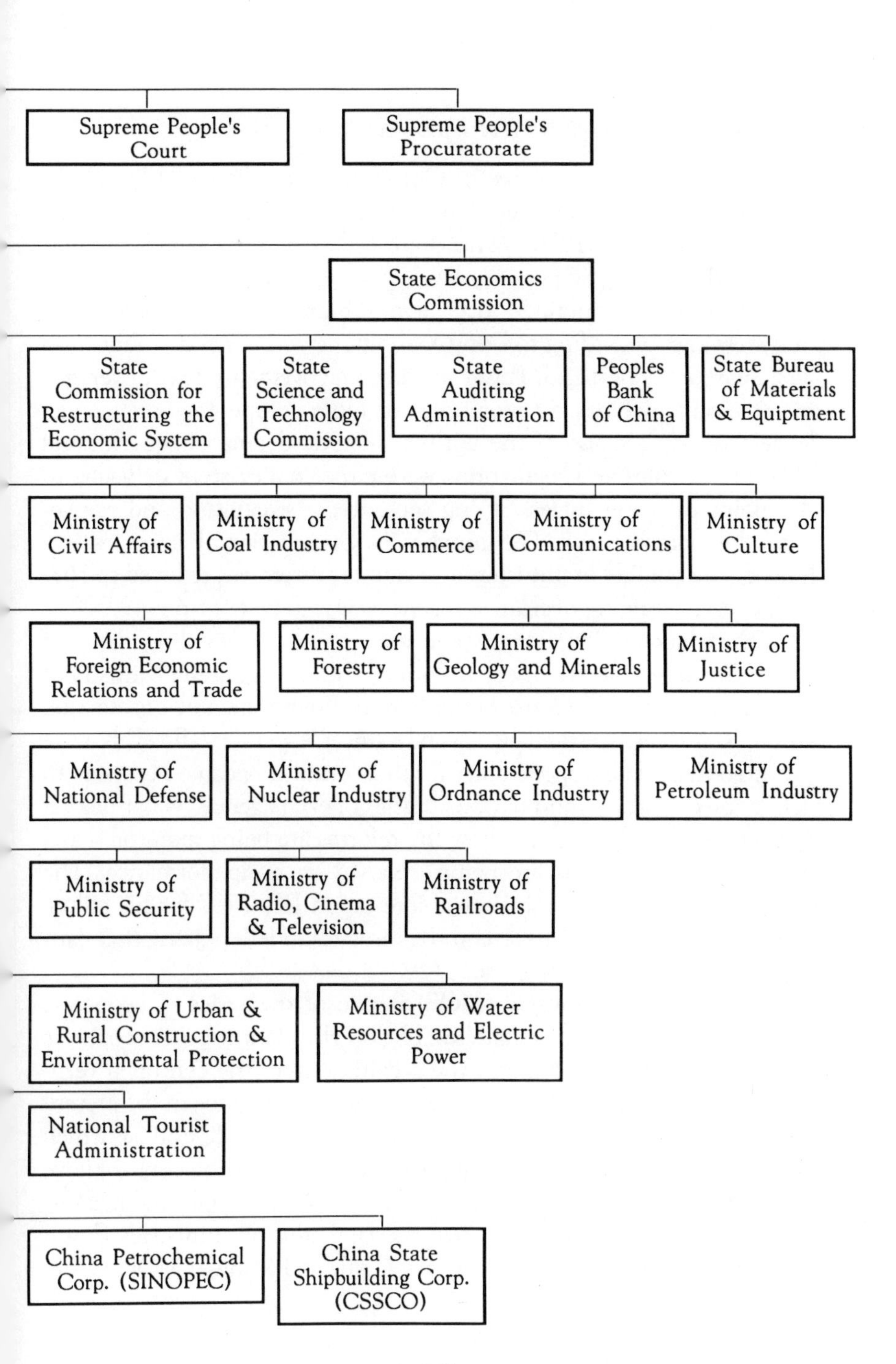
Supreme People's Court
Supreme People's Procuratorate
State Economics Commission
State Commission for Restructuring the Economic System
State Science and Technology Commission
State Auditing Administration
Peoples Bank of China
State Bureau of Materials & Equiptment
Ministry of Civil Affairs
Ministry of Coal Industry
Ministry of Commerce
Ministry of Communications
Ministry of Culture
Ministry of Foreign Economic Relations and Trade
Ministry of Forestry
Ministry of Geology and Minerals
Ministry of Justice
Ministry of National Defense
Ministry of Nuclear Industry
Ministry of Ordnance Industry
Ministry of Petroleum Industry
Ministry of Public Security
Ministry of Radio, Cinema & Television
Ministry of Railroads
Ministry of Urban & Rural Construction & Environmental Protection
Ministry of Water Resources and Electric Power
National Tourist Administration
China Petrochemical Corp. (SINOPEC)
China State Shipbuilding Corp. (CSSCO)

they have yet to emerge as practical concerns in China. Undoubtedly they will do so.

Organization and Materials Allocation

Recent organizational change in China has been discussed by Clarke (1986). He notes the proliferation of state and party organizations that began in the early 1970s. The State Council, for example, almost doubled the number of ministry-level agencies between 1970 and 1982; the government contained over five hundred ministers and vice-ministers and, according to Deng Xiaoping, stood in need of reduction by a quarter to one-third. Power had become overcentralized, overaged to the point of blocking promotion of more competent but younger leaders—and, because of the antibureaucratic dogma of the Cultural Revolution, rules and regulations were either nonexistent or vague to the point of being useless. Deng set out to "streamline and rejuvenate the bureaucracy." The number of State Council agencies was halved, the number of ministers and vice-ministers was reduced to 167, and, overall, staff of government agencies was to be cut from 49,000 to 32,000 (Clarke, 1986). Ministries were merged (including those of Water Conservation and Electric Power), the new umbrella Ministry of Urban and Rural Construction and Environmental Protection (MOURCEP) was established—and, since leaders at all levels were expected to assume responsibility for their units, it became necessary to define the obligations and responsibilities of each agency. What has yet to become clear, however, is how the reforms are being monitored and what kind of sanctions are being imposed on nonperformance. The abrupt dismissal of the minister of forestry in July 1987 for his failure (due to hospitalization) to attend the disastrous Heilongjiang fires cannot be taken as a reliable guide.

The Ministry of Forestry (MOF) is one of the series of economic ministries that function at state level below the planning and economic commissions. Figure 2 presents the organization of the central government as of February 1986. Other ministries of importance in the present context are Foreign Economic Relations and Trade (MOFERT), which develops policy and guides the import and export trading corporations; the Ministry of Commerce, which is responsible for supply and marketing cooperatives in the provinces; the timber-using ministries (Light Industry, Railways, Communications, Coal, and Mines); MOURCEP;

the corporations for shipbuilding and building materials; and the Ministry of Water Resources and Electric Power. Outside this structure is the PLA—a major user of timber but also involved in harvesting, processing, and forest regeneration in remote areas.

Figure 3 outlines the structure of the Ministry of Forestry. The MOF is responsible for research (through the Academy of Forestry), education (at three forestry universities and four colleges), and the administration of state forestry throughout China. The MOF also operates sawmills, plywood mills, particleboard and fiberboard plants, and a number of "integrated wood factories" (most of which are anything but that). It administers nature reserves and "forest farms" covering 50 million hectares (including 4 million hectares of industrial wood production forests).

The provincial organization replicates that at the center, including research and training activities, and is itself replicated—but in a simpler form—at district (prefecture and county) levels. A departure from this pattern is in Heilongjiang, where forests in production are overseen by a quasi-autonomous Forest Industries Bureau, which is also responsible for the general rural administration of the province, including public security—a unique situation.

The MOF inventories the forest resource every five years and has a forest economics research division that is concerned with forecasting supply and demand. The principal field projects under the MOF are the Plains Afforestation Drive (which includes protective and "Four Around" planting of houses, villages, roads, and rivers); the "Three Norths" shelterbelt system—a gargantuan undertaking that aims to cover some 23 million hectares in northern China with trees by the year 2000; and the continued establishment of "forest farms." The latter are mainly in areas where growth is slow and their management is conservative. Under the seventh FYP, the forest farm component is being emphasized and will include the extension of fast-growing plantations in the southern provinces of Fujian, Jiangxi, Guangdong, Guangxi, and Hunan.

Forest production is the responsibility of the forestry bureaus under the MOF except as noted above for some 6.6 million hectares in Heilongjiang. (The forestry bureaus control 4.6 million hectares of forest land—in the Hinggan Mountains, on wastelands, and on farms—but do no harvesting.)

Harvesting from state forests is targeted by the State Planning Com-

FIGURE 3.
STRUCTURE OF THE MINISTRY OF FORESTRY

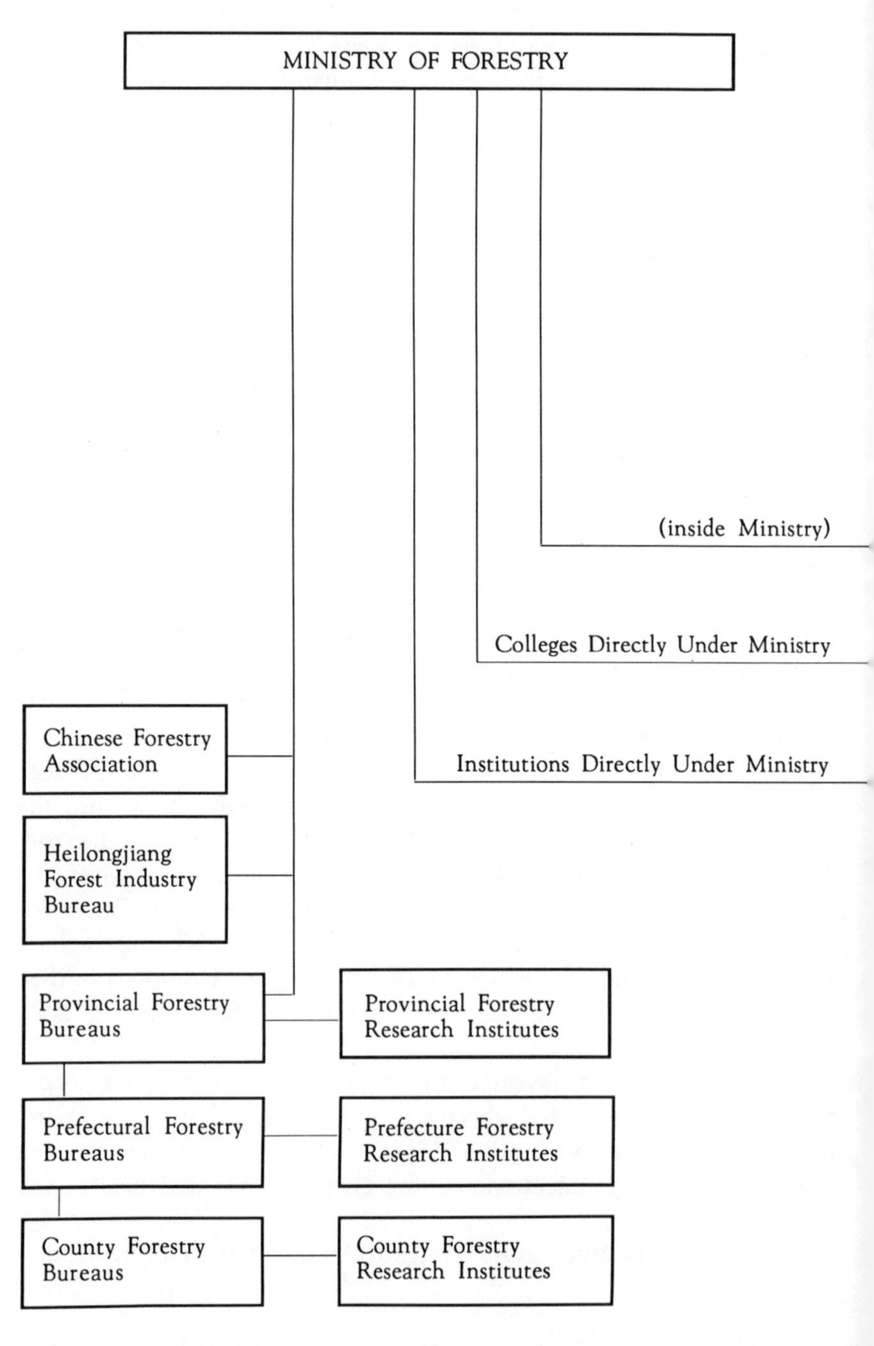

Forest Industry

Silviculture and Management

Natural Resources

Protection of Forests

Security

Planning

Finance

Science and Technology

Education

Propaganda

Foreign Affairs

Personnel Management

"Administration"

Retired Cadre Management

Plan Monitoring

Beijing University of Forestry

Nanjing University of Forestry

Northeast University of Forestry

Middle-South College of Forestry

Southwest College of Forestry

Northwest College of Forestry

College of Forestry Management

Chinese Academy of Forestry

Academy of Forest Planning

Forest Machine Corporation

Material Supply Corporation

Forest Products Corporation

Forest Products Sales Corporation

Forest Seed Corporation

International Cooperation Company

China Forestry Press

Design Institite of Forest Industries

Hinggan Ling Forest Management Bureau

North, Northeast Northwest Shelterbelt Construction Bureau

mission, and production is assembled at logyards—usually at railheads—for grading and pricing by the China Timber Corporation (CTC), which, in effect, provides a consultancy service for the Materials Allocation Bureau under the State Council. The CTC administers supply and distribution but is not responsible for logistics, except in relation to its own distribution centers (at Tianjin, Shanghai, Wuhan, Lanzhou, and Shenyang). The three divisions of CTC are concerned respectively with allocations to user-ministries, allocations to provinces (and municipalities), and the operation of the distribution centers. Most imported logs (80 percent) enter the system via the CTC; the China Native Products Import and Export Corporation (TUHSU) acts as agent. (Only TUHSU can authorize foreign exchange commitments.)

Production from collective or privatized plantations and secondary forest assigned under production responsibility systems is no longer under rigid control. Felling licenses are required for removals greater than 10 m^3 a year, but there are virtually no formal restrictions on the disposal of logs. In the principal industrial wood-producing regions of China (notably the northeast), very little state forest production is assigned outside the state system but inevitably there are leakages. There are also numerous examples of abuse of the system and criticism that too many agencies are involved in the forest industries (Fewsmith, 1985). In the words of one source: "Trees are planted with one hoe but cut with a hundred axes" (FE, 14 March 1981).

The Materials Allocation Bureau (advised by the CTC) is responsible for allocating industrial raw materials, including logs and the products deriving from them (sawnwood for construction, plywood for furniture). Failure of the Chinese (and visitors) to distinguish between logs and lumber has sometimes led to confusion as to how much "wood" is controlled by the center and how much by the provinces. For planning purposes, there are three categories of industrial material in China: Category I comprises strategic "essentials" (including state forest logs and, since 1982, sawn timber), which are allocated centrally under the State Planning Commission. Category II materials are distributed by central government ministries and tend to be specialized goods such as alloys but also including machinery, railway sleepers, and mining timbers (for state mines). Category III materials are those under local (provincial or municipality) allocation. Categories I and II are additionally divided into two parts—one centrally and one regionally allocated. Over recent years, not only have materials been moved from one category to another, but the relative proportions of the allocation have changed. Thus in 1978 some 19 percent of logs were under local control and 81 percent centrally allocated. Since then, 22.7 percent of the "central" component has been turned over to local control.

There are four types of enterprise (Wong, 1986): those under central control (the integrated wood factories, for example), which receive all inputs from the central ministries and make their production available for central allocation (either directly or indirectly); enterprises under local control (province, municipality, and county), which work in a similar self-contained system; those under dual (but primarily central) leadership, in which local inputs are labor and party involvement; and those under dual (but primarily local) control, in which the center provides planning and the local agency the other inputs. The first and third types are "*tongpei*" enterprises, the output of which is predominantly subject to central allocation under the State Plan. (They include the major regional forestry bureaus, though the most important region—Heilongjiang—was transferred in 1983 from type 1 to type 3.)

The material allocation system in China is simpler and less rigid than that in the USSR. Wong (1986) noted that material categories I and II never exceeded 600 items until 1981, compared with 65,000 in the Soviet Union. Moreover, the Chinese product groups are broader (logs and timber, for example, form only two categories) and disaggregation into specific products and contract prices is left to "materials ordering" conferences between users and producers. Perhaps because of this simplicity, product grading (not pricing) appears to be more complicated than necessary. Prybla (1986b) points out that in Russia in 1967 several *million* international wholesale prices were recalculated—by hand—and became instantly obsolete! China is trying to avoid such folly by allowing market signals to determine prices and influence allocations. At the same time, until there is greater availability of supplies, caution must be exercised—and the inexperience of cadres in a market economy (indeed, in some cases, ideological hostility to the reform policies) demands tentative approaches that may result in stop-go tactics.

FOREST POLICY

It is convenient to consider forest policy developments before and after 1975—when the Cultural Revolution ran out of steam and the rule of Mao Zedong came to an effective end. This section also examines policy implementation and annotates the major national tree-planting programs mounted from Beijing.

DEVELOPMENTS FROM 1949 TO 1975

As soon as the Communists achieved national control in 1949, all forest land (except that attached to certain Buddhist temples, which was not taken over until 1951) was nationalized and a Ministry for National Development drew up interim utilization plans. An immediate aim was to rehabilitate and expand sawmilling plants in Manchuria (which before the war were in Japanese hands and had then been largely dismantled by the Russians after 1945) in order to survive the initial shock of a virtual cessation of timber imports from 1950 onward. At the same time, Liang Hsi, professor of forestry at Nanjing University and later to become the first minister of forestry, was invited to prepare a series of guidelines to serve as a basis for a forest policy. His program acknowledged the value of protection afforestation in soil conservation, the need to regulate felling in the Manchurian forests, and the potential role of short-rotation tree crops in the southern provinces. (His scheme was very close to that in place now.)

Little in the way of a formal forest policy emerged before 1952, however, when the MOF was fully established and drew up a "Directive on Mass Afforestation, Cultivation of Forests, and Protection of Forests" that was promulgated by the State Council in September 1953. This document called for the establishment of organizations to introduce fire control in the natural forests and for "vigorous afforestation" of hills and wasteland. It never became a formal part of the First FYP, though the Chinese once claimed to have planted more than 10 million hectares under that plan. It was directed primarily at protective planting by collectives rather than state-sponsored industrial production planting.

Collectivization of forest holdings was carried out very quickly. By 1954 (Ross, 1983), three-quarters of all afforestation was being done by cooperatives—in some cases financed by windfall profits from cutting (and overcutting) existing timber lands. There was a long tradition in Chinese agriculture of "mutual aid," and in many regions it was customary for peasants to pool their labor in planting trees. The "tree associations" were jointly owned by the landholders (whose input was valued at 50 percent of the total) and peasants, who put in labor and tree seedlings. In 1953 the mutual-aid teams were styled "elementary agricultural cooperatives." They have been described as "semisocialist in nature" (Xu, 1986). Few such cooperatives were engaged in industrial forestry, but there were many in which horticultural crops (food and fodder trees) were integrated into the agricultural pattern. In October 1955, the Central Committee of the CCP issued a "Resolution on Agricultural Collectivization" to the effect that trees owned by members of cooperatives should remain under their own care; if they were put under the overall management of the cooperative, there should not necessarily be a change in ownership. Attempts were also made to extend collectivization to hill country, and a "model constitution of agricultural cooperatives" was adopted by the Standing Committee of the NPC.

The second stage of collectivization (formation of the "higher agricultural producer cooperatives") involved groups of 200 or 300 households. The assumption of collective ownership of the means of production tended to confuse the ownership of trees, however, since cadres did not always draw a distinction between naturally regenerated and planted trees. The model constitution of the higher agricultural production cooperatives, adopted in June 1956, provided that in general young forests, nurseries, and blocks of "economic" tree and industrial timber species should be cooperatively owned and, where appropriate, compensation paid for them. By 1957, some 96 percent of China's rural households had reportedly joined advanced agricultural cooperatives and collectivization (including forestry) was more or less complete (Xu, 1986). There were immediate problems. According to Xu, many peasant farmers were undercompensated: In some cases, members of the cooperatives were given the option of drawing dividends as an alternative to compensation, but in many instances cadres failed to implement the regulations in the interests of the peasants. This was a cause of so-called illegal felling in later years.

The collective forest farms established after 1957–1958 were more successful. By May 1958, some 1455 such farms had been set up in China and by September 1960 there were 80,000 with a total labor force of close to 1 million (Xu, 1986). They were said to be more successful because of improved management, realism with respect to what might be achieved, and the lack of "superficial mass movement." That they did not continue as originally envisaged was a result of economic problems created by the Great Leap and calls for self-sufficiency in grain, which led to timber forests and hill country being cleared and cultivated. Throughout the 1960s and early 1970s, there were problems stemming from the inexperience (and incompetence) of cadres.

Developments Since 1975

Since the mid-1970s the development of collective forestry has been closely linked with the identification of industrial forest production bases. In counties where such bases have been recognized, subsidies are available for forest regeneration. But again, there has been dissatisfaction with the way in which the proceeds from cutting are shared. The more flexible "adjustment and reform" policies began to be felt in forestry in the early 1980s. The collective forests were renamed "village forest farms" when the communes were effectively abolished in 1982. Guidelines for production responsibility systems were introduced and attempts made to protect the rights of usufruct. Since then, "multiutilization" of forests has developed. By 1983 there were 175,000 forest farms run by villages—managing 16.7 million hectares of hill and forest, of which 10.9 million was in forest.

A shareholder and contract system has been widely introduced that is characterized by joint management according to the input value of shares. The ownership remains collective, but the individual trees are converted into shares and accorded a money value that is then allocated among the owners of the usufruct. Management may be contracted to a number of households and the income distributed proportionately. To introduce such schemes, rights of usufruct were divided into two types of share: the basic share, or "old" share, was converted from the existing collective forest (including both natural forest and plantations) and "new" shares were issued in return for the investment of labor or other inputs. The basic share is privately owned and can be inherited or transferred; it attracts a dividend, but it cannot be withdrawn. The new share, on the other hand, can be withdrawn, transferred, or inherited. Shareholding certificates are issued to all members. The shareholders become members of an organization that elects a board of directors and is responsible for overall management, the preparation of annual work plans, financial accounting, and the imposition of "economic punishments" to lawbreakers who interfere with the forest management plans.

As in other fields, there is a wide spectrum of multiform entities concerned with forestry development. Xu (1986) cites examples of individuals or households that have specialized in forestry operations undertaking contracts which may cover individual activities (raising seedlings, thinning, felling) or embrace the overall management of a forest area. There are joint regional/household and county/household ventures, and there are enterprise/forestry department ventures. The Chang Jiang Paper Mill of Yibin City in Sichuan, for example, has an agreement with the local forestry bureau to establish some 3000 ha of bamboo forest for use by the enterprise. There are joint-management ventures involving timber production and timber processing units. In Bailuan, Fujian province, for example, the state sawmill has a ten-year joint development and management contract with the township whereby the mill uses the growing stock from some 700 ha of hill land and is responsible for road construction, logging, timber processing, and marketing. The township supplies labor and receives an agreed price for the logs. The sawmill will—it is hoped—return the regenerated hills to the township within five years of cutting.

Since 1981, major users of state wood (the coal, railways, and construction industries, for example) have been required to contribute financially to reforestation. In many areas, they find it preferable to enter into a joint venture with landholders to raise trees rather than pay a levy to Beijing.

Private forest protection companies have been set up to control pests and diseases (and undertaking to pay compensation for losses that exceed agreed limits—if they result from "errors in their work"). Similarly, following practices in agricultural extension, there are joint forest technology and management ventures that undertake technical contracts with production enterprises and households specializing in forestry. They provide nursery services, for example, charging a 3 percent service fee for raising seedlings and, again, paying compensation in the event of failure to deliver the contracted goods and services. There are also associations of specialist forestry producers—organizations representing the specialized households and providing infrastructural facilities (bookkeeping, secretarial, and office services, and so forth). They may also operate nurseries and provide access to bulk purchases of fertilizer, seeds, and the like.

In all parts of the economy, the specialized households represent a significant departure from past policies—and, as noted earlier, could become as powerful a lever for forestry development as the *Baogan Daohu* (contract responsibility) system. A recent paper (Hou and Wang, 1986) gives examples of their operation in remote areas where individual families were unable to combine forestry and agriculture (because of the distance between jobs). One such relates to a family of three members that was given 16.7 ha of land for an annual rent of 200 yuan: From 20,000 timber trees and 500 "economic" trees, the net income was reportedly 2100 yuan. The household paid 500 yuan for seedlings and other inputs and retained the balance.

According to Hou and Wang, there are twelve requirements for the successful settlement of specialized households in hill country:

1. Housing
2. Provision of grain
3. Provision of short-term benefits and an assured income (for example, from "economic" tree crops)
4. Road construction (by collectives)
5. Water supply
6. Schooling
7. Medical services
8. Grain milling facilities
9. Lighting
10. Technical assistance
11. Finance
12. Radio and television

These are all aspects that require outside support if people are to be persuaded to settle and work in hill country in isolation. Hou and Wang's report confirms that there are really no problems in persuading farmers to grow trees where they can see an income within a year or two and where there is security of tenure. There are references in their study to rights of usufruct in hill forest fixed for fifty or eighty years (in Qinshui county of Shanxi province) under regulations determined by the county administration. The usufruct can be passed down to the next generation; there is no deduction or agricultural tax taken from the output of the original cultivated fields and hills; no state purchase quotas are allocated; and the original trees are evaluated and regarded as capital. The added value belongs entirely to the individual family. The household has priority in distribution of seed, fertilizer, and pesticide and in the provision of low-interest loans, technical aid in production, access to schooling, and employment. Since subsidies are payable only on seedlings that survive, reportedly the rate of establishment has greatly improved.

Hou and Wang also cite a contract in highly eroded loess country in Shanxi. A six-member family contracted 16.7 ha in a drainage basin, of which 1.6 ha had been cultivated and on which there was a single tree. The household had to build dams and reforest the slopes, planting 25,000 industrial trees, 4000 fruit trees, and 0.3 ha of mulberry. Again, benefits were generated within three years from the fruit trees. The collective contributed a fund of 3000 yuan, and the benefits were divided 30 percent to the collective and 70 percent to the household (with a 40:60 ratio for the industrial trees).

Since 1984, collectives have been encouraged to recruit retired cadres and forest workers, as well as unemployed middle-school leavers, to carry out reforestation of hill lands under contract. Examples are given of sand dune afforestation, but always in conjunction with usable grassland. Such lands are allotted "as much as possible" to rural households for contracted control; shares in the increased yields of both grass and crops (as well as timber) are proportional to the input of the contract. Similarly, allocation of cut-over forest land allows for household and organization to have shares in whatever growing stock may be present on the site—to provide an incentive for supplementary planting.

In general it is argued that 10 to 20 ha of forest land is appropriate for a single specialized household and 70 ha is an absolute maximum. There is a need for improvement in management and particularly systems of internal responsibility and financial accounting. Some households are indeed specializing in these fields.

Financial assistance for afforestation in China has not been consistent. An afforestation subsidy was provided in 1963 by the state to communes and production brigades, but it came to an end in 1966 for the duration of the Cultural Revolution. With the establishment of the timber forest bases in South China in 1974, subsidies were available for "key production units." Beginning with 20 million yuan in 1976, funds rose to 32.44 million yuan in 1979 and played a bridging role to the introduction of subsidized loans that began in 1986. The subsidy averaged 225 yuan per hectare (APFC, 1987). There are now loans available for planting "economic" species at a standard rate of 7.92 percent. The loans are for seven to ten years (and may be extended to fifteen years in special cases). In 1986, some 300 million yuan was made available through the Agriculture Bank of China. Loans are also available for tending and management of existing economic crops and for diversifying forest production. During the Seventh FYP, an annual investment of 200–300 million yuan is to be allocated for tending. There are special allocations for "key" forestry projects. In the first phase of the Three Norths project, for example, 3 billion yuan was expended. (The benefits are projected at 1 billion yuan annually.) Finance is also provided by local funding and provincial government allocations, mainly for nursery development. There is a forest tending fund deriving from a levy on thinning and felling, bamboo sales, and other sources, of 15 yuan per cubic meter of timber—or per 100 bamboo poles—and taken from state-owned forests. A toll is also taken from collective forests—8 yuan per cubic meter of timber cut by the state, or per 100 bamboo poles, and 2 yuan per cubic meter levied on the growers of forest produce. The levy goes mainly to reforest state-owned cut-over land and to regenerate collective-owned land. It also serves to pay interest on bank loans provided for reforestation.

There is a further stumpage tax collected from collective-owned forests. Before 1985 it was prescribed—39 yuan for Chinese fir, 22 yuan for pines, 25 yuan for "Grade B," lesser known species. When timber markets were freed in South China and prices became negotiable, a general percentage of 20 to 40 percent of stumpage was to be retained for reforestation.

According to Xu (1986) the outlook for collective and joint-venture forestry in China is bright. Its strength lies in the diversity of contractual arrangements possible for forest establishment and management, the introduction of marketization (with the abolition of state purchase of timber produced in collectively owned forests in 1985), and the resolve to integrate timber production with consumption. In the words of Xu (1986:36):

> We do not favor the formulation of a national cooperative forestry economy now, or the compulsory introduction of a certain fixed mode of management from above down to the grass roots level. We are of the view that the independent formulation of constitutions by the organizations themselves in the light of varying local conditions is a wiser approach. It is imperative to encourage people to make their own choices, pool their collective wisdom and create appropriate constitutions and regulations. This approach of respecting the initiative of people and drawing on their wisdom avoids harmful

> consequences caused by rigid and uniform administrative order and facilitates the development of a cooperative forestry economy.
>
> Under the prevailing conditions, the household contracting form which turns producer and manager into one and links responsibility with rights and interests is most capable of being accepted by peasants. . . . The household contracting system is a component part of the cooperative economy. Herein lies the vitality of dual level management in the cooperative forest economy.

The functions of management are seen as monitoring contracts, protecting the signatories from abuse, coordination of various forestry activities (such as fire control), training, and the introduction of new technology. There is an acknowledged need to strengthen the legal system in order to protect rights to usufruct and ensure that the newly created plantations are not subjected to malicious damage.

Policies with respect to state production forests receive much less publicity than those concerned with reforestation. In *A Brief Account of China's Forestry,* published in English (Anon., 1984), the main problems in the state-owned forest areas were claimed to be lack of forest roads, disorganized production layout, inaccessibility of overmature forests in remote areas and excessive felling in accessible areas, and failure to utilize forest and mill residues. These problems were addressed in 1980 with a resolve to stabilize the annual cut, reduce production targets, and, when necessary, import logs to make up the shortfall between demand and supply. A forest security bureau was established in the Ministry of Forestry, and various foreign economic cooperation and technological exchanges were set in train. The Twelfth Party Congress reaffirmed the strategic objective to raise the forest cover to 20 percent of the total land area, to double the timber output, and to quadruple the total output value of the forest industry in 1980 by the end of the century. It was assumed that forest depletion would be lowered to the level of annual increment and timber shortages would be alleviated by that time. As noted in Chapter 3, the target for demand and supply balance has perforce had to be revised to the year 2040.

In 1986, the State Council agreed to raise the price of timber from state forests and to allow enterprises to retain most of their earnings. Income and regulatory tax exemption dates from 1987, and it is expected that a proportion of earnings will be invested in reforestation. In the state industrial enterprises, the incentive does not amount to very much, in view of their poor profit record. An adjunct, therefore, is to encourage mergers between them and commercial organizations. According to *China Agriculture* (Li Tang, 1987) there are "forestry–industry–commerce" combinations in 120 counties through twenty provinces. Problems of coordination and traditional rivalries are acknowledged, though, and it is too soon to judge whether or not these combinations will succeed. Their objectives—like so many—are entirely commendable.

Policy Implementation

Ross (1988) has analyzed China's forest policy in relation to the three landholding categories—state, collective, and private. He emphasizes the inadequacy of early investment in forestry—partly because of the regime's preference for industry over agriculture and partly because the investment was inefficient—pointing out that "total investment in forest management since 1949 barely exceeds the amount invested in water conservancy in any given year in the late 1970's." He argues that water conservancy and forestry both fulfill the functions of conservation and flood protection, but this complementarity did not receive serious attention at the highest levels of leadership until after the Chang River floods in 1981. It is doubtful, in fact, whether the relationship was not recognized—indeed references can be cited in great number to indicate the contrary—but the criticism of unbalanced investment may be valid. The recent merger of the Ministries of Water Conservation and Electric Power Generation exemplifies traditional engineering rather than biological relationships. It has to be noted, however, that protection and production forestry in China, as in most countries, have so far been quite separate—both geographically and technically. Most of the forest being logged for production, in northeastern China, is not protection forest and it is doubtful whether to have left it unlogged would have had any effect on flooding. Certainly land use leaves much to be desired and enormous areas of forest are annually affected by fire and insect damage. It can be argued that this represents poor management, and foresters in all countries would protest the inadequate investment in management, protection, and regeneration.

The three forest sectors in China correspond with policy implementation systems. Ross (1988:17) characterizes the arrangement in these words: "The state system is essentially bureaucratic and largely relies on hierarchy commands to secure compliance, the collective sector features campaigns alongside plans, and the private sector largely relies on material incentives organized through market-type mechanisms. The prominence of each at any given point in time is largely a reflection of broader political dynamics." Though it is impossible to quantify the contribution of each component to the forest economy, there is no doubt that in the near future the state system (harvesting the natural forest and controlling timber imports) will continue to dominate the "planned" industrial wood supply, with the possible exception of pulpwood. The state sector represents some 62 million hectares of forested land, rather more than 50 percent of the total. The state, however, is directly responsible for less than 15 percent of reforestation—mostly in remote areas where there is little permanent settlement.

Ross cited another shortcoming within the state sector as well—encroachment by local populations on forest land and the treatment of forest as common property. This is not a major problem in the northeast (though even here there is evidence of cultivation within regenerated areas close to settlements), but it is undoubtedly important in the south and in the poverty-stricken northwest. Ross cites the case of the Ziwuling Mountains in eastern Gansu where the forest (which is within the "Three Norths" area) has retreated by more than 6 miles since the establishment of the PRC—due to land hunger and exploitation for fuel. Other examples of illegal felling have been cited earlier, and even in the sanctum of the strict panda reserves in Sichuan there is evidence of illegal logging and cultivation (see Chapter 3). The state's response has been to tighten the forest law and (for the first time ever) to prescribe penalties of reparation and restitution for damage to forests, illegal cutting, burning, and theft. Ross contends that, despite legislation and despite increasing the ranks of the forestry police, China's new market orientation is more likely to have increased forest destruction than to have contained it.

Ross is also critical of collective forestry operations, analyzing them in terms of timber prices (which are kept artificially low through a "compulsory and monopsonistic procurement process") and inadequate subsidies. There has been much discussion in the Chinese technical press with respect to stumpage and profitability of both collective and private forests. There have been complaints of maladministration within some collectives and, in areas where there are alternative labor demands, tree farms have had difficulty in maintaining the work force. All the socio-economic problems associated with forestry the world over have been raised in the context of China. Nevertheless, people who grow trees are not always motivated primarily by market opportunities.

Ross has explored at some length the roles of "exhortation and environmental ethics" in China's forest economy and concludes—perhaps too hastily—that the failure has been substantial. It may be that fluctuations in policies and instabilities in the economy provided signals of uncertainty that were communicated to collective members—though, as noted earlier, the rural sector generally (and forestry in particular) tend to be insensitive to political change. Nor is the sector amenable to centralized controls. Certainly there is evidence that the "Great Iron Ricebowl" syndrome was well entrenched in the communes, and it was more damaging in forestry than in agriculture because of the long-term production process and the rigidity of underpricing. There were few material incentives to participate in collective forestry work.

The inadequate technical base of the forestry profession in China provides another reason for failure (which is not, of course, restricted to the collective sector). Mao Zedong's early policies of urging scientists to learn from peasants and celebrating "science for the people" culminated in the denigration of academic achievement (indeed, education itself). Its acceptance by so many intelligent people remains perhaps the biggest enigma in an enigmatic country. The fact remains that collectives were urged to respond to slogans based on the experience of localized models ("Learn from Dazhai," for example) entirely inappropriate in a wider context. Farmers growing annual crops recognize the importance of microenvironments and know that a successful crop in one field can fail utterly a short distance away—they can demonstrate it quickly. Forestry, with its long gestation period, requires much longer and the damage is more far reaching. There has always been a lack of a scientific base in both forestry and agriculture, of course, and practices owe much to empiricism. (Even in highly developed countries, agricultural practices are not entirely dissociated from astrology.) Farmers do not need to know why things work, and their ministrations are usually a combination of faith, hope, and experience. But in many parts of China where collective afforestation was being urged, there was little experience, hopes were soon dashed, and faith was void. Again, it is notable that in the traditional tree crop areas of the North China Plain, the massive technical problems associated with collective forest farming were not experienced.

Problems were exacerbated by poor leadership. In a collective system, leadership quality is clearly of vital importance. Ross argues that grass-roots leadership may have weakened in time and that the importance given to the production of grain relegated forestry cadres to a low level in the pecking order. Certainly the forestry profession has little status in China. Where there is an element of career choice (as in scientific research), the low regard for forestry is quite evident.

The absence of proper tending of trees once planted—a notable feature of China in 1963—was particularly evident in commune afforestation schemes. The single-minded concentration on establishment was acknowledged by foresters at the time but not by the political leadership. Later visitors (including FAO missions) considered that much had been done to overcome the initial neglect. Yet even in 1986 plantations on the more difficult sites—especially those that were aerially seeded—show very poor stocking. Although they are mainly in areas forming part of the state forest farms, it would be a mistake to blame state implementation. The poor showing is as much due to the intractability of site and climate and lack of experience. Thus in the traditional tree farming provinces (such as Henan) plantations have been established on poor sandy soils subject to water and wind erosion and on hostile mountain sites by both state and collective teams. Problems increase as the sites available for reforestation become more and more marginal.

In the private sector, Ross (1988) notes the destruction of small woodlots and orchards during the Cultural Revolution and the virtual disappearance of private forest plots other than the "Four Around" plantings. This situation has changed dramatically with the extension of contract responsibility to forestry and, reportedly, the endorsement of afforestation as a flood control measure by the former party general secretary, Hu Yaobang.

The speed of implementing the new reform policies is illustrated by the fact that the new forestry law—adopted provisionally in 1979—lagged behind practice. The distribution of collective land to individuals, for example, was quite outside the spirit of the new law (as originally drafted), and the prescription of specific penalties within the regulation was not contemplated. These are notable features of the law eventually ratified.

PROGRAMS

The first specific forestry program in China clearly based on a resource assessment was set out in the National Plan for Agriculture, which was to run from 1956 to 1968; it arose from a national conference on forestry held in December 1955. The program assumed a total existing volumetric resource of 5400 million cubic meters (derived from figures compiled by the Nationalist government before the war and agreed by Soviet consultants), which would last about thirty-five years. (Average annual requirements over the period 1960–1990 were considered to be 150 million cubic meters.) In order to meet requirements when the resource ran out, some 105 million hectares was to be afforested within the twelve-year period—giving a forest cover of 20 percent of the land surface. (This figure was revised from a target of 18 percent set in 1953.) As in the previous program, the bulk of afforestation was to be in the form of protection and shelter plantations in difficult country. According to the minister of forestry in 1956, "The National Conference on Afforestation . . . demanded that we plant 100 million hectares of barren lands and barren mountains within the next twelve years. If we can overcome all kinds of difficulties and implement such ambitious plans on schedule, we will have written a new page in the forestry history of China, and scored a success unprecedented in the forestry history of the world. Is this not glorious? . . . But China has the greatest number of barren hills in the world. . . . They are salty, alkaline, or sandy lands presenting enormous technical problems."

In the twelve-year plan, emphasis was also placed on "economic crops and fast-growing species." As time went on (and, particularly, after the formation of the communes in 1958), the latter became increasingly prominent in governmental pronouncements, and present policy is to devote about 50 percent of the planting area to such crops. In 1958, also, the decision was made to concentrate the greater part of production afforestation (85 percent) in areas south of the Chang.

At this stage the government also became concerned about standards of protection in the hill forests and called for a more rational land-use policy: "How can we prevent the indiscriminate reclamation of hilly land and the loss of water and soil? As far as we can see, only when we are able to make out a proper production plan for the high country—particularly a plan for using the land rationally and solving the conflicting needs of agriculture, forestry, and pasturage—can we point out to the masses the bright future, and can we prevent them from reclaiming the hills recklessly. . . . Fire is the greatest enemy of forests. We must resolutely and thoroughly eliminate it; we must stop shouting slogans; we must take effective measures" (Richardson, 1966:58).

This clarion call from the minister of forestry was followed by a series of measures designed to improve forest protection and the coordination of land-use policies. As is usual in China—and despite the minister's injunction—massive propaganda campaigns formed the first step: Newspaper articles, lectures, theater performances, and ubiquitous cartoons and posters exhorted the populace to protect the new forests. Regulations limiting access to afforestation areas and prohibiting fuel-gathering were issued, and more than 3000 State Forest Service Centers were established to police them. In 1957, thirty-seven weather and fire danger forecasting stations were set up, aerial reconnaissance was begun, and the closure of forest land to grazing was attempted. An elaborate organization for collecting information about tree diseases and insect pests was also established, along much the same lines as that for agriculture (see Buchanan, 1960), and chemical and biological control of forest insects was introduced.

In attempts to rationalize and coordinate land-use policy, the Academia Sinica, reorganized and strengthened in 1955, and the Scientific Planning Committee (established by the State Planning Commission in 1956 and amalgamated with the State Technological Commission to become the Scientific and Technological Planning Commission in 1958) played increasingly important roles. The Academia Sinica organized expeditions to the remote parts of China (including plant resource surveys), established research centers in desert regions (see Chu, 1959b), and began the preparation of technical handbooks of soil and water conservation. The Scientific Planning Committee set up a series of regional land-use committees, advisory to the provincial people's councils, and several national bodies such as the Working Committee for the Harnessing of Deserts.

The 1960s and 1970s were much taken up with the Great Green Wall project (renamed the Three Norths project in 1978 and given further injections of finance) and the establishment of the "Southern Forest Production Base"—an attempt to transfer the major reforestation effort to the provinces south of the Chang River (where growth rates and species range are more favorable than in the northeast). As noted earlier, this was a decision dating from 1958. Publicity was also given to fuelwood plantings and "rural self-sufficiency" in order to reduce the misuse of shelter and industrial plantations for firewood. Despite claims of nearly 4 million hectares of fuelwood plots by 1976, there is still constant exhortation to replace wood fuel by coal in all areas where it is available. Fuelwood policies are reviewed by FAO (1982).

Recent MOF publications emphasize three programs; the first is the Obligatory Tree Planting Program, the second covers the Great Plains Program, and the third concerns the Three Norths shelterbelt system. The Obligatory Tree Planting Program has been misleadingly translated as "compulsory." (It is, in fact, obligatory in the sense of being an obligation to the community.) It was formally adopted in 1981 and provides that "where conditions exist, with the exception of the aged, the disabled, and people weak in constitution, all citizens of the PRC above the age of 11 should each year plant three to five trees in light of actual conditions or do other work of afforestation such as raising the seedlings and caring for young growth with equivalent amount of labor" (Ross, 1983). Since 1982, it is claimed that one billion trees have been planted annually under this program—many of them as part of the Great Plains and Three Norths programs.

The Great Plains total nearly 100 million hectares and make up nearly 10 percent of China's total land area. The more important of them are the Sanjiang Plain, the Northeast Plain, the North China Plain, the Plains of the Middle and Lower Reaches of the Chang, and the Pearl River Delta. They support 350 million people and contain 40 million hectares of farmland—some 40 percent of the national total. Since the plains are the agricultural production centers of China, it is at first surprising that they are targeted for forestry. Many visitors to China will attest, however, that the integration of trees with agricultural production is the success story of the Chinese forest economy. There are *five* components of agroforestry in China: farm shelterbelts to serve as windbreaks, constructed either as narrow belts or as networks around agricultural land; the "Four Around" planting—around houses and villages and along roads and rivers; dune forests to fix moving sands; timber-producing woodlots along rivers and barren hills; and widely spaced trees intercropped with various cereal and vegetable species. It is difficult to quantify production from these plantations (since they are usually described in terms of numbers of trees or lengths of roadway afforested rather than area), but they serve many purposes. The production of fruit and nuts, the growing of fodder for livestock, fuelwood, green manure (mulches), as well as specialty timber species (such as *Paulownia*) and fiber plants for weaving (such as false indigo), are traditional farm activities.

The shelterbelt system is spread over nearly 4 million square kilometers in twelve provinces, autonomous regions, or municipalities. Originally the Great Green Wall, it was to cover some 2 million hectares stretching from Heilongjiang to Junggar Pendi. The original objective was to provide an ecological screen against the northern deserts and to bring into agricultural production a large area of what is at present extensively managed pastoral land and advancing sand dunes. The climates are extreme and the soils deteriorating through desertification as well as salinization. One section of the system was to run for 1200 km from Heilongjiang to Liaoning (NCNA, 1 June 1961); another through northern Hebei and Shanxi (NCNA, 1 December 1961); a third for 600 km from Shaanxi into Gansu (SCMP, 1960); another bordering the Tengger Desert (Hsu, 1959); with supplementary arid zone plantings from northern Xinjiang (NCNA, 8 August 1962) to Junggar (NCNA, 21 November 1962).

Edaphically and climatically, the overall project covers a wide range of conditions and it might be expected that many diverse species would be involved. Surprisingly, this is not so—probably because the limiting factor for tree growth, whether in Manchuria or Xinjiang, is lack of moisture. The principal species are the hardy *Populus simonii, Ulmus pumila, Salix matsudana,* and *Eleagnus angustifolia* with, as secondary species, *Populus ussuriensis* in Manchuria, *P. diversifolia* and *P. pseudosimonii* in the north central provinces, and *Tamarix chinensis* and *Hippophae rhamnoides* in the saline soils of Xinjiang. The advance of the desert is graphically annotated in Chinese literature. A major problem of course is the silting of rivers, canals, and reservoirs caused by moving sand and loess.

The Three Norths project involves the establishment of mixed forest and grassland, large-scale shelterbelts and networks; the creation of shelter and fuelwood forests, economic plantations, and commercial timber forests; and the closure of areas of mountain and desert to grazing in the hope that natural revegetation will occur. In inaccessible areas the program is being implemented by the PLA through aerial seeding.

It is difficult to judge the success of the Three Norths project. The target for the first eight years was to establish 6 million hectares of forest (12.8 percent shelterbelts, 23.2 percent dune-fixing forests, 55.34 percent "soil and water conservation" forests, and 8.7 percent other types). Again, however, areas have been reclassified within the Three Norths project area; indeed, some entire counties in Liaoning province have been assigned to the project area and their existing shelter plantations counted as part of the Three Norths target. Although this is an imaginative and exciting project, it is difficult to envisage success on the scale postulated without massive investment by the state and the use of the huge labor force provided by the PLA.

An interesting development in 1985 was the establishment of the "Green China Fund," which accepts contributions from overseas Chinese and visitors, as well as from domestic sources, to be used exclusively for afforestation projects. It is administered by a high-powered foundation (the Afforestation Foundation of China) and makes grants or concessionary loans for approved projects, principally in afforestation.

FOREST LAW

Forest policy is expressed through the Basic Forest Law—a document that has undergone many changes since its first proclamation in 1953. The most recent ratification was effected in 1986. For the first time, regulations qualifying the law invoke criminal penalties.

Development

The forest law in China is more a statement of policy and exhortation than it is prescriptive. The first such statement was the "directive on mass afforestation, cultivation of forests, and protection of forests" promulgated by the State Council in September 1953. This document called for the establishment of organizations to introduce fire control in the natural forests and for vigorous afforestation of hills and wastelands. Many such "campaign documents" and regulations limiting access to forest areas, prohibiting fuel gathering, urging fire protection, and so forth, followed during the next decade. During my 1963 visit to China, regulations governing the protection of forests were issued by the State Council on 27 May and accompanied by a *People's Daily* editorial emphasizing their importance. The regulations were published as an appendix to my previous book (Richardson, 1966). During my 1986 visit, a further set of regulations on forest protection was issued—this time qualifying the forest law adopted by the Seventh Session of the Standing Committee of the Sixth NPC. Its history is rather different from that of the 1963 document.

The provisional forest law was in fact adopted at the February 1979 meeting of the Standing Committee of the NPC and was one of the first items of formal legislation enacted during the post-Mao regime. The law was very soon overtaken by events—its emphasis was on the indivisibility of state and collective property and, as Ross (1988) points out, the assignment of collective forests to individuals was virtually prohibited. The law was not ratified, however, and it was followed in 1981 by a State Council decision (FE, 14 March 1981) that has had far-reaching implications for forestry in China. This document points out that there are too many departments dealing with timber and, moreover, their systems are confused. It is critical of lax application of discipline toward forestry, indiscriminate felling, and other abuses and points out uncertainties regarding rights to forest production and to the use of wasteland as major constraints upon further reforestation. The document called for a five-year plan to be prepared within a year by all provinces relating to afforestation and increased forest protection. The forest law was declared to be for trial implementation only and urged the eradication of leftist ideology in forestry development. What became known as the Three Fixes or "Unchangeable Things" were formalized as a result of this document. The needs to be "fixed" were the rapid resolution of forest rights of ownership and usufruct, the demarcation of wasteland for household reforestation, and the establishment of contract responsibility systems for collective forestry operations. Reportedly some two million property rights cases were resolved within the next two and a half years.

Ratification

The basic forestry law was reconsidered by the Sixth NPC in September 1984. (An official English translation is presented in Appendix C.) Article 23 gives the effect of law to individual and collective ownership of forest resources. The law also imposes stronger penalties for indiscriminate felling of trees and attempts to establish closer links between harvesting and reforestation. The detailed regulations for the implementation of this law did not emerge until 1986 and were published in the *Chinese Economic Daily* on 15 May. The length of time elapsing between the original drafting of the law and its implementation is a measure of the change in thinking that was occurring in China during the post-Mao period—involving almost continuous amendment to the provisions of the law. The regulations in fact redefine the forest resource base and establish new targets for reforestation.

Of particular interest is the prescription of penalties for offenses against the law. These require reparations as well as punitive fines. There are punishments for the falsification of felling certificates and transportation permits and for the disregard of regulations governing fire protection. In general, the proceeds of fines are to be paid to the owners of the trees. Introducing the regulations, a vice-minister of forestry, Dong Zhiyong, pointed out that stressing the detail of penalties in such a law was unusual. It follows the adoption of a Criminal Law code in 1984, which also contains penalty provisions for illegal felling.

CONCLUSIONS

It is too early to judge the success of new forest policies and, in particular, contract responsibility systems and the quasi-privatization policies. Success in agriculture seems undoubted (see Walker, 1984), however, and the policies can be expected to work with economic crops offering short-term returns. In the traditional tree-planting provinces they are likely to be successful even in the long run because of the regard for forests and the prestige associated with woodlot ownership. But whether forestry will hold attractions in areas where it is a new pursuit—and where financial returns are long removed from the investment—remains to be seen. Success may well depend on the defense of rights of usufruct against claims by other households or indeed state organizations. There have already been reports of newly privatized forests being destroyed by neighbors of the new owners; the intensification of interpersonal competition that the new policies involve is the subject of academic studies (Gold, 1985; Zweig, 1986). Conflicts within communities are exacerbated in China by the inability of people to change their domicile, while the sheer size of the country poses problems of governability that no other country in the world has to face. It is easy to call for devolution of authority in decision-making, and there is ample evidence that the economy has in the past suffered from excessive control. It is less simple to prescribe the optimum degree of decentralization, given the variety of outcomes that may emerge. (See, for example, the Naughton and Wong studies in *CELT 2000.*)

There is, as yet, no evidence that China has solved the tenurial (management) problems of difficult sites. Even the examples cited by Hou and Wang (1986) and by Xu (1986) do not cover regions where trees have not been grown before, nor do they cover terrain that is so steep as to require aerial seeding. (Significant areas of the Three Norths shelter plantings come into this category.) There is a tendency, too, in China, to describe "trials" as though they were proven practices. Many of the examples given in these reports are experimental—however promising they are, it would be unwise to extrapolate far beyond them yet.

In 1983 forestry received the strong support of a new minister of forestry, Yang Zhong, including the extension of privatization to established forests (though he drew the line at inheritance). The forest law, after six years of provisional status, was finally enacted and regulations were issued in 1986 (Appendix C). Revolutionary changes in price regulation were enacted (including relief for some 225 counties from timber procurement and delivery quotas to enable them to obtain higher prices). Logging controls are to be more closely regulated (under the slogan "strict regulation in the mountains, relaxation in the lowlands"). But apart from the control of cutting, the forestry sector is virtually free. It remains to be seen whether China's forestry households (and cadres) have the technical expertise—and freedom from official corruption—to enable their new role.

Ross (1988) makes the point that "fear of policy instability is the most critical impediment" to sustained forestry in China. In the past the law has not always been applied consistently, and there have been differences in ad hoc interpretation. If the newly established property rights are not respected by authority, it is unlikely that they will be observed by the people. Changing policies with respect to "common" land and rights to its usufruct have—worldwide—led to overexploitation (whether for grazing or for fuelwood) and destruction of values held in trust. China's most pressing needs in forestry are for certainty, continuity, and common sense with respect to forest policy and law.

A realistic policy for the forest industry has yet to emerge. There is agreement that the Soviet Union's guidelines for integration are inappropriate, and there is growing concern to develop a policy for the pulp and paper industry that enables some control of pollution without commitment to the operational scales of North America and Scandinavia. Much will depend on forest products pricing policies—an area of particular concern to the World Bank (Delfs, 1987). That prices are a factor in this discussion is a measure of change in China since 1963.

NOTES

1. This is a convention of both capitalist and socialist economics. Under the former, natural resources have no value unless and until they are marketed; to socialists they are valueless until there is a labor input.

6

PRODUCTION PRACTICES

IT IS IMPOSSIBLE TO NOTE—let alone annotate—production forestry practices in so vast and variable an area as China in one chapter of a book. In my 1966 survey two chapters treated this subject—and even then summarily—and a separate chapter discussed the role of Michurinist biology in forestry in China. Less extensive treatment is warranted now because—since the early 1960s—there have been several published reports that outline silvicultural practices in selected parts of China. (See FAO, 1978, 1982; Dickerman, 1980; Luukkanen, 1980; Matthews, 1980; Turnbull, 1981; Krugman et al., 1983.) Moreover, Michurinism is no longer a topic of interest among forest scientists. (Many, indeed, have never heard of Michurin—or Lysenko.) This chapter, therefore, concentrates on changes evident in field practices since the 1960s.

As with timber production, I devoted considerable discussion in 1966 to statistical claims and counterclaims relating to areas afforested and seedling survival. During my air and rail travel in China, I tried to estimate tree cover by spot observations at fixed intervals. My reasons for what proved to be a very tedious exercise (the data did not compensate for the sleep sacrificed in collecting them) were twofold. First, there were big discrepancies between claims published in China of areas reforested annually since 1949; second, cadres invariably claimed survival rates in excess of 70 percent irrespective of species or site conditions. Figures for areas reportedly afforested were greatly distorted by cadre claims during the Great Leap—at one stage, the New China News

Agency was reporting plantings of more than 3 million hectares *per month*—while only in the northeast and in experimental plantings of the research institutes could the survival rates claimed be confirmed.

MOF officials now accept that the initial failure was widespread; indeed, over substantial areas there is little evidence remaining of afforestation. The figures presented in Table 13 are what Chinese planners believe—and the fact that there is variation downward as well as upward promotes confidence that the statistics are more reliably based than those of the 1960s. The question of survival, however, has not been resolved: In "Four Around" forestry and in timber forests in the northeast, there is clear evidence that greater attention is now paid to tending new plantations than before and that it has paid dividends. Plantings on "bare hills," reclaimed desert, and areas that have been aerially seeded, however, are less convincing.

Nor have problems of reforestation of newly harvested forests in the plateaus of the southwest been tackled successfully. Insolation is extreme (the diurnal temperature range can exceed 25°C), and forests that should be regenerated naturally under a shelterwood are being clear-felled and replanted or reseeded. Successful establishment in these conditions is improbable.

NATURAL FOREST MANAGEMENT

Systematic management of China's natural forests is almost nonexistent: The primary forests are becoming increasingly inaccessible and the cut-over secondary forests are rapidly changing to artificial—and increasingly species-poor—mixtures with exotics.

PRIMARY FORESTS

Management in the primary nonproduction forests is token only: They are protected by inaccessibility, and intervention is limited to the control of hunting and gathering and to fire prevention where possible. Chinese statistics acknowledge an annual loss in excess of 1 million hectares (from fire, insect damage, and disease)—which is not excessive compared with similar forests in other countries (Canada, say, or the USSR) but much more than wood-starved China can afford. Moreover, only part of this loss is in the overmature (and therefore relatively unproductive) forests of the northeast; in the hills of the southwest, there is still shifting cultivation (swidden) and forest clearance for settled agriculture. In the dry northern pastoral zones, fire remains a tool of the grazier. Some Chinese scientists are also concerned about the entrepreneurial activities of the PLA in the border forests of the northeast and along the Mekong River in the south. Forest losses from these sources cannot be quantified, but they must be substantial. Over 2000 forest protection stations have been established across China and are manned by a forest police force of more than 30,000. In the northeast, there is an Aviation Forest Protection Bureau (at Jiagdaqi), and aerial patrols allegedly cover more than 33 million hectares. As the June 1987 fires in the northeast revealed, however, there are weaknesses in the speed of response to the outbreak of fire and in the ability to mobilize the firefighting forces (Richardson and Salem, 1987).

Secondary Forests

The management of cut-over primary forest is perhaps the issue in most need of attention in China's forestry. In most provinces there are a few remaining areas of secondary forest. Their significance is to some extent hidden by the redefinition of forest resources noted in Chapter 3, but the 1976 Japanese assessment (Shiraishi, 1983) estimated some 50 million hectares (50 percent of the total timber forest area) to be secondary forest. Significant areas will doubtless be made available for village or household management (and if the existing trees become part of the shareholding, they will probably be felled to finance reforestation). If they remain under provincial control, management will be restricted to enrichment (probably with exotic species) and selection logging at a set rotation age or diameter.

Henan (in many ways a model province for production and protection forestry plantations) has a natural forest reserve of some 60 million cubic meters (of *Pinus tabulaeformis, P. massoniana, Quercus acutissima, Q. glauca, Cunninghamia lanceolata,* and other species) yielding 500,000 m^3 annually. The "rotation" is said to be thirty to fifty years to give logs of 25 cm diameter. Thus management consists of cutting all trees in excess of that diameter annually, with enrichment planting when the opportunity arises. The species planted are mainly *Cunninghamia lanceolata* and *Pinus massoniana,* with *Populus tomentosa* and

Paulownia fortunei at the forest edges. The introduced clumps may be thinned—beginning at age 8 for *Cunninghamia* and age 10 for pine. Thus, eventually, the species-rich mixed forest will become coniferous plantation—higher yielding, perhaps, but questionably more effective in watershed protection.

Liaoning is a province with over 1 million hectares of natural forest—virtually all of it secondary and comprising *Quercus mongolica, Betula* spp., *Tilia* spp., *Fraxinus mongolica, Phellodendron amurense,* and other species. The favored plantation species, however, are *Populus* spp., *Robinia pseudoacacia, Pinus tabulaeformis,* and *Larix olgensis.* The cutting limit here is 40 cm and geared to the requirements of hardwood furniture and veneer logs. Pine and larch are used for enrichment at spacings of 1.75 × 1.75 m with the intention of thinning at three- or four-year intervals from age 15. The larch will be cut at forty to fifty years, the pine at sixty to eighty years; both are expected to yield 170 to 180 m^3/ha. As in most provinces, the silvicultural regime is a statement of intent rather than established practice. Pine and larch are both light-demanding species—well suited to plantation growth but a surprising choice for introduction into the shade of existing forest. Nonetheless, the deciduous canopy enables their survival and, again, the rich deciduous broad-leaved forest will eventually be transformed into coniferous monoculture.

Several species of larch are being planted, including *Larix japonica* (the fastest growing but susceptible to frost) and *L. dahurica* (photoperiodically better suited to the north), as well as *L. olgensis.* The preferred pine species are *Pinus koraiensis* and *P. sylvestris* var. *mongolica* (on sandy soils), but *P. tabulaeformis* appears to survive better under shade.

In the northeast, the intent of management is similar to that in Henan, but the cut-overs are more recent and the diameter limits less rigidly enforced. Natural regeneration favors the low-value *Betula, Quercus,* and *Acer* species, which grow much faster than the furniture hardwoods (*Fraxinus, Ulmus, Tilia* spp.); enrichment planting, therefore, is again with pines (*Pinus koraiensis; P. sylvestris* var. *mongolica*) and the larches. In this case, although the original mixed coniferous/broad-leaved forest type is changing to coniferous plantation, oak coppice survives as an understory. In the north, the optimum rotation for Korean pine is between 80 and 120 years, while larch degenerates beyond 60 years. The silvicultural regime calls for the oak to be thinned to favor the conifers. In the northeast, there is great reluctance to thin larch.

In the southwest where natural forest is being harvested (in the mountains of Sichuan and Yunnan, for example), land with agricultural potential is generally converted and there is no enrichment planting. In the northwest, the natural forest edges are grazed but the climate determines the carrying capacity (and settlement density) so the forest is probably intact.

Provinces where there are natural forest remnants are urged to "close the hills" so that they may recover. Where this is done—and the closures are policed—it is successful in enabling regrowth, which at least provides firewood. Selection logging, however, is dysgenic and, over the centuries, trees with good form and growth rates have disappeared. Such hill areas are now being allocated to specialized households for management under contract responsibility schemes.

The secondary forest areas in Yunnan (*Keteleeria evelyniana* forest on West Mountain near Kunming) and Heilongjiang (Liang Sui) that I visited and photographed in both 1963 and 1986 show marked differences. The southwest forests—accessible to the city—have been overgrazed and foliage has been collected to feed rabbits: There is little evidence here of diameter growth. The Liang Sui forest resembles an even-aged plantation of pine and larch, with occasional individual *Pinus* and *Abies* trees. To be sure, it needs thinning and it is doubtful whether the increment exceeds 5 m^3/ha, but the contrast highlights the effects of population pressures. In sum, then, the fact of dysgenic selection logging over many years underlines the importance of genetic improvement programs to service China's reforestation campaigns.

SEED COLLECTION AND NURSERY PRACTICE

In the 1960s, China's tree nurseries—especially those serving noncommercial plantings (urban and environmental needs)—were impressive. They are still competent, but less innovative. Nor have they yet begun to apply the results of genetic research to field practice. The following paragraphs describe seed collection methods and the techniques employed in a range of nurseries.

SEED COLLECTION

The mass reforestation campaigns beginning in the 1950s called for huge quantities of tree seed—which was collected by all means possible, with little regard for quality or provenance. Seed centers were established by the MOF in about 1961 and attempts have been made to improve culling standards but with limited success. The "creaming" of *Pinus koraiensis* seed—an edible species that is sorted for separation of the biggest seeds—was noted in 1963 and again in 1986. A U.S. mission of tree improvement specialists (Krugman et al., 1983) observed that tree improvement programs begun in the 1950s were abandoned during the Cultural Revolution; the mission was critical of seed selection intensity and found fault with attempts to manage orchards for both seed and timber production.

Though the vast bulk of seed used in China is still collected without much regard for genetic quality, there are some 10,000 ha of seed orchards and stands. The most advanced province in this regard remains Guangdong—which in 1962 introduced regulations prohibiting communes from using tree seed other than that supplied by the MOF. Guangdong was heavily involved in direct seeding establishment and pioneered the introduction of both *Eucalyptus* spp. and exotic *Pinus* spp. Seed orchards are predominantly for *Cunninghamia lanceolata* and *Pinus* spp. in central and southern China; *Larix olgensis, Pinus koraiensis,* and *P. sylvestris* are the primary seed species in the northeast. Orchards are established by grafting, but frequently they are open-pollinated. Krugman et al. (1983) note that the objectives of seed orchards—and progeny testing—are not always understood.

The use of mother tree seed stands—plantations (perhaps of unknown origin) established and treated from the start for seed production—is favored in China over the designation of seed stands in existing forest. (The reason may be not unconnected with the foresters' reluctance to thin heavily and to risk compromising timber production.) Yet it might be expected that, certainly from the conifers, greater production of good seed would be assured by this method. For shelter and special-purpose species, of course, individual tree collections are more appropriate.

Tree seeds are used extensively in China (because of aerial seeding programs and other direct seeding establishment) but less so than in the 1950s when "nest sowing" of seed was widespread. It is still not uncommon to put two seeds into each dibble-hole, but attempts to produce eight to ten seedlings from the same spot are no longer in vogue.

Visitors to any country can be misled by the enthusiasms of their hosts. Tree improvement is a fashionable field in forestry, and China's forest scientists are keen to be in the forefront. Moreover, it is a research area that requires little by way of capital investment and uses scientifically "elegant" techniques (such as anther and tissue culture). Researchers in China, however, take shortcuts and are apt to move so far ahead of practice as to be out of touch. The major species used in China are easy to propagate and do not need elaborate techniques. Similarly, hybridization programs may be premature in light of what still has to be done in the way of genotype selection.

Nursery Practice

Major changes in nursery practice since 1963 relate to mechanization and fertilizer regimes, the latter fairly recent in origin. The State Forest nursery at Dailing remains typical for the northeast: It has increased in area from 20 to 27 ha, but the annual planting program has not significantly enlarged—the extra space is used to produce flowers and seedlings (of a variety of ornamental and horticultural species) for sale. The principal tree species remain *Pinus koraiensis, Larix dahurica, Pinus sylvestris* var. *mongolica,* and certain hardwoods—*Fraxinus mandshurica, Tilia amurensis, Phellodendron amurense, Populus maximowiczii* (not recorded in 1963), and *Juglans mandshurica.* The conifers make up 70 percent of production of around 8 million seedlings annually, with some 16 to 25 million seedlings and transplants held in stock.

Seed is now drill-sown (in 1963 it was broadcast) at densities rather less than 0.5 kg/m^2 for larch and Scotch pine and about 1 kg/m^2 for Korean pine. The aim is to obtain 280 to 320 seedlings per square meter. Germination varies with seed quality. Seedlings are held in the nursery until 15–20 cm tall (for pine) and 30–40 cm (for larch). They are three-year plants—either two years, undercut, plus one year or one year, transplanted, plus two years, depending on the initial density. Undercutting is a new mechanized development since 1963. Hardwood stocks for forest planting have now standardized on 1 + 1 or 2 + 1 transplants, and the poplars are raised from cuttings. The Dailing nursery still has problems of uneven germination, which ranges from 60 percent to "better than 95 percent."

The fertilizer regime remains a mixture of organic (night soil, town refuse, and forest litter) and artificial urea and phosphate with periodic green cropping. Artificial fertilizer use is being reduced because a number of factories are closing—under the new competitive regime, they are uneconomic. And the old standby of forest nurseries the world over—spent hops—has virtually disappeared, as chemicals now substitute for hops in breweries. In some forest nurseries they have been replaced by biogas sludge, but in Heilongjiang green cropping with soybeans and the foliage of *Robinia pseudoacacia* is used to prepare seedbeds.

Irrigation facilities were installed in 1965 and operate from May to August. Insecticide may be used in the irrigation system, but knapsack sprayers are still in use. Weeding continues to be done by hand and provides seasonal employment for middle-school leavers. During winter the seedlings are buried with soil, which is raked off in spring.

Apart from variable seed quality (a comment applying to seed orchard production as well as to forest collections) the Dailing nursery is well run and the plants generally well grown. There is a general reluctance to cull (and to destroy culled material) throughout China. If it were done more vigorously, it might be possible to reduce the planting age of at least some species. A nursery at Linkou (Heilongjiang) is experimenting with a huge plastic greenhouse in an attempt to speed the development of containerized pine seedlings: It is very much a demonstration nursery and includes electronically controlled mist irrigation from an automatic, traveling boom system. Simpler means more generally applicable would be to use cloches or polyethylene tunnels.

There are at least 8000 formal forest nurseries in China (MOF, 1984) producing 25 billion seedlings from 400,000 ha. The Dailing nursery by this reckoning is only half the national average in area but is above average—despite its northern location—in productivity. Certainly there is great variation in size and quality. Since Dailing is in an agricultural area with a fairly heavy soil, it is not particularly suitable for pines. The provincial nursery at Zhanggutai (Liaoning), by contrast, is on a sandy soil and produces plants to be used in sand fixation and shelter trials. It comprises 10 ha devoted 50 percent to research production and 50 percent to horticultural production (of which half is sold for the benefit of the nursery staff and half allocated to the adjoining research institute for its own use). The lighter soils suit *Pinus sylvestris* var. *mongolica*, the principal species used in this project. It is drill-sown at a density of 400 seeds per square meter with straw laid between the drills to lay dust and serve as a mulch. The fertilizer regime here is heavily organic (more than 20 metric tons/ha) including wood ash and night soil. Lines of tree seedlings may be intercropped—cabbage is used as a windbreak! The horticultural part of the nursery is a mixture of forest trees, vines (including a species of kiwifruit—*Actinidia argula*—which was formerly known as Chinese gooseberry), peanuts, and ornamental trees for urban planting. Surprisingly there is very little containerization except for the ornamentals.

Farther south, in Henan, nurseries provide excellent-quality stock of *Paulownia* (4 m tall one year after sowing 1 × 1 m in the nursery), *Cunninghamia lanceolata* (being grown for cuttings), and a great variety of species raised for planting on the edge of the loess plateau—in highly saline soils (of pH 8–8.5) and on precipitous slopes. In one nursery (near Mongshan) over 600 species have been grown, of which 400 survived and 10 were selected for regular planting. (They include *Cupressus, Ailanthus, Larix, Amorpha fruticosa, Populus tomentosa, Gleditschia, Salix matsudana,* and the exotic *Robinia pseudoacacia.*) Henan is a pivotal province for forestry in China. Its rivers divide North and South China, and it is the meeting point of tropical and temperate species. Climatically it provides unrivaled opportunities to assess species on their geographic margins.

High standards of nursery practice characterize the Hangzhou Forest Research Institute in Zhejiang (especially for oil-producing trees and introduced southern pines—see Krugman et al., 1983); Guangdong Forest Research Institute (container-grown *Eucalyptus* spp., bamboos); Kunming (Yunnan) Botanical Institute (ornamentals and medicinal plants); and Xishuangbanna (Yunnan) for introduced tropical species, including balsa. The botanical gardens (those that survived the Cultural Revolution intact) also maintain a good impression despite their focus now on sideline production for sale. Nursery practices are well tried, and technicians are both knowledgeable and conscientious. Standards in the old, established nurseries have been well maintained and, as in 1963, they are impressive. The botanical garden nurseries serve taxonomic arboretums, orchards, and demonstration and trial plantings, and they are training grounds for forest nurserymen as well as orchardists and, sometimes, exponents of the art of *p'an tsai* (miniature tree cultivation).

In the south, the present system of preparing beds is to plow and harrow, dig to a depth of about 15 cm, and then fill with a conventional mixture of loam, calcium/magnesium/phosphate (equivalent to 87 kg/ha of phosphorus), and 1 to 2 tons/ha equivalent of organic manure. Turnbull (1981) describes current procedures for raising *Eucalyptus* at Dongmen (Guangxi) and Leizhou (Guangdong). Prepared beds are watered, cut into blocks 8 × 8 × 12 cm, and then covered with organic ash. The soil may be sterilized with formaldehyde and the seed sown in mixture with fine, dry soil; it is covered with sieved organic ash, a layer of pine (or *Casuarina*) needles, and irrigated. Seedlings are pricked out into the earth blocks (one per block), then fertilized twice weekly with a urea solution. The aim is to produce 20–30 cm seedlings in three months. At Leizhou, seed is sown directly in the soil blocks and the seedlings reduced to two per block at the two-leaf-pair stage; seedling growth is similar to that at Dongmen.

The nursery and associated trial gardens of the Botanical Institute at Kunming were impressive in 1963 and still maintain high standards. Because of its success in phytochemical research, the institute is well funded (from Japan and Canada) and the nursery reflects this support. It serves the extensive *Camellia* and *Rhododendron* gardens as well as raising medicinal and ornamental plants and has modern seed storage and propagation facilities.

The smaller rural nurseries serving ad hoc reforestation projects are, as in 1963, much less noteworthy. They may be roughly cleared, sloping hillsides—apparently begun as temporary "acclimatization" sites before planting and kept going, because of seedling demand, as sidelines that have grown into "specialized household" projects. The Changchun Municipality (Jilin) nursery began to raise larch and pine seedlings for planting around a reservoir, but the staff's ability and enthusiasm have turned it into a major tree improvement research focus, a biological control center, and a pioneer in contract research and the production of genetically superior seed for sale. The An Qian nursery (Heilongjiang) was established in 1952 as a vegetable garden for a management training institute; the institute subsequently set up a forest farm and turned the garden into a tree nursery; it now runs a ginseng garden under a contract responsibility system (with seventy-six forest workers) and produces 6 metric tons of fresh root annually.

There are nurseries run by the PLA and at least one operated by the commercial arm of a trade union. There are household nurseries raising seedlings at fixed prices under contract; there are backyard nurseries no bigger than a few square meters raising trees from which to take cuttings for sale; and, in the cities, there are balconies and window boxes in high-rise apartment blocks being used to grow tree seedlings for urban planting. Despite these initiatives, there appears to be less innovation in practice and in experimentation now. In 1963, every nursery offered its own organic fertilizer recipe—some of them sounding more like the contents of a witch's caldron than plant nutrients—and there was greater variation in sowing density, shading, and watering.

PLANTATION FORESTRY

Despite the wealth of tree species available to China's foresters, the problem of what to plant is not an easy one. After all, the Chinese are attempting to establish forests in areas where there is no residual vegetation to serve as a guide to likely performance or where secondary vegetation may be fire-induced and misleading. Except for Guangdong, no province of China had any significant experience of extensive plantation forestry before 1950; since then, there have been numerous failures and it is not surprising perhaps that foresters appear conservative with respect to choice of species—but adventurous when it comes to extending known species beyond their natural range.

Site Preparation and Planting

The improvement in plantation survival so noticeable since the 1960s is due largely to more intensive site preparation and early tending. Except in the southwest, there is no burning off and wherever possible new plantations are established one year after complete plowing to 30 or 40 cm and harrowing. Where this is not feasible, trench or pit planting is adopted; the trenches may be 50 cm deep and lined with green manure. Pits are much bigger than in most countries and filled prior to planting with a mixture of soil and wood ash or compost. The quantity of fertilizer is substantial—Turnbull (1981) notes applications of 3 or 4 kg of organic manure (animal or vegetable) to each hole prior to planting *Eucalyptus*. Trenches are believed to conserve moisture and to provide early shelter. As with direct seeding, it is not uncommon to plant two seedlings per hole.

In Manchuria, on wet or peaty sites, turf planting is still carried out—

using turfs 2 m long × 1 m broad × 30 cm deep; up to ten trees are planted on each turf. Nowadays it is stressed that multiple plantings of this kind are not related to the eight-point charter for agriculture urged by Mao Zedong, which included "close planting"; rather, it is to enable early selection for stem form and vigor. Whatever the rationale, it is not a practice that appeals to visiting *plantation* foresters—for most species, spacings appear much too close and the effects are exacerbated by a reluctance to thin stands until they yield a usable return. (With *Eucalyptus* or poplar in South China, even a one-year plant has value, but in the northeast plantations remain densely stocked for too long. There is a similar reluctance to discard culls in nurseries—a practice that leads to overstocking the plantations.)

After planting, intercropping may be practiced on accessible sites and is a reason for deep planting. The interrow area may be shallow-plowed or hand-dug and organic manure buried, sometimes to a depth of nearly 1 m. On hill country, contour furrowing is usual; if the slope exceeds 20 degrees, planting terraces and trenches may be formed and interplanted with (usually) a legume to stabilize the site. The plants are older than on flat country, though generally smaller planting stock than in the 1960s seems to be used in timber production forest.

The following pages outline silvicultural requirements and practices adopted in China for some of the more important commercial production species. Establishment of environmental and protection plantings is described in Chapter 7.

Major Timber Species

My 1966 survey included a table of 175 principal species used in afforestation projects in China, according to region and physical site, including multiple-purpose, shelter, and wood production species. Although there are few additions to be made to that list today, there are significant changes in their relative importance—especially of the so-called economic species. The increased attention given to agroforestry species can be traced to its espousal by international organizations (FAO, ICRAF, and others) and to the fact that, since the 1970s, most visiting foresters in China have been steered to the traditional tree-planting provinces, where they could not fail to become enthusiastic about the rich variety of multiple-use species, well grown and well maintained. Chinese foresters, too, have become more knowledgeable about exotic species (especially *Eucalyptus*, *Robinia*, some pines and poplars) and, in new afforestation areas, have promoted them.

Despite the emphasis on multiple-purpose species, two-thirds of current planting in China is for wood production (Hsuing, 1980). A further 15 percent is of bamboos, which substitute for many purposes for which timber is used in industrialized countries. The species annotated here are the same as in the earlier publication, except for the omission of *Pinus massoniana* and *Populus yunnanensis.* The former—once the most extensively planted conifer—is no longer regarded as a major production forestry species. (It is still used in aerial seeding because it tolerates a wide range of soil types and inhospitable, frosty, or dry sites, but it is a survivor rather than a high-yield species.)[1] Its place in southern China is being taken by the southern pines of North America. Yunnan poplar is still being planted, but as a "Four Around" species more than in block plantations.

Large-scale afforestation for timber production is dominated by the following species: Manchuria—*Pinus koraiensis, Larix dahurica,* and *Pinus sylvestris* var. *mongolica;* North China Plain—*P. tabulaeformis;* Changjiang provinces and South China—*Cunninghamia lanceolata;* South China—*Eucalyptus* spp.; southwestern China—*Pinus yunnanensis.* Throughout China, poplars are being used and *Eucalyptus* spp., *Pinus* spp., and *Robinia pseudoacacia,* all introduced, are increasingly apparent in the south and southeast. Other high-yielding species such as *Paulownia* are extensively grown in "Four Around" and shelter plantings, but the "bread and butter" indigenous timber species are all coniferous.

PINUS KORAIENSIS (RED PINE, KOREAN PINE) *Pinus koraiensis* is endemic to Manchuria, northern Korea, the Russian Maritimes, and Japan. It is a major component of both the Mixed Coniferous and Deciduous Broad-Leaved and the Deciduous Broad-Leaved formations. It is unusually shade-tolerant for a pine and has obvious silvicultural affinities with *Pinus strobus,* its counterpart in the Mixed Northern Hardwood forest of North America. In natural stands it reaches heights of 45 to 55 m with diameters up to 1.5 m; its natural life is some 600 years. Tree form is good; there are clean cylindrical boles and well-shaped, ornamental crowns. Timber quality is high: The wood is light (density 0.4), soft, straight-grained, and easy to work. Strength properties are good, it seasons and finishes well, and it is generally free of defects. Korean pine is the most valuable general-purpose softwood in North China (the northern equivalent of Chinese fir) and is the species against which other plantation species are measured. The timber is exported to all

China's provinces (though large sizes have become very scarce) and is used in construction, in the round (transmission poles), in joinery, furniture, and veneer production. Residues are used in particleboard and groundwood pulp; it is the preferred raw material for the new medium-density fiberboard (MDF) plants being installed in the northeast. There used to be oleoresin production from trees from south-facing slopes, but this practice is now rare. The seed of Korean pine is collected in large quantities for sale—it is a large edible seed similar to the Mediterranean stone pine (and in fact is sometimes called Korean stone pine).

In plantations *Pinus koraiensis* grows on a range of sites, but it is difficult to establish on waterlogged soils and does not do well on shallow, stony sites. It will, however, tolerate acid soils and—unlike *P. strobus*—it is deep-rooting, thick-barked, and consequently fire-resistant. It is frost-hardy and can withstand very low winter temperatures (down to −45°C) and low rainfall (around 50 cm annually), but it needs a growing season of at least 100 frost-free days. Since it is prone to forking in plantations, there is a move to increase the initial planting density (from 3300/ha to over 6000)—in contrast to the trend in other species. The minimum rotation is now considered to be sixty years (in 1963 it was fifty) and, more generally, seventy to eighty. The oldest continuously measured stands are in forty-year-old plantations in Liaoning that (according to FAO, 1982) had a mean height of 15.6 m, a mean diameter of 20.9 cm, and a volume of 256 m^3/ha. The density at the time of measurement was 1020 stems per hectare. (It is unlikely that these values are representative of most plantations.)

Silvicultural schedules—as with all plantations in China—are eclectic and, as presented to visitors, are usually the tentative recommendations of researchers in Beijing. In the field, there are differences of opinion regarding the relative merits of Korean pine and larch. The latter grows more rapidly for the first fifteen or twenty years but is then overtaken by the pine; on north-facing slopes, pine moves ahead at an earlier age. Its timber quality is generally agreed to be higher—but so is the rotation required to grow high-quality wood. There are also differences of opinion as to whether it should be grown in pure stands or in mixture with broad-leaved trees (birch or oak coppice). As in many countries, the views of foresters are often colored by experiences far removed from the problem on hand. There is evidence in some parts of China that the growth rate of *Cunninghamia lanceolata* falls after the first rotation if it is grown in monoculture. There is a tendency to assume that this will occur with all coniferous species and that mixed coniferous/broad-leaved plantations are more desirable. Moreover, mixtures are expected to suffer less damage from pests than pure coniferous stands. Such assumptions are applied to pines in Manchuria, where there is no evidence one way or the other. The fact that there is argument about such issues represents a change from 1963.

In contrast with other major species, Korean pine has not been planted outside its natural range to any extent. It is of interest, however, that a few plants established from seed in New Zealand in 1964 have survived; they have not grown well, but they are prolific seeders.

LARIX DAHURICA (= *L. GMELINI*) (DAHURIAN LARCH) Larch species are widely distributed in China from Manchuria to Xinjiang, but Dahurian larch of the Hinggan Mountains of northeastern China is the most extensive and economically important. At altitudes over 900 m it forms dense pure stands, reaching a height of 30 to 40 m and a diameter in excess of 1 m with a branch-free straight bole. It is the main constituent of the Northern Coniferous Forest. At lower altitudes in the Hsiao Hinggan Ling it is found in mixture with pine, spruce, fir, and birch. Nowadays a taxonomic distinction is made between *L. dahurica* and *L. olgensis* (= *L. gmelini* var. *olgensis*); the latter is the major component of the Hsiao Hinggan Ling and is thought to date from volcanic eruptions some 180 to 200 years ago. In Chinese field forestry, however, the difference is not recognized—though some foresters argue that the best seed sources for plantation larch are from the *L. olgensis* region.

The life span of *Larix dahurica* is at least 250 years; in plantations, however, it suffers heartrot after 50 or 60 years and the rotation is determined by this feature. The young timber is valued for its natural durability (poles in ground contact have a service life north of the Huang River of 20 to 25 years) and its strength. It is a low-density timber (0.32) and does not season well; nor can it be easily treated; and while it can be pulped, its resin content causes excessive foaming. Except for use in the round (poles, piles, and mine props), it is less favored than Korean pine and is too coarse-grained to peel satisfactorily. The utilization of larch is regarded as the biggest problem facing foresters in Jilin province and ranks second only to that of shrinking forest resources in Heilongjiang. Despite these difficulties, *Larix dahurica* makes up nearly half the existing plantations in Manchuria and, together with the closely related *L. sibirica*, is the major imported species from the USSR. Much timber in China is used as roundwood (including roof trusses and other house construction).

Larch in northeastern plantations may be a mixture of species—certainly of races. It is nonetheless very uniform and can be grown on a wider range of sites than Korean pine. It is easier to establish on wet sites and tolerates a wider pH range.

It is difficult to understand Chinese foresters' reluctance to thin larch. (In the case of pine and hardwoods the lack of thinning may be due to the absence of demand in the sparsely populated forest areas, but larch poles find a ready use in construction.) There is ample evidence that wider initial spacing and heavier thinning give higher increments per hectare (and per tree); yet there is little change in the appearance of larch plantations from 1963. In Jilin, eighteen-year-old *L. olgensis* recorded a mean diameter of 6.1 cm, when planted at a density of 10,000/ha, and 8.0 cm at a density of 4400/ha. An older stand (thirty-eight years) recorded 14.8 cm diameter, dominant height 16.3 m, total volume 161 m^3/ha, and density 1550 stems per hectare. At Liang Sui in Heilongjiang, spacing trials established in 1953 and ranging from 12,500 to 2500 stems per hectare showed (in 1963) the merits of wider spacing. These plots were thinned in 1964, 1974, and 1984. Measured in 1986, the plot data for the widest spacing were as follows:

Unthinned	
Volume	304 m^3/ha
Mean height	19.7 m
Mean diameter	16.2 cm
Thinned 20 percent	
Volume	271 m^3/ha
Mean height	19.8 m
Mean diameter	18.4 cm
Thinned 30 percent	
Volume	247 m^3/ha
Mean height	20.5 m
Mean diameter	19.1 cm
Thinned 50 percent	
Volume	174 m^3/ha
Mean height	21.1 m
Mean diameter	18.8 cm

Despite the evidence of piece size, in Chinese eyes the measured volume per hectare overrides other features. Nonetheless, the older practice of planting two seedlings in each hole at 1 × 1 m spacing is no longer much in vogue, even in the northeast. In the southern provinces, wider spacing is general and thinning is much heavier irrespective of species. This practice doubtless stems from the demand for wood—of any size or shape.

In 1963 I concluded that "a silvicultural schedule for *Larix dahurica* is taking shape, though it is doubtful whether it is generally applied yet." That conclusion must remain, but the schedule now developing is different from that envisaged earlier—which involved pruning at ages 5 and 10, then at two-year intervals, with the disposal of pruned branches and thinnings for fuel, pulpwood, or stakes. In 1986, pruning was not mentioned.

PINUS SYLVESTRIS var. *MONGOLICA* (SCOTCH PINE) Scotch pine is a principal constitutent of the northern coniferous forests around the world from Siberia to Alaska. The Chinese variety is much more restricted in natural forests, but it is a major production species on dry sites and in shelterbelts. It occurs spontaneously on stabilized sand dunes of the Hailar steppes, on the northwestern slopes of the Hinggan Range (with *P. pumila*), and in Jilin (with *Larix* and *Betula*). It is both hardy (surviving temperatures down to −50°C) and drought-resistant and is able to adjust its growing cycle to the variable frost-free season. This property results from the rapid completion of height growth and earlywood formation (within four to six weeks of bud-break) and flexibility with respect to latewood development.

On the Hailar sands (and in Outer Mongolia—see Richardson, 1987b) Scotch pine does not grow quickly or densely; rather, it resembles a savanna woodland species. The stem form is excellent, however, though the branches are coarse and the crowns broad. In China, where it is being planted, it should properly be regarded as an exotic species. Growth is relatively rapid. According to FAO (1982) it has reached a mean height of 8.5 m and a diameter of 12.7 cm at age 20 on Liaoning sand. It has some tolerance to salinization and—on nutrient-poor sites—responds well to admixture with legumes. In the northeast it is planted at higher altitudes than Korean pine or larch and, though light-demanding, has been successfully group-planted (a practice stemming from the Michurinist thesis that competition is less fierce within species than between species). On other than the poorest sites, Scotch pine is sensitive to weed competition and, on cut-over forest, it has suffered from *Armillaria mellea* infection. If it is to be successfully processed, it will need pruning; branches are more persistent than is the case with *P. koraiensis*.

Plantations administered by a municipal research institute in Jilin (around a reservoir) show Scotch pine to be better than *P. koraiensis*, *P. tabulaeformis*, or *Larix olgensis*. Planted at 6600 stems per hectare in 1957, stands have been thinned and designated seed orchards; at 600 stems per hectare, the height averages 15.4 m. In a shelter planting in Liaoning (a block 1000 × 400 m not differing markedly from a production unit), the spacing was 1.5 × 2 m, thinned to 4 × 3 m at age 15; thinning was scheduled at three- to four-year intervals and the expected yield at maturity (sixty years) was 180 m^3/ha. A mean annual increment of 3 m^3/ha may be low compared with, for example, southern pines on good soils, but it is acceptable on an arid site in the northeast.

Not much is known of the timber properties of *P. sylvestris* var. *mongolica* from sites where it is being grown in China. It is regarded as a coarse timber species compared with Korean pine, but mature logs imported from Outer Mongolia have been well received in Beijing.

PINUS TABULAEFORMIS (CHINESE PINE; BLACK PINE; HORSETAIL PINE) The natural distribution of *Pinus tabulaeformis* ranges from the Deciduous–Broad-Leaved formation, through the Mixed Coniferous and Deciduous–Broad-Leaved forests (on the drier sites), the Mixed Deciduous and Evergreen Broad-Leaved ecotypes at high altitudes (1000 m), and the Evergreen Broad-Leaved zone. It varies in habit from a stunted, flat-crowned tree on high ridges to well-grown 20–30 m trees of reasonable form on well-drained slopes in northern China. Ecologically it may be a pioneer, a preclimax species in a broad-leaved seral stage, or a climax component (with *Biota orientalis* and *Juniperus rigida*) on the arid, steep slopes of the hills bordering the North China Plain. Because of its site tolerance, it has also been planted outside its natural range and can now be found from Jilin province to Yunnan.

The timber is coarse-grained but works well and would be easy to treat; it resembles Scandinavian red deal (*P. sylvestris* or North American *P. resinosa*) and is used for much the same purposes (joinery, construction, sleepers, boxwood). It is also tapped for resin.

In China's northern provinces, *P. tabulaeformis* has been widely planted on dry, eroding hill sites, where it tolerates a range of pH from slightly alkaline to acid. It is a major protection species of the loess regions, established on terraces, troughs, or fish-scales and sometimes intercropped. As a timber species it has been planted in the northeast, and statistics cited from Jilin are as follows (Luukkanen, 1980): age—38 years; dominant height—12.8 m; mean diameter—15.3 cm; volume—172 m^3/ha; density—1450 stems per hectare. On this particular site, the yield exceeded that of larch. It also grows well in Liaoning, but the stem form is poor. Farther south, it is an acceptable substitute for *Pinus massoniana* and in Yunnan is being used in place of *P. yunnanensis.*

It is not possible to estimate the extent of *P. tabulaeformis* planting: In no province is it the most common species, but it is a significant component of many. In 1963 it did not commend itself—survival was poor and performance disappointing. With improved planting practices and tending, the prognosis is brighter and it offers a realistic alternative to *P. massoniana* on dry sites.

CUNNINGHAMIA LANCEOLATA (= *C. SINENSIS*) (CHINESE FIR) Chinese fir is an attractive tree with a long history of plantation forestry in China. Its natural distribution is throughout the provinces bordering the Chang River and south into Guizhou and Hunan. In natural forest it grows to 50 m with a straight, cylindrical bole to 2 m diameter and is often clear of branches for 18 to 20 m. In stem form, it resembles its close relatives *Araucaria* and *Agathis* spp. Like *Pinus tabulaeformis,* it has been planted well outside its natural range and because of its coppicing ability—after fire as well as cutting—it has become naturalized (in the south). Menzies (1989) records plantations of Chinese fir dating from the ninth century and used to convert scrub forest. It is sensitive to early frost but has the capacity to shed frost-killed branch tips and regenerate from an adventitious bud. It grows best on deep, well-drained, sandy soils with high temperatures and humidity during the growing season. The coppice shoots can easily be rooted—the usual way of regenerating this species. According to Menzies, the earliest agricultural treatise describing the cultivation of Chinese fir dates from 1273; in the sixteenth century, a system of measurement and valuation was developed for it—an indication of its importance.

The timber and its traditional uses have been extensively described by Wilson (1913) and documented by Menzies (1989). Its durability—and resistance to termites—have made it a classic coffin wood and it is a traditional boat-building timber. The wood is light (0.45), easily worked, fine, and straight-grained, and peels well. It is in demand for plywood, furniture, and ornamental chests. By-products include several essential oils and medicinal bases; the bark is used as a roofing material. Individual trees will grow to a great size, but as early as 1725, in Zhejiang, rotations were suggested as short as twenty years—ranging to eighty years depending on the soil and timber market prices (Menzies, 1989).

Menzies also annotates the early intercropping of *Cunninghamia* and its use in attempts to settle shifting cultivators. Settlers (in 1780) could contract to clear and plant land (with trees and oil species—thea, or tea, and tung) and the proceeds were divided between landlord and tenant "on such a day as the *Cunninghamia* shall be deemed to have been grown to maturity." Menzies suggests that this may have been the forerunner of the Burmese system of establishing teak by "*taungya.*" (This system is widely used to persuade farmers to establish trees: They are permitted to raise food crops in forest areas on condition that they also raise seedlings, moving onto new ground when the tree canopy closes. The Burmese term *taungya* means "hill cultivation.")

Nowadays Chinese fir is a major reforestation species in provinces from Taiwan to Yunnan, but most extensively in Zhejiang, Fujian, northern Guangdong, Guangxi, Hunan, and Guizhou. It has a wide altitudinal range from 200 m above sea level in the eastern and northern regions to 2000–3000 m in western Sichuan and northeast Yunnan (FAO, 1982); obviously the range is related to rainfall, but there are geographic variants according to temperature. It needs good drainage but tolerates acid soils (pH of 4.5–6.5). In 1963, the standard spacing (for cuttings) was 1 × 1 m in well-cultivated soil that was kept clean, sometimes by intercropping. Now there appears to be a wider range of initial spacing (according to site quality, slope, and elevation).

FAO (1982) provides data on growth rates but their origin is not referenced. They imply a preferred density of 2000–2500 stems per hectare and mean annual increment to age 20 of 12–17 m^3/ha. Other visitors (NZ Mission, 1987) postulate a "usable yield" of 75 m^3/ha at age 18 and 150 m^3/ha at age 20 (mean annual increments of 3.2 and 10 m^3 respectively). There is little doubt that intensively managed Chinese fir on suitable sites has the potential to reach mean annual increments of 20 to 25 m^3/ha. There appear to be no serious diseases or insect pests, but there is evidence of reduced growth in second-rotation stands. This issue forms a major focus of research at the Institute of Forestry and Soil Science of the CAS at Shenyang. The reduction in growth is reportedly from a mean annual increment at age 20 of 15 m^3 to 10 m^3. The institute recommends the introduction of broad-leaved trees (*Castanopsis*), but intercropping with agricultural legumes may be a more profitable alternative. There is some mixed planting of *Cunninghamia* with *Robinia pseudoacacia* and *Sassafras tzuma* in Henan and, farther south, ditches are dug between the rows and filled with green manure. In Henan the initial spacing is 1.5 × 1.5 m, but it is still too close. Nonetheless, the plantations of *Cunninghamia lanceolata* are an impressive sight and a pleasing contrast to their state in 1963.

Cunninghamia is a species with potential as an exotic in other parts of the world and its timber would have a high value. It has, of course, been grown as an ornamental tree but should do well as a production species in, say, Mediterranean countries and subtropical regions of the Southern Hemisphere.

Every forester has his own favorite species, for whatever reason. Of the conifers, Chinese foresters from the northeast might choose Korean pine; but many of the others as well as most visitors would opt for *Cunninghamia lanceolata.*

EUCALYPTUS SPP. *Eucalyptus* spp. were introduced into Guangdong province in the late nineteenth century from Italy and have since been planted in Yunnan, Guangdong, Guangxi, Fujian, and Zhejiang as production species. In 1963, *E. globulus* (Chinese blue gum) and *E. citriodora* (lemon-scented gum) were annotated as particularly promising, but *E. camaldulensis* and the cold-resistant *E. cinerea* and *E. rubida* (from the USSR) were also seen. Since the 1970s, under the guidance of FAO, selections have been introduced from Australia. Turnbull (1981) claims that more than 300 "species, varieties, and hybrids" have been taken into China from various sources. Some 200 species are said to be still cultivated, but many are wrongly named or their records of origin have been lost. In 1980, there were estimated to be over 330,000 ha of *Eucalyptus* plantations in China—excluding the many miles of linear and "Four Around" plantings. Since 1980, the planting rate has increased and the range has been extended.

In Yunnan, the preferred species are *E. globulus* and *E. maidenii* at higher elevations (over 800 m) and *E. robusta* and *E. camaldulensis* lower down—in recent years, the two latter species have suffered damage from unseasonal frosts. In Sichuan, *E. botryoides, E. camaldulensis,* and *E. viminalis* have been planted. Turnbull (1981) cites ten-year-old *E. botryoides* in northern Sichuan reaching 19 m in height and 23 cm in diameter; at Chengdu, *E. viminalis* has grown to 27 m and 31 cm diameter at age 17. Many species have been grown in Guangdong (especially at the Leizhou forest farm); *E. robusta, E. citriodora, E. camaldulensis,* and *E. tereticornis* were the first species planted extensively, but nowadays *E. exserta* and *E. citriodora* are the preferred species. (They tolerate poor alluvial soils and are easy to propagate.) *Eucalyptus exserta* includes the "*E. Leizhou* No. 1" hybrid—the exact origin of which is not known though it was a deliberate cross. It has grown in plantations to a height of 6 m and 4.5 cm diameter at two years of age. Mean annual increments of 15 m^3/ha have been recorded for *E. exserta* and *E. citriodora* with a maximum of 25 m^3/ha. *Eucalyptus cloeziana* and *E. urophylla* have also given high yields. Several visitors to China (Krugman et al., 1983; Turnbull, 1981) have argued that China has been premature in setting up hybridization programs without adequate selection. Turnbull notes that this practice probably stems from self-reliance policies and making the best of gene stock available within China; but it may be due in part to the apparent reluctance of Chinese researchers to communicate with each other. Recently an Australian aid project was established at Dongmen (Guangxi autonomous region) to develop a *Eucalyptus* improvement program. It may be expected that China will select from Australian—and, perhaps, Brazilian—populations.

Most *Eucalyptus* species in China are grown on a fifteen- to twenty-year rotation for mine props, poles, fuelwood, and the like, but some may be grown to twenty-five years to yield saw-timber. Thinning is prescribed at age 3–5, 7, and 10–12 (Turnbull, 1981); the first thinning reduces the initial stocking from about 4000 to 2000 stems per hectare (*E. citriodora*) or 3000 (*E. exserta*). The final stockings will be, respectively, 900 and 1500 stems per hectare. Thinning yields on very good sites may be on the order of 14 m^3/ha (FAO, 1982).

Eucalyptus exserta is a coppicing species and, for roundwood and fuel plantations, regeneration follows clear-cutting; *E. citriodora,* however, must be replanted. Both species are regarded as general-purpose hardwoods in China, but *E. citriodora* is used for the production of citronella oil. It is also planted around living quarters as a mosquito repellent.

Some Chinese foresters have recently become aware of the environmental controversy surrounding the extended use of *Eucalyptus* spp. It is argued that such plantations may create a host of undesirable consequences for soil and water conservation and for people. In India, it has been claimed to lead to desertification; in Portugal, *Eucalyptus* is described—somewhat extravagantly—as a "fascist" tree because it supposedly impoverishes farmers; in Thailand, there have been attempts to have it banned. The truth, of course, is that the genus is damned by success: Because of its fast growth, it does use a lot of water and nutrients; and if trees are needed for purposes other than biomass production, in most situations *Eucalyptus* spp. are inappropriate. Undoubtedly, there are many examples of its misuse worldwide (and it has sometimes been planted in unwise locations in China)—but these are arguments in favor of defining objectives and improving the selection of species and seed sources, rather than banning its planting.

PINUS YUNNANENSIS (YUNNAN PINE) Yunnan pine is very like *P. tabulaeformis* and may in fact be a variety of it. Ecologically it is the western equivalent of *P. massoniana* in the evergreen broad-leaved forests of Yunnan, Guizhou, and Guangxi. It is a pioneer species following fire—and together with *P. kesiya* it is also found in Myanmar, Thailand, and Vietnam, wherever there is shifting cultivation. In natural stands it is not high-yielding, reaching heights of only 20 to 30 m and diameters to 50 cm before giving way to seral hardwoods; it has excellent stem form, however, in contrast to *P. massoniana*, and has not suffered the dysgenic selection of the latter. Its good form also gives it an advantage over *P. tabulaeformis* as a pole timber.

Pinus yunnanensis prefers moister sites than the other species, but this characteristic is reportedly due to its intolerance of low atmospheric humidity rather than soil moisture. It is invariably planted on steep slopes, often very closely spaced on cultivated contour strips. In 1963, in Yunnan, it was the only species for which specific blanking prescriptions were given. (Where the initial "take" fell below 70 percent, dead seedlings were replaced for two consecutive years.) Nowadays blanking is general; indeed, *P. yunnanensis* is being used to replace failed *P. massoniana*.

The timber is similar to red deal and has a specific gravity of 0.45. It is a useful general-purpose softwood used in the round for house construction. It is also tapped for resin, and resin-impregnated splints are used for lighting. Most of the plantations have been established by collectives (former communes), sometimes in mixture with evergreen oaks—reportedly to provide alternative fodder for sheep and goats. It is unlikely that *P. yunnanensis* will rival the other plantation conifers of China but, in a region lacking softwoods, it may have a role to play.

It is the most extensively planted conifer in both Sichuan and Yunnan provinces and is a memorable feature of the railway journey between Chengdu and Kunming—many thousands of hectares have been planted on the precipitous sides of the canyons through which the line runs. Though the soils are dry, the hills are frequently shrouded in fog, so the humidity is high. (It is said that the dogs in Sichuan bark only when they see the sun!) In Yunnan, seed of *P. yunnanensis* (and *P. kesiya*) is sometimes broadcast in mixture with dry rice in a form of *taungya* cultivation. Farmers are paid 25 yuan/*mu*—provided the trees are successfully established (85 percent) within three years. The sites are degraded *Castanopsis*, *Schima*, and *Betula* scrub.

Other Production Species

POPLARS AND BAMBOO Of the species I identified in 1963 as significant for timber production (as distinct from fuelwood and multiple-purpose species), poplars and bamboo are not annotated here because of the impossibility of doing justice to them. There are specialist works dealing with both groups, and the major species are discussed by FAO (1982). Poplars are noted in Krugman et al. (1983) and Matthews (1980). IDRC (1987) has published a monograph on bamboos in Asia and the Pacific. There is also an extensive Japanese literature on bamboo, and it is perhaps rightly described as the most useful plant in the world—the Chinese claim more than twelve hundred ways of using it.

Over 140 species and varieties of poplar are being grown at the Chao Bai forest farm near Beijing, and the genus is a major focus for hybridization. Krugman et al. (1983) note that *Populus euramericana* (*P. deltoides* × *P. nigra*) from East Germany is producing nearly three times the volume of the more widely grown *P. canadensis*. The commonest native species grown in plantations are *P. simonii* (several varieties), *P. cathayana*, and the hybrid between them (*P. pseudosimonii*). The aspen poplars (such as *P. tremula* var. *davidiana*) are common colonizers of cut-over and burned areas of the northeast, and the desert poplars (such as

P. euphratica, P. pruinosa, P. alba) are important shelter species. Hybrid poplars are more important in linear plantings (the ubiquitous *P. tomentosa* in northern and central China, for example, and *P. yunnanensis* in the south). But with new emphasis on quick returns from forest plantations it seems likely that the poplars will increase in importance—perhaps in mixture. In 1963 (Richardson, 1966:96) I recorded that "all Chinese poplars appear to be free of the major pests and diseases to which the genus is susceptible in temperate regions." Unfortunately that statement is no longer true.

There are more than 300 kinds of bamboo in China, but easily the most commonly planted is *Phyllostachys pubescens,* accounting for 70 percent of plantation production. (It was introduced into Japan in the eighteenth century and now accounts for 50 percent of Japan's production.) There are three major bamboo regions in China stretching between latitudes 18° and 35°N across the country. The northern species are single-culm bamboos and mostly confined to sheltered valleys; *P. bambusoides* is particularly hardy. In the south, most species are multiple-culmed (such as *Sinocalamus* spp.); in the central belt, they are mixed (such as *Indocalamus* spp.). *Phyllostachys pubescens* plantations are concentrated in the provinces of Zhejiang, Jiangxi, and Hunan; there remain huge areas of mountain bamboo at elevations up to 3500 m in Yunnan and Sichuan. FAO (1982) notes that bamboo management in China is very conservative.

EXOTIC SPECIES Apart from *Eucalyptus* spp. and some poplars, the important plantation introductions into China are the southern pines, *Taxodium distichum,* and *Robinia pseudoacacia.* Krugman and Kellison (1986) estimate that some 1.25 million hectares of southern pines (*Pinus taeda* and *P. elliottii*) have been successfully established in China and the area is increasing by 40,000 ha a year. Moreover, pilot field trials are under way with *P. caribaea, P. palustris,* and *P. serotina* and smaller-scale tests of more than twenty other species of pine are recorded at the Hangzhou Forest Research Institute, Zhejiang. The USDA is involved in a cooperative tree improvement program with China, using agreed study designs at twenty-five locations. The results will be of worldwide interest.

Pinus elliottii was first introduced from the United States in 1946, and most earlier plantings are of seed from this initial arboretum trial. *Pinus taeda* came much later from unknown sources. The two species are believed to offer the best prospects of any exotics for large-scale plantation forestry in southeastern China. They have been established in fourteen provinces and frequently outgrow the *Pinus massoniana*, which they will probably replace.

Taxodium distichum var. *nutans* is also a common exotic species in South China. Introduced in the early years of the century, it is of good form and growth rate and is being grown in mixture with the deciduous conifer *Metasequoia glyptostroboides*. *Robinia pseudoacacia* is another North American species introduced at the same time. It grows on a range of sites in China but is particularly useful on poor sands. It has been tried in virtually all provinces of China and is successful over some 25 degrees of latitude—as a street tree, a soil improver, a fodder tree, in windbreaks, and in plantations. It may be grown in mixture with high-volume producers such as poplar. *Robinia pseudoacacia* provides a good example of China's eclectic approach to choice of species for new forest areas—in marked contrast to the country's conservatism in the traditional forest production regions and with respect to silvicultural schedules.

OTHERS Of the other timber species listed in 1966, the following justify a higher ranking now than they were previously afforded:

Cupressus funebris
Melia azedarach
Cryptomeria japonica (introduced)
Metasequoia glyptostroboides
Quercus mongolica
Sassafras spp.
Pinus armandi
Salix matsudana
Picea likiangensis
Fraxinus mandshurica

Other tree species (*Paulownia*, *Casuarina*) are more important but are properly regarded as special or multiple-purpose species. And some of the species noted in the 1966 list no longer warrant the attention given them (*Salix wilsonii*, *Keteleeria* spp., *Abies delavayi*).

Special and Multipurpose Species

Trees have always been important in China for purposes other than industrial timber production. Marco Polo is often quoted nowadays in support of China's ancient record of tree planting to "afford much comfort to travelers." In fact, it was the Mongolian leader Genghis Khan, who in the thirteenth century ruled a goodly part of China, who should be given credit for the early special-purpose plantings. (His empire, reckoned as a proportion of the known world, has never been exceeded; he administered it with the aid of fast Mongolian horses traveling routes marked by trees.) It may be, however, that tree planting for fuelwood predates Genghis Khan. According to Pei (1985), the Dai people living in Xishuangbanna (Yunnan province) have raised *Cassia siamea* for fuel and house-poles for nearly 2000 years; it is coppiced on a four-year cycle and a rotation reportedly of up to 200 years. Thus the tradition survives and, as noted earlier, the espousal of agroforestry by the international agencies has focused particular attention on what used to be translated as "economic" or "industrial" species. These include trees grown for fruits, medicines, oil, fodder, and organic manure as well as for fuel and timber. (For present purposes trees planted primarily for amenity and protection are treated in Chapter 7.)

Special-purpose species identified in 1963 are listed in Table 27. The list even then laid no claim to be comprehensive. Species emphasized by subsequent visitors include *Acacia* spp. (for tanning); *Dalbergia* spp. (yielding wax); several introduced species (*Amorpha fruticosa*—forage; *Hevea brasiliensis*—rubber; *Carya illinoensis*—nuts; *Anarcardium occidentalis*—nuts); *Mangifera indica* (fruit); and *Salix* spp. (osiers). The lists do not include the huge number of medicinal plants.

Species from Southwest China were inadequately represented in my earlier work; since the 1960s, significant ethnobotanical work has been done on the tropical species of Yunnan and the high-altitude trees of the Tibetan border, and several multipurpose, oil-bearing species have been identified. Trees underemphasized in 1963 include the *Paulownia* species, which have long been grown for cabinet woods and can also be used as fodder. FAO (1978) presents a note on the Changsha municipal nursery, which raises *Cedrus deodara, Pinus massoniana,* and *Cunninghamia lanceolata* for roadside planting as air pollution detectors. The same report lists the palm *Livistona chinensis,* grown for its fibrous leaves and for erosion control.

TABLE 27
SPECIAL AND MULTIPURPOSE SPECIES NOTED IN 1963

Region	Low Altitude ("Easy" Country)	High Altitude	Problem Sites
			Sand Dunes
NW Nei Monggol	*Prunus armeniaca*		*Haloxylon ammodendron*
NW Xinjiang			*Hovenia dulcis*
N Gansu			*Tamarix pentandra*
N Qinghai			*Ailanthus altissima*
			Crataegus pinnatifida
			Sand Dunes
E Xinjiang	*Morus alba*	*Juglans regia*	*Sophora japonica*
SE Gansu	*Sophora japonica*	*Pyrus malus*	*Elaeagnus angustifolia*
S Nei Monggol	*Catalpa bungei*	*P. communis*	*Populus simonii*
N Hebei		*Diospyros kaki*	*P. pseudosimonii*
Liaoning		*Zizyphus jujuba*	*P. diversifolia*
N Shanxi		*Prunus armeniaca*	*Ulmus pumila*
		Sophora japonica var. *pendula*	*U. laciniata*
			Saline Soils
			Tamarix chinensis
			Hippophae rhamnoides
			Sand Dunes
NE Nei Monggol	*Morus alba*	*Phellodendron amurense*	*Salix matsudana*
Heilongjiang	*Juglans regia*	*Tilia mandshurica*	*S. mongolica*
Jilin	*J. mandshurica*	*T. taquetii*	*Elaeagnus angustifolia*
			Pinus sylvestris var. *mongolicum*
			Mountain Sites
Shandong	*Morus alba*	*Phyllostachys edulis*	*Pinus tabulaeformis*
S Hebei	*Quercus acutissima*	*Sassafras tsuma*	*Cryptomerica japonica*
S Shanxi	*Cinnamomum camphora*	*Quercus variabilis*	*Pseudolarix amabilis*
Shaanxi	*C. kanahirai*	*Q. acutissima*	
Henan	*C. cassia*	*Castanea henryi*	
Jiangsu	*Zelkova schneideriana*	*C. mollisima*	
Hubei	*Thea sinensis*	*Thea oleosa*	
Anhui	*Dendrocalamus strictus*	*T. sinensis*	
Zhejiang	*Phyllostachys* spp.	*Aesculus chinensis*	
S Guizhou	*Arundinaria* spp.	*Gingko biloba*	
N Hunan	*Pterocarya stenoptera*	*Bischoffia javanica*	
N Jiangxi	*Sapium sebiferum*		
NW Fujian			

Region	Low Altitude ("Easy" Country)	High Altitude	Problem Sites
			Mountain Sites
Sichuan	*Phoebe nanmu*	*Phoebe nanmu*	*Pseudotsuga sinensis*
N Guizhou	*Cinnamomum camphora*	*Juglans regia*	*Castanopsis platycantha*
SE Qinghai	*C. cassia*	*Thea oleosa*	*Tsuga yunnanensis*
	Sinocalamus affinis	*T. sinensis*	*Lithocarpus cleistocarpus*
	Morus alba	*Aleurites fordii*	
	Juglans cathayana	*Phyllostachys edulis*	
	Pterocarya stenoptera	*P. bambusoides*	
	Alnus cremastogyne	*P. puberula*	
	Gingko biloba	*Citrus* spp.	
E Xizang	*Juglans regia*		
Qamdo			
NW Yunnan			
SW Sichuan			
Yunnan	*Catalpa duclouxii*	*Cupressus funebris*	
SW Guizhou	*Lindera communis*	*Prunus persica*	
W Jiangxi	*Cupressus torulosa*	*Pyrus communis*	
	Rauwolfia spp.		
	Adina racemosa		
	Sapium sebiferum		
			Sand Dunes
Jiangxi	*Cinnamomum camphora*	*Citrus* spp.	*Casuarina equisetifolia*
Guangdong	*C. cassia*	*Canarium album*	*Pandanus odoratissimus*
Hunan	*Eucalyptus citriodora*	*C. pimela*	var. *sinensis*
Jiangxi	*Bambusa stenostachys*	*Euphoria longana*	
Fujian	*Dendrocalamus latiflorus*	*Litchi chinensis*	Swamps
	D. giganteus	*Cinnamomum camphora*	*Rhizophora mucronata*
	Arundinaria amabilis	*C. kanahirai*	
	Citrus spp.	*C. micranthum*	
	Morus alba	*Zizyphus spinosa*	
	Encommia ulmoides	*Ficus lacor*	
	Bischoffia trifoliata	*Aleurites cordata*	
	Sapium sebiferum		
	Litchi chinensis		

SOURCE: Richardson (1966).

The special-purpose and multiple-purpose species feature in Four Around plantings: Usually they are widely spaced (initially 2–5 m) in line plantings, but in shelterbelts they may be much closer. Where possible they are pruned for fuel and thinned. (Even bamboos are pruned.) They are more intensively managed than plantation trees, of course, and are intercropped. Yields have been measured for crops from *Cunninghamia* plantations—the output in a county of Hubei province was reportedly 6.5 metric tons/ha of grain, watermelon, and vegetables, yielding 50 percent of total income. In Hunan, intercropping has produced 20 metric tons/ha of green manure. These yields are exceptionally high, but the merits of intercropping are not in question. Obviously, cultural practices vary with site and there are notable differences in the same area. In general, practices appear more pragmatic than in the case of plantation forestry.

The *Paulownia* species merit particular mention—not only because of the antiquity of cultural tradition (the earliest monograph on *Paulownia* dates from 1049) but because of the burgeoning interest in the species as an exotic introduction in other parts of the world. Canada (through IDRC) is supporting research on *Paulownia* at six centers in China and has sponsored publication of a monograph on *Paulownia* in China (IDRC, 1986).

Seven species of *Paulownia* are planted: *P. elongata, P. catalpifolia, P. tomentosa, P. glabrata, P. fortunei, P. fargesii,* and *P. kawakamii.* They cover a wide range of climates and elevations to 1600 m. They need deep, well-drained soils and are ideal for wide spacing and intercropping. The most favored species in Henan are *P. elongata, P. catalpifolia, P. fargesii,* and *P. kawakamii. Paulownia* can be grown from cuttings, but in Henan it is raised from seeds sown at a spacing of 1 × 1 m. If the one-year seedlings do not reach 4 m in height, they are cut back and allowed to sprout. Plantation spacing is 6 × 6 m (intercropped) and the rotation fifteen years, which is believed to coincide with the maximum mean annual increment; merchantable trees can be cut from thinnings at age 5. The timber has a high strength/weight ratio and, unlike many fast-growing species, is decorative; easily seasoned and worked, it is used for cabinet making and furniture. It has been exported to Japan (from Taiwan as well as from mainland provinces) for many decades. Where there is adequate moisture, the mean height growth of *Paulownia* exceeds 2 m annually, and a five-year-old tree can reach 30 cm in diameter. A ten-year-old tree can produce some 30 kg of dry foliage and 400 kg of branch wood (Chin and Toomey, 1986).

Paulownia is said to have a beneficial effect on microclimate, and in the Huang River basin it can significantly increase grain yields. Like all "miracle" trees, it suffers from pests—notably gypsy moth and a microplasmic witches'-broom—and is prone to flood damage. Nonetheless, it is a well-proven species in China and presently grows on more than 1.5 million hectares of intercropped land (more than double the area in the 1970s). It has even attracted the attention of Deng Xiaoping, who visited the Cheng Guan forest farm in Shandong province in 1983 to view an acclaimed "treasure" (Chin and Toomey, 1986).

Another species highlighted by FAO (1982), but limited so far to Hunan and Hubei provinces, is *Sassafras.* It is intensively managed—planted in 1 × 1 m holes at 2 × 3.3 m spacing (1500 seedlings per hectare) and intercropped with *Amorpha fruticosa* for green manure. It is grown on a fifteen-year rotation and can yield 30 to 35 m^3/ha annually. It is listed in Table 27 but was not a particularly noteworthy species in 1963. (*Casuarina* is noted in Chapter 7 since it is primarily a shelter species for coastal dune protection.)

FOREST HEALTH AND PROTECTION

The most obvious agents of damage to trees in China are insects. Perhaps because of the expansion of line planting along roads and railways—and very low bird populations—insect infestations appeared more numerous in 1986 than in 1963. Since the late 1950s, Chinese agricultural scientists have been concerned to monitor pathogens and foresters have followed their lead.

According to Wagner (1987) the intensity of valley agriculture in China traditionally enabled cultural and physical control of pests there. The hills were neglected, however, and as deforestation proceeded under increasing population pressures, bird and insect populations began to change. In the late 1950s, cultural and labor-intensive methods in agriculture were derogated. Chemical control—together with some resistance breeding—was urged as a scientific alternative but proved to be an expensive one. From 1958 to 1970, chemical pest control dominated other methods and, despite short-term successes, led to significant reductions in populations of parasites and predators, as well as the target pests. According to Wagner, the caterpillar *Dendrolimus punctatus* (*pini*) was first subjected to spraying with benzene hexachloride (BHC) in 1955: Within ten generations, its chemical resistance had increased fifty to a hundredfold and populations of the beneficial parasitic wasp *Trichogramma* had been decimated; by 1975, 5 million hectares of forest were affected by *Dendrolimus.* Since 1972, there has been a return to cultural and biological control at the local level, but doubts have been expressed about the likely success of "integrated pest management" following the demise of the communes.

A notable feature of China's plantations in 1963 was the absence of bird life—a result of earlier campaigns aimed at reducing the population of sparrows, a war against birds that became indiscriminate. The scarcity of birds is still a feature of forestry in China, but efforts to encourage insect-eaters are apparent in many areas. Households raise insectivorous birds (such as magpies) that are rented out for control purposes. Nesting boxes are a common feature now—at least in research plots. Insect damage appears to be a particular problem with poplars in line plantings and in neglected pine plantings on poor sites.

An interesting development is the use of abandoned nursery sites to monitor insect pests and diseases. Where trees have been raised for special projects—such as ad hoc roadside or shelter plantings—nursery stock is left for periodic monitoring of insect pests. It sounds like a more valuable practice than it is: Not only are the abandoned seedlings unrepresentative, but also foresters are not well versed in insect identification. Nonetheless, the nurseries could be organized to serve this function more effectively. Wagner (1987) discusses integrated pest management in China and emphasizes the importance of introducing new technology through existing political structures and cultural practices.

Apart from *Dendrolimus*—for which the wasp *Trichogramma*, the fungus *Beauveria bassiana*, and the bacterium *Bacillus thuringiensis* are used as parasites—*Tortrix* and gypsy moth (*Lymantria*) are serious pests in conifer plantations. China has sought international assistance for insect control on poplars (Morris, 1985). The problem insects are:

Wood Borers
- *Anaplophora glabripennis*
- *Trirachys orientalis*
- *Saperda populnea*
- *Paranthrene tabaniformis*
- *Aegeria apiformis*
- *Cryptorhynchus lapathi*

Defoliators
- *Clostera anachoreta*
- *Stilpnotia salicis*
- *Lymantria dispar*
- *Apocheima cinerarius*

Others
Chaitophora sp. (poplar aphis)
Anacampsis populella (leaf roller)

Twenty provinces in China reportedly have serious problems of poplar damage from insects. Morris was impressed by the remedial action under way in Shanxi province. Based on his observations there, he drew up a project of pilot studies of integrated control measures involving chemical spray, selection and breeding of fast-growing genotypes (insect infestation is frequently associated with poor growth), and biological controls.

Some 200 insect species are associated with poplars in China; 20 of them do serious damage. One-third of the poplars planted are said to be damaged over an area of 7 million hectares (Morris, 1985). Earlier visitors to China (among them Dickerman, 1980) expressed concern at what may be excessive dependence on particular clones of poplars (especially *P. tomentosa* in Beijing). Attention given to clonal propagation of poplars, *Eucalyptus,* and *Paulownia* stands in contrast to the haphazard seed collections for conifers.

Apart from the poplars, the most obvious problems relate to the large area of poorly maintained *Pinus massoniana* in the south. Formerly a major timber production species, most of it can now be harvested only for firewood. It is chronically attacked by *Dendrolimus* and *Matsucoccus matsumurae.* The prescribed treatment is to thin out the worst affected trees and to interplant other species—*Eucalyptus, Camellia, Cunninghamia,* and *Sassafras.*

Insects damaging other tree species include *Polychrosis cunninghamiacola* (a leaf roller) on Chinese fir; *Algedonia coclesalis* and *Ceracris kiangsu* on bamboos; the bark beetles *Ips subelongatus* and *Blastophagus piniperda;* the larch case-bearer, *Coleophora laricella; Dendrolimus superans* and *D. spectabilis* on pine and larch; *Dioryctria abietella, D. splendidella,* and *D. mendacella* (cone and shoot destroyers); the termite *Odontotermes formosanus;* and *Biston marginata,* a defoliator of *Camellia.* The usual run of fungal diseases is recorded—damping-off, rusts (including blister rust), *Lophodermium pinastri, Armillaria mellea, Fusarium* wilt (affecting tung trees), *Endothia parasitica* on chestnut, and Dutch elm disease. *Paulownia* is susceptible to a witches'-broom and *Larix* to heartrot. In general, however, fungal disease is less noticeable than insect damage.

As in other fields, there is a wide gap between research and practice in forest protection. In the laboratory, researchers are isolating viruses and experimenting with pheromones; in the field, control is overwhelmingly manual. The highway bureaus, in particular, mobilize large numbers of schoolchildren to crush the eggs of borers and capture the adult beetles. Nonetheless, there have been major changes in field programs since the 1960s. Insectaries are common, but they are not always used; where they are, the results are impressive. In the *Trichogramma* programs, for example, host eggs from the oak silk moth are employed and in one province alone (Zhejiang) more than 100,000 kg of cocoons are used: Eggs are glued onto cards and set to attract *Trichogramma;* the cards, each carrying some 9000 *Trichogramma,* are then distributed according to the density of *Dendrolimus.* The method is adapted from agricultural pest control. It is not universally acclaimed—in Guangdong foresters use plastic containers and rolled leaves containing the parasite. It is claimed that *Trichogramma* plus two microbial insecticides are used on more than 600,000 ha of forest annually (compared with over 1 million hectares chemically treated). It is not clear that these figures refer to closed forests.

Problems of rodents in grasslands and shrub forest areas have been noted in Chapter 2. In Yunnan, a very different pest is the spread of the creeper *Eupatorium odoratum* after burning. It is a recent introduction and Chinese agriculturists have no answer to it; it takes eight to ten years for a tree cover (*Trema orientalis*) to come through it. It may be beneficial, though, in that it enforces a rest period from cultivation on impoverished soils and is helping to reduce burning.

CONCLUSIONS

The most striking difference in production forestry practices from 1963 is the attention given to early tending: Plantations are weeded two or three times in the first two years and often again in the third and fourth years. The slogan "70 percent is in the management" dates from the 1970s. It has paid dividends. Apart from tending, plantation forestry silviculture in China remains conservative. Initial spacings in production forests are close (except for poplars) and thinning—even of light-demanding pioneer species—strikes the visitor as timid. Moreover, it is not clear that thinning schedules are applied routinely. Visitors are taken to see experimental plots in several provinces, but there is little evidence that the spacing and thinning prescriptions are being followed in the field, even in the northeast. Moreover, some young plantations visited in 1963 have been clear-felled in advance of rotation age. There may be very good reasons for this cutting, but it suggests that silvicultural schedules are—at least—flexible.

Foresters today are prepared to argue the pros and cons of thinning intensity, and this would not have happened in 1963. Other controversial issues are freely discussed—notably the timber supply/demand crisis and what should be done about it; pest control (biological or chemical); choice of species in agroforestry; even the accuracy of the national forest inventory. (At least one province does not accept Beijing's interpretation of the 1981 sampling and continues to use its own 1976 data.)

In the traditional tree planting provinces of the North China Plain and in the south, planters are more adventurous in their readiness to try new species (often outside their natural range) in new agroforestry combinations, though they still rely on traditional species (such as *Pinus massoniana*) for aerial seeding and on poor sites. Possibly it is the farmers, rather than foresters, who are responsible for this impression. Certainly farm extension services seem more dynamic and innovative than in forestry, and farmers have had economic incentives for a longer time.

The effects of the contract responsibility schemes on silvicultural practices will be of interest and, perhaps, crucial. Free markets should increase prices for larger logs more than for pulpwood sizes; and there should be financial incentives to thin more heavily. The standard of pruning (for fuelwood) should improve, as well, when trees are identified by households instead of being considered communal property. Pruning is still not viewed as a means of improving log quality, however, even by researchers—it is simply a source of fuelwood.

There remains a gap between research and practice that needs to be bridged (see Chapter 8). Except in forest entomology and pathology, there is little evidence that day-to-day field problems influence research programs. The rift is most glaring with respect to wood processing, but it exists in forest management, too. Thus researchers are well aware of the importance of seed selection and tree improvement—the U.S. Forestry Exchange Program has served China well in this regard—but the large majority of seed used in reforestation programs is still randomly collected in the field from unselected parent material. Perhaps the most serious deficiency in forestry (not only in China) is failure to involve forest managers in designing research programs and setting priorities. This task is left to the producers—rather than the users—of research.

NOTES

1. The Society of American Foresters took part in an exchange visit with China in 1980. Their report (SAF, 1980) suggests that the story of *P. massoniana* might serve as a warning to all foresters of the dysgenic effects of overcutting. Over much of its natural range (covering ten provinces in central and southern China) it has been eliminated, and the only *natural* stands remaining are remote and inaccessible. Dysgenic selection resulted in the poor quality of trees in the large-scale plantations of *P. massoniana* that were a notable feature in the 1960s. The SAF team concluded that while the species has not been totally lost, it will be many years before trees of acceptable quality will be available.

7

ENVIRONMENTAL AND PROTECTION FORESTRY

EVEN VISITORS WHO TRAVEL no further than Beijing and the Ming tombs are impressed with the evident success of environmental tree planting in China. The multiple-row avenue of poplars, birches, willows, planes, and pines (and other species) along the 30-km route from airport to city center—especially in a summer drought—is astonishing. Much of the rest of the capital city's planting lives up to its promise. Elsewhere in China, too, the greenery of the cities and towns offers compensation for the cramped and meager living conditions inside the high-rise beehives. The boulevards are reminiscent of Washington and Canberra—but more impressive than either because of the speed with which greening keeps pace with construction activity—involving transplantation of "instant" trees, sometimes 8 to 10 m tall. China's urban forestry expertise—and the gardening traditions from which it derives—provide hope that other kinds of forestry may eventually become as successful.

The challenges that face protection forestry, however, are, in the light of history, more daunting. In past centuries it might be argued that China's failure to protect its great watersheds was due to ignorance. In fact, this is unlikely; the role of trees in water conservation and erosion control has long been understood. The effects of forest removal and slope cultivation, as well as recovery following famine and depopulation, have been extensively documented. According to Deng (1927), the grim lessons to be learned from China's experience were well known in the reign of the Emperor Chen during the Chou dynasty (1122–256 B.C.). Many decrees were aimed at protecting forests—including the ultimate sanction: "Persons that failed to plant trees were not allowed to have coffins." But despite the vision of men like Mencius, who warned of the need for conservation, feudal warfare brought the dynasty to an end and, with it, the end of the "golden period of the forest history of China." The following "dark period" lasted, according to Deng, until 1911. Dynasties came and went but the forests did not: During prosperous times, agriculture expanded at the expense of the hill forests; then came floods, drought, famine—and insurrection, the ultimate contributor to forest destruction. As early as 1844 Huc (cited by Lowdermilk, 1926) described the degradation of soils that followed forest removal, and both Sowerby (1924a; 1924b) and Lowdermilk (1926) provide further documentation. Lowdermilk published early photographs (from 1924 and 1925) of wasteful exploitation, steepland cultivation, and subsequent denudation. Even then, he believed that watershed protection (an end to cultivation and grazing) would enable recovery.

More recent prophets are less sanguine. Smil (1982) has assembled a bibliography of 249 publications on environmental aspects of economic development in China and concludes that "deforestation appears to be the country's most critical environmental problem." Lampton (1986), too, documents a sorry story of continued overcutting and resource deterioration; some Chinese believe that "an unprecedented breakdown of their ecosystem is a real threat" (Smil, 1982). This chapter outlines recent developments in urban environmental forestry and the setting up of natural (or seminatural) forest reserves. The continuing problems of water conservation, desertification, and shelter are discussed, together with the use of trees in what are truly herculean endeavors to mitigate them.

ENVIRONMENTAL FORESTRY

My 1966 survey devoted a chapter to arboriculture, describing roadside and city tree planting, temple trees, and *p'an tsai* (miniature tree culture). Tree cultivation in these contexts is an art form in China that has captured the enthusiasm of many visitors, and there have been several recent accounts of urban forestry, botanical parks and gardens, and the imaginative use of trees in landscape. (See Moss, 1965; Ovington, 1975; Pollard, 1977; Mathews, 1980; Keswick, 1978.) The pleasing contrasts between the symmetry of palace or temple gardens and the busy informality of "parks for people" remain.

Even the authors of the FAO (1982) report were moved to lyricism by the Chinese garden and what is described as "the ultimate illusion of the natural"—the Confucian virtues of "simplicity, antiquity, and harmony." It is doubtful whether many of the present-day users of parks and gardens in China are aware that they are reflecting a Taoist "search for unity with the universe" or that their fascination with climbing high places (whether mountains or fire towers) is a manifestation of Buddhist reverence. But the gardens are nevertheless surprisingly free of vandalism and generally escaped the ravages of the Cultural Revolution. These green spaces are undoubtedly China's greatest recreational asset—ranging from extravagant and ancient imperial gardens to modern extrovert playgrounds populated by gaudy concrete animals in remote rural villages.

Urban Forestry

Equally successful are the urban forests of China. A major change from the 1960s is the increase in air pollution that is the inevitable consequence of urbanization. City trees—selected in the 1950s for their powers of recovery from browsing by draft animals (which included camels in Beijing)—have had to be replaced with species resistant to leaded gasoline and smoke. But the scale of planting and the combination of environmental improvement and utility are still prominent features. Trees may be planted in single rows or belts, but they are usually close-spaced (2–4 m) and established as standards when they are anything from 3 to 8 m in height. They are tended meticulously, pruned regularly, and irrigated. In many towns, they are treated with sewage. The foliage is usually collected and used to feed municipal livestock or for mulching.

Species vary but are selected with an eye to usefulness as well as ornament. In the southern cities, many more conspicuously flowering species will grow (such as *Jacaranda mimosifolia*, *Hibiscus* spp., and *Lagerstroemia speciosa* in Guangzhou; *Melia azedarach*, *Michelia alba*,

Diospyros kaki, Delonix regia, in Kunming); but oil trees (*Aleurites moluccana; Melaleuca leucodendron*) and other "economic" species (*Casuarina equisetifolia; Bombax malabaricum*) are not neglected. In provinces farther north, the common urban species are high-volume clones of agroforestry species, especially poplars, willows, plane trees (the road from the airport to the city of Shanghai is a notable avenue of *Platanus* × *acerifolia*), and, increasingly, *Metasequoia glyptostroboides.*

The following species are reported to be resistant to air pollution (see Pollard, 1977):

Gingko biloba	*Gleditschia japonica*
Ailanthus altissima	*Maackia amurensis*
Persica (Prunus) davidiana	*Populus canadensis*
Robinia pseudoacacia	*Salix matsudana*
Sophora japonica	*Tamarix chinensis*
Ulmus pumila	*Zizyphus jujuba*

The fast-growing *Paulownia* is used wherever it will grow, as are the ubiquitous mulberry (*Morus alba*) and *Sapium sebiferum, Catalpa obovata, Cinnamomum camphora,* and *Citrus* spp.

In every town and city of China, industrial eyesores (and slums) have now been more or less hidden by greenery. Again, the evident lack of vandalism is noteworthy. Urban forestry is the preserve of the municipal and town governments, most of which maintain their own nurseries, run by horticultural bureaus. Beijing has several nurseries—one on the visitor's route is at Xinanjiao in the southwest suburb. It covers 32 ha, employs 150 staff, and produces 150,000 trees annually (about 6 percent of Beijing's requirement). It has over 300 species or varieties, some of which are "sideline" production for sale. The success of China's urban planting may be due in part to the practice of transplanting with a large ball of soil around the roots; the volume of soil lost from the nurseries is substantial and is made good by replacement from city building sites.

Roadside and railside planting outside the towns is the responsibility of highway bureaus and the Ministry of Railways. The quality of planting is not as uniformly high as in the urban areas, however, and the selection of species is uncompromisingly utilitarian. Poplars, *Paulownia, Cunninghamia, Sassafras, Eucalyptus, Larix,* and *Pinus* are frequent—often in several rows and closely spaced. In the south, *Casuarina* and even *Grevillea robusta* are not uncommon. A feature of roadside plantings, however, is their even height: In some species (poplar, for example, and *Platanus*) this may be the result of close planting, but it is equally noticeable in *Larix* in the northeast. It was a notable feature of railway plantings in 1963.

Forest Preserves

The authors of the FAO (1982) report argue that the ancient view of nature reflected by traditional gardeners is perpetuated in the design and development of "wildland resources." The creation of "forest preserves" is recent: Though the earliest of them (Dinghushan) dates from 1950, it was the 1972 Stockholm Conference on the Human Environment that focused China's attention on "integrated environmental management" (Qu and Li, 1981). A national environmental conference was held in 1973, and the 1978 constitution included explicit reference to the state's responsibility for environmental protection. Legislation was enacted in 1979, and by 1980 nature preserves in twenty-one provinces had been announced. Some 130 reserves covering 8 million hectares have now been designated: All are in primary forest, but many of those intended to protect wildlife and endangered plant species are in seminatural ecotypes.

The first natural reserves to become part of the UNESCO MAB network—at Dinghu (Guangzhou), Changbai (Jilin), and Wolong (Sichuan)—are modified from their pristine state. Dinghu is the site of an ancient Buddhist temple and two-thirds of its 1200 ha has been planted with pine or otherwise modified; Changbai is a former hunting forest and the site of a sanatorium; Wolong—home of the giant panda—has been extensively planted with *Larix* (at the expense of the natural bamboo needed by the pandas). Other preserves include Jianfengling (on Hainan Island) with a much reduced remnant of tropical rain forest on the mountaintop; Huaping (Guangxi), noted for its specimens of *Cathaya argyrophylla* (silver pine or silver fir); "Friendship Park" in the Altay Mountains, "frequented by herdsmen and scientists"; Jingpo Lake; and Qising Lizi (Heilongjiang)—which is as close to primary forest as it is possible to find in China and which provides habitat for the few remaining Siberian tigers.

The native preserves are annotated by Enderton (1985); she has sympathy with the problems of maintaining reserves in a country such as China where, obviously, protection of wildlife cannot be given a high priority, even within the environmental protection policy. She notes (1985:117): "There is a considerable gap between China's articulated conservation policy and the implementation and enforcement of that policy." Like other observers, she is aware of the conflict in China over responsibility and jurisdiction. Thus the State Council has established a committee for Natural Environment Protection, which includes wildlife. It has to work through the MOF in forest areas, however, and foresters are charged with maximizing income from *all* resources under their control. Similarly, trade in skins is regulated by the Ministry of Trade, which sets quotas to be culled—which, of course, creates a market and encourages poaching. Even though China is a signatory to the CITES convention giving all wildcat species protected status, the ministry maintains its quotas. Other abuses—including the offering of endangered species on the menus of gourmet restaurants—have been publicized.

Natural protected areas in China are intended to serve four purposes: conservation, scientific research, production, and tourism. Since it is also part of the policy that "such nature conservation as is undertaken should relate closely to productive needs" (Wang, 1980:10), there is conflict and compromise. It is acknowledged that protected areas are inadequate and that, even in managed reserves, the contradiction between conservation and production has not been resolved. The objectives are commendable but they have yet to be reached. Although there are no universal rules of management, in most protected areas there is an attempt to zone uses. A core area, representing the original vegetation, is to be used for ecological research and to enable monitoring. The core is surrounded by a buffer zone that may be "a semiexploitative place" for management of the vegetation and wildlife but in a way that can be strictly measured. Then there is a so-called experimental zone that is in fact exploited and may be deliberately transformed into artificial ecosystems to demonstrate what production function the particular environment may serve.

A feature of most reserves in China is the provision of facilities for visitors. Whether a few hectares of forest or a "tourist monastery," there are almost invariably physical facilities to provide a focus: a hall for accommodating schoolchildren; a lookout; and a collection of stuffed animals, often along a trail as if in drunken imitation of the famous avenue leading to the Ming tombs. Such facilities attract litter—as noted earlier, a sad but striking feature of China revisited.

The success of nature reserve management, as in all countries, depends to a large extent on the establishment of good relations between the conservators and local people who may have traditionally used the reserve. At Dinghu, for example, provision has been made for alternative fuel sources (access to the pine forest and the establishment of fast-growing species of *Acacia, Cassia, Leucaena,* and others); there have been no serious conflicts. At Wolong reserve, on the other hand, attempts to relocate farmers in nontraditional housing in villages have been resisted. (Despite the charisma surrounding pandas in the West, not all scientists—and few Wolong farmers—see great merit in perpetuating a species which, like the homosexual whooping cranes of North America, feel no urge to reproduce their species.)

PROTECTION FORESTRY

In my 1966 survey, the chapter on water conservation and protection forestry emphasized the Siamese-twin nature of their relationship throughout the world—attempts to separate them are very risky. Prodigious problems face China with respect to river control, land reclamation and erosion, desert stabilization, shelter, and the rehabilitation of overexploited protection forests. Much publicity was being given at that time to the mass of small-scale water conservation projects—especially for irrigation and power generation—in contrast to the vast scale of the earlier land reclamation projects under the guise of arable farms, which dispossessed pastoralists (and, in turn, led to overgrazing of the remaining grassland and pressures on the scarce forest resources to provide animal fodder and fuelwood). It was concluded, however, that land reclamation was a field in which the government was applying lessons learned from past mistakes relatively rapidly. In all aspects, foresters appeared to be closely associated with water conservation and the protection of farmland. I noted in 1966 (p. 129): "The Chinese recognize deficiencies and appear to be making creditable attempts, within the limits of the politico-social structure, to avoid them in the future. Doubtless they will commit new errors and they still face massive problems with finance, organization, technology and biology. Any prognosis, of course, needs to be hedged around with qualifications but, on present evidence, it is the biological problems that may prove to be the most intractable."

Twenty-five years later, that conclusion must be modified. The more recent impression is that, technically, China has made great strides and there is a wealth of information deriving from experience relating to water conservation, desert stabilization, and protection forestry. The bureaucratic problems, however, remain and if anything may have been exacerbated by the post-Mao reforms. Emphasis has shifted again to large-scale engineering works; the problems are political and financial.

Water Conservation and Soil Erosion

A perceptive study by Lampton (1986:387) on water politics and economic change in China starts with the recognition that water politics revolves around "the twin issues of who gets water when there is not enough and who must take it when there is too much." The horizontal cleavages within the administrative system are simply not geared to decisions that cut across bureaucratic boundaries—decisions in which controlled land management, the maintenance of cooperative irrigation projects and facilities, and the like clash head-on with the perceived self-interest of individuals whose entrepreneurial instincts are being encouraged so enthusiastically by the new reforms.

The challenges are clear-cut: Industrial water consumption in China in the year 2000 will be three to four times higher than it is today; and while there have been impressive achievements in irrigation and in the construction of reservoirs, dikes, and embankments, there have also been disastrous floods (in 1981 and 1983) of the Chang—which runs through the geographic focal centers of China's economic development. Lampton notes Chinese estimates that if a flood similar to that which occurred in 1954 were to hit the Chang Basin now, as many as 7 million people could be displaced and economic losses could reach 20 billion yuan. The much smaller 1981 flood caused 2 billion yuan worth of damage and left 1.13 million people homeless, 2600 factories wiped out, 98,000 ha of farmland covered with debris, and over 20,000 ha of reclaimed land simply swept away. The July 1983 floods affected eighty counties and 8 million people in Sichuan province alone—with estimated income losses there of 200 million yuan. (These are Lampton's figures from Chinese sources.)

Interestingly, these two floods emphasized the relations between water conservation, land use, and forestry and led to an acknowledgment in Beijing that it was not enough merely to report the fulfillment of unspecified tree planting targets. Attention was drawn to the political problems of forestry discussed in Chapter 5. A source quoted by Lampton records that "the crux of the land problems is the lack of unified scientific controls. Scientific land control includes legislation, investigation, planning, registration, statistics, and approval and supervision for its use." (Such problems of interagency integration are not unknown in other countries of the world.)

For the first time, forest destruction since 1949 has been acknowledged and publicized. A reduction in overall forest cover was reported and 430,000 km^2 of the loess plateau classified as affected by erosion (Betke and Kuchler, 1987, citing Chinese sources). Despite "preliminary control" of 17.5 percent of this area, the sediment yield of the Huang Basin upstream from the Sanmen Xia is said to have increased by 32 percent since 1949. The dramatic statements regarding reductions in forest area in many counties in the Upper Chang Basin date from soon after the 1981 floods. There was debate over the extent to which deforestation was responsible for those floods: The Ministry of Forestry argued that more investment in forestry was needed whereas the Ministry of Water Conservation and Electric Power favored greater expenditure on storage dams (Ross, 1988). Even though investment in reforestation was stepped up, much of it went into new "economic" crops rather than protective plantings.

There is also a problem in China of uneven distribution of water resources. Lampton records that every hectare of cultivated land in the Chang Basin has more than nine times as much water available to it as in the Huang Basin. There are ambitious plans for interregional water transfers, but experts are agreed that the waste of water is phenomenal. Much of it results from inability to rationalize distribution—which again stems in large part from bureaucratic stovepipe vision. The idea of unified cooperative management of a widespread resource such as a river is beyond the system. Lampton cites the disparity in productivity of reservoirs stemming from arguments about jurisdiction.

Apart from the Chang and Huang catchments, areas of southern China and the northeast are reported to be endangered by soil erosion. In Jiangxi, the affected area has allegedly increased from 4 percent of the province in the 1950s to 20 percent at present; in Guangdong, 18 percent of the provincial area remains affected (Betke and Kuchler, 1987). To the northeast, varying percentages from 15.4 (Jilin) to 50 (Heilongjiang) are reportedly losing soil. It is difficult to evaluate such claims: The fact that they are made in Chinese publications does not make them realistic. (The Chinese environmental movement—as well as that in the West—has its profiteers of doom, and official statistics can be used to foster alarm.) If the figures for the northeast are correct to within a few orders of magnitude, however, soil loss is a serious matter because it is not clearly evident. Indeed, crop yields have never been higher (Walker, 1984), and if good soil is disappearing at the rate implied by Qian Zhengying (the source quoted by Betke and Kuchler)—50 cm of topsoil over ten to twenty years—dire trouble is imminent. Smil (1982:8) refers to "previously unknown serious droughts and dust storms in Heilongjiang, due to over-cutting the forests and grassland destruction." He quotes Chinese sources to the effect that, throughout China, 15 percent of the land surface suffers serious erosion.

In 1982, the Ministry of Water Conservation merged with the Ministry of Electric Power. One reason postulated for the merger is that the Ministry of Electric Power is a large revenue producer whereas water is underpriced and produces little revenue. A more logical merger in China might have been between water conservation and forestry, had there been no financial constraints. But "many cadres are reluctant to relinquish the security of the system they have come to understand for the vagaries of a market place they cannot control and with which they have no experience" (Lampton, 1986:405).

In contrast with the bureaucratic problems (and the few signs of their solution), there is a wealth of information on techniques of conservation land management and protection forestry. Increasingly, protective planting is being decentralized; by 1985, over 3 million households were reported to be engaged in planting trees for soil and water conservation under contract responsibility, and the time is envisaged when 50 million households will be so involved (Hou and Wang, 1986). In the river catchments, the main species are willows and poplars, with fast-growing multipurpose trees along canal and irrigation ditches. Along the Chang and Huang, literally thousands of miles of trees (including *Paulownia* and *Robinia pseudoacacia*) have been planted—with considerable success, as may be seen from aerial observation.[1] The technical problems are minor along the rivers; the challenges are in the hills.

The FAO study (1982) outlines the development and use of hydrological data bases along the Huang River and attempts to define a land-use classification. The restoration of forestry to lands over 15 degrees in slope—and the conversion of marginal and steeply sloping cultivated lands to forest or pastoral use—have priority. Many examples of conservation farming and erosion control techniques have attracted favorable comment from visitors: The planting methods—using contour planting, terracing, fish scales, level ditching and steps (with or without live breastwork planting), ditch/pit combinations, double furrows, and so forth—were illustrated in my earlier survey (Richardson, 1966) and have not changed much. They aim to combine moisture conservation with soil stabilization. Where this can be achieved, a wide range of multiple-purpose tree species can be planted including such fruit trees as mulberry, date (*Zizyphus jujuba*), and walnut. Where erosion is more serious, the construction of checkdams and aprons (of soil/cement, woven wattle, willow stakes, and other materials) must precede replanting—which may then be done in patches or rows across gullies and ravines. Species planted tend to be restricted more to the poplars, *Ulmus pumila, Tamarix juniperina, Hippophae rhamnoides,* and the saxaul, *Haloxylon ammodendron;* they are not intended to yield timber but may be managed for fuelwood.

There is little new in the methods and details of erosion-control cultural practices, which are available in any textbook dealing with the subject. The MOF in Beijing issues simple low-cost manuals for the guidance of households. Since the new extension system—using contractors—came into vogue, there has been a massive increase in publication of such handbooks. The CAS, too, continues to publish extensive surveys of water conservation practices, which are of interest for the meticulous detail in which they describe planting systems and "homemade" manures. The joint publications with the USSR Academy of Sciences on erosion control—which were a feature of the 1950s—are no longer issued.

The steep, eroding loess country of China's watersheds is a target for aerial seeding. As noted earlier, the success of aerial seeding programs is unknown. Frequent references are made in the Chinese press to the activities of both the PLA and China Airlines in aerial seeding—usually in the remote border regions and reportedly to coincide with glacier snowmelt. Few visitors report on these programs and none at first hand. Photographs published in China often show more herb and shrub species (such as *Astragalus adsurgens*) than trees, though southern stands of *Pinus yunnanensis* and *P. massoniana* are featured (MOF, 1984). There is no reason why aerial seeding should not be successful; there are many precedents. But success in the harsh environments in which it is used in China would be a cause for detailed documentation and the extension of the techniques to other parts of the world. A recent report presented at a workshop in Chengdu (Li and Zhang, 1985) notes the aerial seeding of 330,000 ha of "grassland" in Sichuan, resulting in satisfactory stocking of only 133,000 ha, and stresses the need to establish drought-resistant legumes before attempting to sow tree seed. The same report outlines the 1982 Grasslands Management Act, which bans cultivation on slopes in excess of 25 percent and restricts further "land reclamation" for grain farming; at the same time, farming communities in mountain areas were exempted from compulsory sales of grain and from grain taxes for periods of three to five years.

The dramatic results claimed for protective watershed management in reducing runoff in the southwestern mountains strain credibility: Even the work of senior scientists is inadequately replicated and evaluations are sometimes naive.

DESERTIFICATION

Next to soil erosion, the extension of China's great northern deserts is the country's most serious nonurban environmental problem. The CAS Institute of Desert Research (at Lanzhou in Gansu province) estimates China's desert area at 1.3 million square kilometers, of which some 25,000 km^2 have become desert in recent decades; a further 150,000 km^2 of land are at risk of desertification (according to CAS scientists cited by Betke and Kuchler, 1987). Of the causes of desertification, fuelwood collection is allegedly responsible for 32.4 percent, overgrazing for 29.4 percent, and cultivation of marginal lands for 23.3 percent. By contrast, "sand encroachment" is responsible for only 5.5 percent. Compared with erosion, increasing desertification is a minor problem in China, according to these figures. Much depends, however, on definition: Other sources estimate desert increases of 260,000 km^2 since 1949.

China contributed a number of papers to the United Nations Conference on Desertification held in Kenya in 1977. Case studies were published in 1980 and have been drawn upon in the preparation of this account. They obviously predate the Gang of Four and ascribe much of the success in stemming the advance of the deserts to Mao Zedong; nonetheless, they contain material of technical interest and describe tactics and methods of desert control that could be applied elsewhere. Most of the conservation measures combine simple engineering works with tree or shrub planting. They are in sharp contrast to the multi-billion-dollar schemes under consideration for river control in the ministries in Beijing.

In planting for desert control, the major innovations appear to have been the development of new varieties and the extension of species ranges through attention to maintenance, rather than the development of new techniques. A report on combating desertification (Anon., 1977) discusses four tactical approaches:

1. Oasis Protection: The protection of oases through the establishment of forest belts on their periphery—together with forest networks covering the farmland within the oasis, the construction of "contain sand, cultivate grass" areas on the uncultivated fringes, and full utilization of surface water resources
2. Stabilization of Shifting Sands: In sand areas that are partly stabilized, the construction of grass "*kulun*" (literally, enclosures) to develop stock farming through supplementary feeding together with further "contain sand, cultivate grass" areas to increase stocking
3. Road and Railway Protection: Reservoir construction along rivers and in lake basins for water conservation at the time of snowmelt in the mountains, combined with the construction of protective forest belts to reclaim land
4. Oasis Development: Along railways and highways in deserts, engineering methods such as the creation of sand shields to stabilize the surface followed by the planting of sand-tolerant species

The report then lists numerous examples of successful protection works in the desert—many of which are duplications of earlier measures, as of course are the techniques. The paper was written in the days when farmers were being urged to "learn from Dazhai" and perhaps the most valuable lesson was self-reliance.

Methods of combating desertification have not changed much in twenty-five years, but publications describing them in English are more readily available today. The following paragraphs derive from case studies from translated Chinese sources of oasis development, sand stabilization, protection, and reclamation.

OASIS PROTECTION In the northwest oases, the first stage is to establish a forest belt on the fringe and then on low-lying land in between the major dunes. The species comprise mainly poplar (*Populus cuspidata*) and sand date (*Elaeagnus angustifolia*). The idea is to surround the dunes with vegetation to reduce their movement and the effect of wind. In between the shields are planted *Haloxylon ammodendron* and other sand-fixing plants. Where the land is flat, species such as tamarisk (*Tamarix ramosissima*) are planted along the desert face to reduce the sand accumulation within the forest belt. Around oases that fringe the gravel Gobi (or on wind-eroded land, where the danger is from high winds rather than physical movement of sand), the practice is to dig ditches and establish multiple strips of closely planted trees along them, using the ditches for irrigation. The ditches are 1.5 m wide and 4 to 5 m apart with trees closely planted in between. Again, sand dates are much used, together with poplars (*Populus folleana*) and elm (*Ulmus pumila*). On the leeward side, more poplars and mulberry (*Morus alba*) may be grown. Elsewhere *Salix matsudana* and *Populus simonii* are used in mixture.

Within the oases, the network belts are usually 8 to 11 m wide and consist of five or six rows of trees, with secondary belts 6 to 8 m wide with three or four rows of trees. The forest belts occupy 7 percent of the total land area, and as a rule of thumb the distance between the tree strips in the first row of network plots should be fifteen times the height of the plants; for the second row it may be fifteen to seventeen times; and for the third row, about thirty times. The strips should be complemented by shrubs to increase the lower density of the belts. Information has been accumulated on the efficiency of forest belts of various heights; the principles involved are the same as in shelterbelt establishment anywhere in the world.

STABILIZATION OF SHIFTING SANDS The "block at the front, pull from behind" method involves planting the low-lying land at the base of the dunes on the leeward side and then attempting to establish sand willow (*Artemisia ordosica*) to reduce the surface wind velocity in advance of the dune—the idea is to level the top of the dune to the point at which vegetation can be established on it. In general, practices that involve working against the face of the dune run counter to the usual approach in Europe and elsewhere. They are unavoidable in many parts of China because of the barren, arid nature of the sands.

A variant is to establish trees on the low-lying land among the sand dunes and "stabilizing plants" on the dunes themselves—leading to the planting of forest trees once the sands have been fixed. These are likely to be poplar (*Populus pseudosimonii*), indigo (*Amorpha fruticosa*), and *Salix flavida*. Pine species may be added later when the sand has stabilized (*Pinus sylvestris* var. *mongolica*, *P. tabulaeformis*, and others).

Where necessary, the initial vegetation establishment is assisted by irrigation. "*Kulun*" are designed to keep out animals, slow down the movement of sand, and enable the cultivation of grass and other plants within the enclosure. They may be built of wire, mud and wattle, or live fencing (rooted stakes), and their construction may be preceded by the development of groundwater resources by digging wells and waterholes where there is water at depth. Again, there is nothing revolutionary about these structures and, where they have been well constructed, their effects are remarkable—grass yields within the *kulun* have been recorded at nearly twenty times that of unimproved pastureland (Anon., 1976). The 1977 report on combating desertification also gives measures of the effects of enclosure and records pasture production and soil development inside them.

ROAD AND RAILWAY PROTECTION There is nothing new in the methods of protecting trans-desert railways and roads. The slogan is again "contain sand, cultivate grass" with whatever method is locally appropriate. The techniques include sand stabilization through the construction of "shields" of brushwood with stabilizing plants planted inside them. (The shields are in 1 × 1 m squares.) In some instances the vegetation cover has increased from 3 to 30 or 40 percent. Many of these developments have occurred along the Paotow–Lanzhou Railway on the southeastern fringe of the Junggar Desert. Along Gobi areas where there is a danger of sand drift, canals have been built to bring in irrigation water for afforestation. The trees are planted very deep (often 1 m). An alternative is to build shields of stones or even blocks of clay or salt, interspersed with reed checkerboard squares.

OASIS DEVELOPMENT In expanding the arable and pastoral areas of oases it is first necessary to establish a water supply from groundwater or water from snowmelt. Reservoirs may be built on land between sand dunes, on lakes and marshes, or along rivers, to collect and conserve floodwaters and reclaim the desert fringes. Various methods of soil improvement are used—including the introduction of less alkaline sands from outside and the cultivation of leguminous green manure crops. Soil may be introduced into pits ("guest soil") to ameliorate the Gobi and enable the planting of more demanding species; the pits are 1 m deep and about 60 cm long by 40 cm wide. Remarkable improvements in nutrient status have been claimed.

INTEGRATION Another report in the same United Nations volume uses Wushenchao Commune in Inner Mongolia as a case study. (The same commune is cited by FAO 1982 to exemplify problems and their successful solution in Northwest China.) It describes the development of grass *kulun* from enclosures established for growing trees and fodder and then transformed into grazing land. It is an interesting experiment, but there are obvious risks of overcropping and overgrazing. Moreover, it is in an area of temperate-zone dry grassland (rather than desert) that already contains a large number of plant species. Where there is moving sand, the first stage is to establish a plant cover that will reduce the wind velocity and level the dunes. Shrubs are followed by trees and then grass under the protection of the forest.

The sequence is given as "water, grass, forest, fodder—four-element *kulun.*" At Wushenchao there are four types of *kulun* (fodder, grazing, sand control, and the four-element variety). The first two are for the protection and cultivation of natural pastures; the third is for the control of desertification; the fourth is the ultimate objective. The enclosures were originally earthen walls and fences, but nowadays barbed wire is used instead. The development of water resources begins with the creation of wells—either "big pit" wells or "sand drift" wells, with mechanized and artesian wells for groundwater deeper than 100 m. Originally the wells were lined with clay bricks, but plastic sheeting is now used. Spray irrigation has been introduced. "Big pit" wells are in fact reservoirs up to an acre in extent and about 3 m deep. "Sand drift" wells are concrete tubes, 0.5 m in diameter and 20 to 30 m deep, that allow water to seep in but not sand.

The second stage is the establishment of a protective forest belt of trees and shrubs—mainly sand willows at right angles to the wind direction—with fodder shrubs and grasses introduced when the trees are 3 or 4 m high. The belts are leveled and cultivated, and grazing grasses and grain crops are introduced. The grazing grasses include sweet clover (*Melilotus alba*) and the purple alfalfa (*Medicago sativa*). Grain crops are intended as supplementary feed during winter and spring. The final stage is described as "multiple activities" that go beyond the original objectives of the *kulun*—nursery production, fish ponds, the growing of vegetables and fruit.

The report concludes: "You can smash the spiritual shackles of the theory of predestination that deserts are beyond control and reverse the relationship between man and the sand. In the spear/shield situation—either man controls the sand or the sand controls man—man plays the active role. The deserts are limited, while the knowledge and ability of man to combat them are unlimited."

A third report in the United Nations publication is entitled "Tame the Wind, Harness the Sand, and Transform the Gobi," from the Office of Environmental Protection in Xinjiang Uighur autonomous region. It is concerned with Turpan country—a basin in the Tienshan Range. It is an ancient agricultural area and was once a major economic and cultural center. Again, the theme is "contain sand, cultivate grass" on the periphery of the oases with a protective network within them. The first stage is the protection of existing natural vegetation—including species of camel thorn (*Alhagi canescens*), the evocative plump maiden (*Karelinia caspica*), and "deer horn grass" (*Scorzonera divaricata*); where the soil is highly mineralized but with surface groundwater, reeds, tamarisk, and salt ear tree (*Halostachys belangeviana*) are found as well as "mouse melon" (*Capparis spinosa*). Following protection, the next stage is winter irrigation and cultivation of grass. When the vegetation cover reaches 60 percent, measures are taken to allow rotational grazing and fodder collection in strips or plots. The tree networks within the cultivated areas are 150 to 200 m apart and the trees reach a height of 15 to 20 m.

The new features in reclamation are the introduction of selected varieties of crop rather than the techniques of shelter, sand fixation, and water conservation. In particular, sorghum and grapes are relatively new species in the Gobi areas. For grapes, planting ditches are built 5 m apart—2.3 m wide at the top, 40 to 60 cm wide at the bottom, and 40 cm deep. Oblong planting pits 1 m deep, 60 cm long, and 40 cm wide are then filled as before with "guest soil." Each pit is planted with two plants, and other crops (melon and legumes) may be planted alongside the ditches during the first two years of cultivation. A fertilizer furrow is dug 60 cm deep on one side of the pit to take liquid manure.

Salinization and Alkalinization

UNEP (1982) reports that over 350,000 km^2 of China is affected by salinization and alkalinization. Betke and Kuchler (1987) quote estimates of 6.6 million hectares of irrigated land affected in the early 1980s and perhaps over 7 million hectares in 1986. The areas affected are principally in the north—the North China Plain alone accounts for over 3 million hectares.

Salinization is part of the downside of irrigation and water conservation activities, which have increased dramatically since the 1950s. The problem has been tackled around desert oases through the selection of tolerant species and varieties. The former Ministry of Water Conservation claimed that 4.4 million hectares of alkaline and saline farmland were improved between 1949 and 1983 (cited by Betke and Kuchler, 1987), but it is not clear to what point. Certainly saline soil will grow crops—in Saudi Arabia, for example, a salt concentration of over 2000 ppm is not considered unmanageable. ("Fresh" water has 250 ppm and ocean water, 35,000 ppm.) Salt-tolerant varieties of many crop species are known—including barley, cotton, sorghum, wheat, and soybean. Tree crops are generally more tolerant than herbs, and perennial species are less affected than annuals.

In recent years extensive work has been done on salinization and alkalinization—some of it financed by newly rich oil-producing countries—as well as by more traditional research sponsors in, for example, Israel and Australia. Western Australia, in particular, has developed the use of tree species to mitigate the effect of salinization of grazing land. There could be advantage (perhaps mutual) in cooperation between China and these countries in the field of species and varietal selection (and breeding) for salt and drought tolerance.

Other Factors

The loss of agricultural land through erosion and other means of degradation in China is unquestionably serious, though the precise extent of it is unknown. Some official statistics indicate a figure of 99.3 million hectares under cultivation; other data are considerably higher, reflecting (according to Betke and Kuchler) massive transfers of forest land to agriculture. Sometimes forgotten is the continuing loss of agricultural land (usually the most productive) to urban and rural construction. A *China Daily* report dated 19 May 1987 states that the total loss of farmland in 1986 was 600,000 ha; in 1985, it was 1 million hectares. Less than half the loss was compensated by newly developed land. There is a growing political sensitivity to land-use changes, and legal instruments are being created to protect land resources. However successful they may be, the need to increase land productivity will remain—and one of China's major tools to that end is its almost legendary use of shelterbelts.

SHELTER PLANTING

The Three Norths program incorporating the "Great Green Wall" of China was outlined in Chapter 5. The 20-million-hectare reforestation project is easily the most ambitious undertaking anywhere in the world: It involves a vast and complex shelterbelt system extending from Heilongjiang to Xinjiang. The program in fact includes all kinds of forestry within the counties forming part of the system, whether primarily for shelter or for fuel. Other shelter planting (on a more domestic scale) also forms part of the Great Plains program as well as the coastal dune afforestation (the "Great Green Wall of South China") in the southeast.

The World Food Program (WFP) of the UN is involved in several "food for work" projects in China, including the Three Norths project. The program (described as a prototype for 3000 km^2) includes water and soil conservation, farm shelter (windbreaks), sand stabilization, pasture improvement, and provision of range shelter in Ningxhia Hui autonomous region. (Halfway through the project period, the tree planting components were ahead of schedule.) The Three Norths program is in two phases: The first, from 1978 to 1985, was designed to increase the forest cover in 396 counties from 4 to 6 percent of the land surface; the second phase, from 1986 to 1990, aims to tackle a further 460 counties and 7 million hectares. Special emphasis will then be placed on railway shelter. The WFP component involves close planting (4000/ha) of poplar, elm, willow, and other trees and, apparently, multiple planting (5–10 trees/hole). Pasture establishment has been attempted but proved difficult on hill lands perhaps due to drought. The project is to introduce sainfoin (*Onobrychis sativa*) from a WFP project in Turkey and is experimenting with different ground surface treatments to intercept runoff. Evaluation missions have reported favorably on the project—particularly the quality of planting stock.

The structure of the major shelterbelts has been described in FAO (1982). They vary according to function. In areas of high wind, the objective is to moderate its velocity but not to provide an impenetrable barrier; the structure is open, giving a filter effect. Where sand movement is a major problem, belts are narrow but dense at the base.

Regeneration is begun several years in advance of felling the older shelter. On established farms, the shelter plantings are part of the agricultural system, either as lattices or as uniform shelterwood.[2]

There is, of course, considerable variation in the *patterns* of shelter employed. In Manchuria, the major belts are in the form of 50-m-wide strips set 90 degrees to the prevailing wind direction and spaced some 10 km apart; they are cut obliquely by narrower belts—also 10 km apart—thus giving a series of diamond-shaped enclosures, each about 100 km^2 in area. The pattern may be repeated *within* the enclosures at varying intensity, depending on the need for protection. In Nei Mongol, Shanxi, and Gansu, barriers tend to be solid—1.5 and 2.5 km wide—abutting on the steppe or desert. Within the belts the structure again varies, but normally there is a frontal zone of erosion-resistant legumes (herbaceous); then a belt of "survivor" trees (selected for their ability to survive irrespective of growth rate); there may then be a belt of fruit trees (several rows); and finally a zone of perennial herbs and (possibly) medicinal plants. If the shelter planting abuts onto moving sand, the area in front of the belt may need to be fixed (using brushwood, palisades, and the like) or protected by a "sand-sealing" operation—a zone several hundreds of meters wide in which trees are planted in dune hollows and shrubs and grasses on the sides and tops of the dunes; as the sealing takes effect, the shelterbelt may be extended.

The methods of establishing shelterbelts have not changed significantly since the 1960s. The standard prescriptions given in the Lin (1959) handbook of water conservation are still followed. Primary "field protection belts" should be spaced at a distance equal to twenty times the expected height of the belt at twenty to thirty years; secondary belts running with the prevailing winds can be more widely spaced and the trees planted to avoid turbulence. The FAO through the WFP projects has produced a set of standard design criteria for physical treatment of sloping surfaces (see Holder et al., 1985). The small number of tree genera used in shelter planting along the fringes of the sand zones has been noted—*Populus, Salix, Ulmus* (along watercourses), *Elaeagnus, Tamarix, Hippophae, Haloxylon,* and *Artemisia.* On farmland a much wider range of species will grow, and various checklists of shelter species are available. Table 28 presents a list for three regions—North, Northwest, and South China—from a recently published silviculture textbook. The familiar genera from the north are represented, and several introduced species are included for the south.

TABLE 28
PRINCIPAL SHELTER SPECIES IN NORTH, SOUTH, AND NORTHWEST CHINA

North

Populus simonii	*Populus* × *euramericana* cv. *Sacrau 79*
P. pseudosimonii	*Elaeagnus angustifolia*
P. lasiocarpa	*Salix gordejevii*
Armeniaca sibirica	*Hippophae rhamnoides*
Gleditschia horrida	*Alnus japonica*
Crataegus pinnatifida	*Salix matsudana*
Prunus amygdalus	*S. purpurea*
Populus berolinensis	*Zanthoxylum bungeanum*
Fraxinus mandshurica	*Hedysarum scoparium*
Glyptostrobus pensilis	*Zizyphus jujuba*
Larix leptolepis	*Salix babylonica*
Populus tomentosa	*Populus cathayana*
Caragana korschinsky	*Pinus tabulaeformis*
Xanthoceras sorbifolia	*Castanea mollissima*
Ulmus pumila	*Paulownia elongata*
Populus baichehensis	*Platanus* × *acerifolia*
Fraxinus chinensis	*Sophora japonica*
Populus pekinensis	*Robinia pseudoacacia*
P. canadensis	*Diospyros kaki*
Larix olgensis	*Caragana microphylla*
L. principis-ruprechtii	*Lycium barbarum*
L. gmelini	*Lespedeza bicolor*
L. sibirica	*Juglans mandshurica*
Picea koraiensis	*Tamarix chinensis*
Gleditschia chinensis	*Euonymus bungeana*
Platycladus orientalis	*Juniperus chinensis*
Juglans regia	*Populus nigra* var. *italica*
Malus prunifolia	*Ulmus macrocarpa*
Ailanthus altissima	*Phellodendron amurense*
Quercus acutissima	*Populus alba*
Pinus thunbergii	*Amorpha fruticosa*
Quercus mongolica	*Salix mongolica*
Populus simonii × *nigra*	*Populus simonii* × *Salix matsudana*
Corylus heterophylla	*Pinus sylvestris*
Populus nigra var. *thevestina*	*Carya illinoensis*
Morus alba	*Prunus persica*

South

Duabanga grandiflora	*Metasequoia glyptostroboides*
Eucalyptus robusta	*Zelkova schneideriana*
Pinus massoniana	*Casuarina equisetifolia*

TABLE 28 (*Continued*)
PRINCIPAL SHELTER SPECIES IN NORTH, SOUTH, AND NORTHWEST CHINA

South	
Pinus caribaea	*Artocarpus heterophylla*
Anthocephalus chinensis	*Eucalyptus globulus*
Acacia confusa	*Taxodium ascendens*
Eucommia ulmoides	*Cunninghamia lanceolata*
Pterocarya stenoptera	*Camellia oleifera*
Acer negundo	*Platanus* × *acerifolia*
Fraxinus americana	*Cryptomeria fortunei*
Sapium sebiferum	*Eucalyptus citriodora*
Pistacia chinensis	*Morus alba*
Cocos nucifera	*Hibiscus tiliaceus*
Pinus elliottii	*Taxodium distichum*
Calophyllum inophyllum	*Camptotheca acuminata*
Eucalyptus exserta	*Cinnamomum camphora*
Koelreuteria paniculata	*Pandanus odoratissimus*
Pinus taeda	*Toona sinensis*
Northwest	
Populus nigra var. *thevestina* × *simonii*	*Populus sinopyramidalis* cv. *Nanlin*
Fraxinus sogdiana	*P. bolleana*
Ulmus laevis	*Caragana intermedia*
Haloxylon persicum	*Tamarix ramosissima*
H. ammodendron	
Populus euphratica	

SOURCE: Chao Xinshen (1983).

In 1988, I prepared the shelterbelt component for a WFP irrigation project in the Lhasa River Valley of Tibet. The river is fed by snowmelt and the floodplains are stony and dry; parts are highly saline. Based on trees planted in the project area the proposed species were poplars, willows, and nitrogen-fixing *Hippophae* spp., *Amorpha fruticosa,* and *Sophora* spp. A major problem will be to persuade farmers to protect the trees (there is no tree-planting tradition in Tibet) and foresters to accept fuelwood production as having a higher priority than industrial timber. As elsewhere in China, there is extreme reluctance to thin stands of trees; the notion that shelterbelts might be regenerated other than by clear-felling is also foreign to Tibetans.

Considerably less publicized than the northern shelter plantings of

China are the protective belts of trees established along the southern coast and fronting the South China Sea to stabilize sand dunes and form a typhoon barrier (as well as supplying timber and fuelwood). The principal species is *Casuarina equisetifolia,* which is salt- and drought-resistant. The shelterbelt system is mainly in Guangdong province but links with belts along the coast of Guangxi autonomous region and Fujian. Its total length is more than 3000 km and its area over 1 million hectares.

Early plantings in Guangdong in 1954 were a failure, with less than 3 percent survival—reportedly because a typhoon occurred within weeks of the initial planting. Techniques involved bare-rooted seedlings that were pit-planted in specially prepared mud. This method gave way to the use of containers and, eventually, to bamboo baskets containing seedlings 1 to 1.5 m in height. The baskets are expensive to produce, but the costs are said to be covered by the increased agricultural yields. Rice production is reported at 7 tons/ha, and one particular coastal strip (14 km × 110 m) yields 4000 m^3 of timber plus 750 tons of fuel (see Zhu, 1981).

CONCLUSIONS

Environmental forestry and protection forestry in China provide a contrast. Trees established primarily for environmental improvement—in parks, gardens, and streets—are meticulously cared for to the extent that the crowns are sometimes pruned for no other reason than to give them symmetry. The protection forests, by way of contrast, are often less than 50 percent successful and look untidy; as was the case twenty-five years ago, they are carelessly pruned for firewood and (with exceptions) are not nearly so impressive. Similarly—and expectedly—more care is lavished on the natural preserves that provide a focus for visitors than on the steepland protection forests.

Smil (1982), discussing China's agro-ecosystems, notes that there may be a total food deficit of about 5 percent—this shortfall does not imply nationwide shortages but regional disparities leading to periods of malnutrition in provinces affected by flood and drought. Agriculture is vulnerable because of its dependence on human and animal energy. Organic manures are still more important in contributing nutrients to agriculture than chemical fertilizers, and traditional biomass fuels (wood and agricultural residues) are far more important than modern fuels. Despite the large number of small hydropower stations (more than 90,000), the per capita use of electricity is less than 40 kwh annually (not enough for one light bulb continuously); coal is of poor quality and inefficient; and the biogas program has run into snags. The majority of Chinese still depend overwhelmingly on wood fuel and, when none is available, on animal dung. Thus the food/fuel dilemma of India is a problem in China too. Smil adds convincing support to Field's (1982) claim that China is an energy-constrained economy. Moreover, the country's fuel conversion efficiency is low (28 percent compared with 40 to 50 percent in North America and 60 percent in Japan), a situation that exacerbates pollution (Smil, 1981).

As noted earlier, it is the hill country management that must be improved if food production is to increase: Water conservation and its efficient use are crucial. Techniques are known and the problems appreciated. The major concerns are whether—now that the communes have been discredited—there is a focus for mobilization of people and resources and whether the new market philosophy permits long-term need to dominate immediate benefits. The experience of the WFP is encouraging. Environmental protection is seldom—perhaps never—a decisive factor in implementing policy in any country. Rather, environmental concerns provide supplementary support for decisions that are seen to be more cost-effective (in the short term) than alternatives and that conserve long-term interests (or at least keep open the option). Glaeser (1987) distinguishes "traditional" environmental problems such as flood and drought (with which China has coped well) and "modern" problems that are often side effects of technology—such as the insect plagues following the introduction of high-yielding cereal varieties or the water and air pollution that follow urbanization. These modern problems require information systems and research if they are to be as effectively tackled as the traditional problems.

But China is now very much aware of what was once called the zero-sum game. There is environmental legislation, it is being used, and problems are widely discussed. The fact that Smil's critical review of the environmental aspects of China's economic development (Smil, 1982) relied almost entirely on Chinese sources is—as he himself points out—an encouraging sign. The will, as well as the goodwill, may be forthcoming. If the wish of Qu Geping, chief of the Office of Environmental

Protection, to double expenditure on environmental management to 1 percent of GNP by the year 1990 (Qu, pers. comm., 1987) could be fulfilled—and if protection forestry in the hills can emulate the success of environmental forestry in the cities—there is no need for undue alarm.

NOTES

1. It is not easy to take photographs from airplanes in China. Some excellent pictures of riverbank (and coastal) protection plantings, however, are reproduced in MOF (1984).
2. Over the years, China has built up much experience of farm shelter planting and has published many data on microclimatic effects and the improvement of crop yields. Few of these trials, however, are properly replicated and statistically valid. Even those established at experiment stations are often inadequately designed and analyzed.

8

EDUCATION, TRAINING, AND RESEARCH

THE ACADEMIC REVISITING CHINA after twenty-three years rapidly becomes aware of changes in his own attitudes as well as changes in China. In the 1960s most Westerners with sufficient curiosity to visit that country were—if not idealistic—certainly hopeful that they might find an educational philosophy more socially relevant and persuasive than that which characterized the academic institutions of Western Europe and North America—and to which the student generation of both regions was then reacting. And, indeed, China became a focus of admiration for the dedication with which overworked teachers set about their task of combining productive work and study in training, though there was an element of despair in contemplation of the magnitude of that task. By 1986, there could only be sadness at the appalling waste of the decade of the Cultural Revolution.

My guide and mentor in 1963 was Cheng Wanchun, a former chairman of the Forestry Department of Nanking University, vice-director of the Academy of Forest Science, and one of few foresters ever to be elected to the Academia Sinica. He was a stimulating and immensely knowledgeable dendrologist who had achieved an international reputation through his discovery and identification of the "living fossil tree," *Metasequoia glyptostroboides*. He was a linguist, a poet, and a historian. There is no one now in forestry in China with the breadth of experience and expertise of Cheng. There are able and knowledgeable foresters, true, but the hiatus of the Cultural Revolution—and the subsequent lack of job mobility—have severely restricted their experience. It seems, too, that forestry's status in the scientific hierarchy has been reduced, possibly because of bureaucratic self-containment and the constriction of outlook.

THE CULTURAL REVOLUTION AND AFTER

According to Lin (1971) the Cultural Revolution was a confrontation between social change and Confucian mandarinate tradition. The directive of Mao Zedong on 7 May 1966 (after which the "May the 7th" Cadre Schools were named) was contained in a letter to Lin Piao and called for the PLA's involvement in teaching political and cultural subjects as well as agriculture and industrial production. Every school was to be linked with a factory or commune and teachers were to be drawn from the PLA, peasants, technicians, and academics. Most universities were closed until 1970, but when they reopened the academics were sent to the countryside to do manual work—often under the "guidance" of the PLA. In the words of one zealot: "We must use our perspiration from labor to wash away our selfishness, and in practical struggle temper a red art to be boundlessly loyal to Chairman Mao."

China's educational policies in the 1980s have been assessed by Shirk and Smith in the volume *China: The 80's Era* (Ginsburg and Lalor, 1984). An effect of the Cultural Revolution noted by Shirk was the greatly increased enrollment at primary and secondary institutions. This wave of students exceeded their capacity and resulted in the switching of resources from the senior high schools and tertiary education system generally, as well as attempts to abolish distinctions between the lower schools. The elite schools were condemned as "little treasure pagodas" and distinctions between specialized technical vocational schools and junior high schools were criticized. Government-financed schools were to be replaced by "people-financed schools." University selection was detached from school performance and based on political attitude. There were political attacks on academics and requirements that college students (and their teachers) undertake manual labor.

By the time the Cultural Revolution was officially terminated, there was massive unemployment among middle and high school graduates—

Chinese estimates run as high as 26 million. Moreover, those who had jobs were ill-equipped to do them. Garside (1981) wrote of the vocational testing of university graduates working in scientific and technical jobs in Shanghai in 1977. Examinations were based on high school course requirements and there was no time limitation. The results: 68 percent failed in mathematics, 70 percent failed in physics, and 76 percent failed in chemistry. The New China News Agency reported that "some people could not answer one question in their own specialty's most basic knowledge. They just handed in blank papers." Garside also compares the number of students in tertiary education per thousand of population at that time in a number of countries. The United States had 50, Canada 36, Japan 21, the USSR 16, Brazil 11, India 5, and China 0.1.

The problem of scientific manpower is not restricted to the lack of educational facilities. The system of compulsory assignment to jobs does not ensure that the best minds go where they are most needed. Zhou Enlai as early as 1956 complained that "some are assigned one task today and another tomorrow, but are never given a job for which they are qualified." This problem was aggravated by the anti-intellectualism propagated during the Cultural Revolution.

Deng Xiaopeng in 1979 reestablished the "key" schools at all levels. They are institutions (including universities) allocated a higher than average share of resources to enable them to provide the best possible education for outstanding students. Stratification applies to institutions, to graduates, and to levels of mobility. At the tertiary level, the task of mitigating the virtual closure of universities was attempted by expanding the lower-status end of higher education—by creating part-time institutions (night schools, correspondence colleges, television colleges) and enlarging the technical-vocational institutions. Inevitably, there is a status hierarchy ranging from the key faculties and universities down to the vocational colleges run by the economic ministries (such as forestry). The key universities have first choice of student, then the colleges in the second level (teachers' colleges, foreign trade institutions, and others); the vocational schools are assigned the remainder.

The reintroduction of academic standards for admission to universities has brought with it opportunities for nepotism and corruption. There have been numerous references in the Chinese press to the leakage of examination papers and to admissions to higher institutions through the "back door." Malpractice is compounded at the postgraduate level, and there have been complaints of irregular admissions to overseas programs—the pinnacle of the tertiary system. Candidates accepted by overseas institutions have reportedly been substituted by nominations from PLA and party officials and qualifications forged.

There is increased mobility for graduates but very little for faculty members, and it is not unusual for able students to remain as teachers for their entire career in the same institution. Promotion comes from within the unit and there is little opportunity to move elsewhere. The administration of the institution is dual—there is an academic system and a political one. The influence of the party appears to be more readily accepted in academic institutions than in industrial enterprises.

Since the Cultural Revolution, the major expansion of tertiary-level institutions has been with respect to single-discipline institutes. Even in universities, however, there is a tendency for students to remain within their discipline and within the narrow confines of a vocational curriculum. There is very little interaction between the humanities and science, and in all fields there is limited intellectual content or analysis (Smith, 1984). A notable feature of guides and interpreters in 1986 was the limited extent of their general knowledge of China—let alone the rest of the world. One, for example, knew of the Japanese war in Manchuria but was unaware of the Russian occupation that preceded it. An ecologist and senior research scientist from Beijing could not recognize any plant species in Sichuan. These examples remain in one's mind because of their contrast with the evident savoir-faire encountered in 1963.

Smith notes the widening rift between groups *within* academic systems: red versus expert, city versus country, science versus humanities. Scientists in particular have argued that there is no need for a political ideology to achieve success. Shirk observes that in the past China could be proud of its egalitarian socialist society despite economic backwardness—but now that egalitarianism is disappearing from educational institutions, there is a need to find a substitute for the political cynicism that is an aftermath of the Cultural Revolution. Today there is an emphasis on "socialist spiritual civilization," which Shirk describes as a synthesis of Confucian and Maoist principles of moral behavior. Such a system will have to coexist with explicit sexism in the education system—streaming of students begins in the middle schools at age 11 and the established ratio is two-thirds male to one-third female for higher levels. (There are very few female senior faculty members or, indeed, senior party officials.) There is also said to be implicit racism in the system, but this bias is in part economic in that China's minorities tend to live in the isolated and socially deprived areas of the country. Demonstrations in 1986 by African students at the University of Tianjin, however, suggest that racist tendencies may be only in part economic.

EDUCATION AND TRAINING

There are six regional forestry colleges (tertiary level) in China under the control of the Ministry of Forestry plus a national college of forest management. In addition, there are five provincial colleges, eighteen forestry departments within agricultural and "communist labor" universities, and forty-three forestry technical schools. Some statistics are presented in Table 29. There are also eighteen training colleges and schools for cadres involved with forestry, more than fifty worker-training institutes (vocational schools), and a radio/television college that specializes in forestry. Some 57,000 students have graduated from the colleges and 69,000 from the technical schools (APFC, 1987). No students were admitted between 1966 and 1972—indeed some of the university departments were transferred en masse to the countryside, including the premier Beijing College of Forestry, which transferred to Yunnan—and from 1972 to 1976 "worker-peasant-soldier" students were admitted with no academic qualifications or entrance examinations. Graduates from this period pose problems.

The forestry curricula are standardized. Usually the forestry departments of agricultural colleges teach only the silviculture component; the syllabus, course handouts, and textbooks are prescribed. There are few electives, and options normally require an additional year of instruction. Options include wood utilization and landscape architecture, and the basic courses are reminiscent of prewar European curricula. Only in the Beijing Forestry University have computer analysis and modeling been introduced into forestry courses (in 1986).

The technical vocational schools provide one-year courses taught through a combination of study and work. This concept is excellent and entirely relevant to China, but it needs good teachers and the study and work components must be structured and integrated. All too often, they are not.

The absence of experienced senior professors (and researchers) is

TABLE 29

BASIC STATISTICS ON FORESTRY EDUCATION: 1984

A. Tertiary Forestry Institutions

Name of Institution	Postgraduates Enrolled	Graduates	New Students Enrolled	Students Enrolled	Teachers, Staff, and Workers	Full-time Teachers
All institutions of forestry	358	2593	4211	12,517	9689	3799
Beijing Institute of Forestry	104	193	325	1,313	1212	540
Northeast Institute of Forestry	143	641	1064	2,668	2051	730
Nanjing Institute of Forestry	65	411	516	1,749	1384	520
Zhongshan Institute of Forestry		189	349	932	966	401
Northwest Institute of Forestry	17	92	151	515	457	189
Southwest Institute of Forestry	16	123	245	723	579	249
Inner Mongolia Institute of Forestry	1	255	442	1,418	650	331
Jilin Institute of Forestry		179	344	868	783	197
Fujian Institute of Forestry	12	208	333	1,001	761	324
Zhejiang Institute of Forestry		154	242	828	385	162
Hebei School of Forestry		148	200	502	461	156

TABLE 29 (*Continued*)

BASIC STATISTICS ON FORESTRY EDUCATION: 1984

B. Full-time Teachers

Name of Institution	Professors	Associate Professors	Lecturers	Instructors	Assistants
All institutions of forestry	36	330	1641	232	1560
Beijing Institute of Forestry	9	73	241	14	203
Northeast Institute of Forestry	7	113	340	20	250
Nanjing Institute of Forestry	6	62	316	1	135
Zhongshan Institute of Forestry	3	23	155	35	185
Northwest Institute of Forestry	5	14	58	11	101
Southwest Institute of Forestry	4	8	87	54	96
Inner Mongolia Institute of Forestry		13	134	6	178
Jilin Institute of Forestry		2	70	67	58
Fujian Institute of Forestry	2	10	112	5	195
Zhejiang Institute of Forestry		3	50	9	100
Hebei School of Forestry		9	78	10	59

C. Vocational Schools: 1984

Discipline	Number of Schools	Graduates	New Students Enrolled	Total Students Enrolled	Teachers, Staff, and Workers	Full-time Teachers
Agriculture	235	22,149	29,602	71,558	29,088	11,334
Forestry	43	4,893	8,047	16,441	6,153	2,077
Land reclamation	14	987	1,437	3,370	1,282	401
Water conservation	43	4,616	5,741	15,905	7,040	2,717

SOURCE: Department of Planning and Finance, Ministry of Education (1986).

notable—they were the elder statesmen of the forestry profession and some had been trained abroad (especially in France and the United States). Many are now broken physically and intellectually and have lacked the resilience to cope with the sustained derision and indignities heaped upon them during the Cultural Revolution. Most have retired, though a few remain as "advisers." There is no one to take their place.

Comparing the Beijing Forestry University in 1986 with its status in 1963 (when it was only a "college"), one finds there have been administrative changes but few with respect to curriculum. The college became a university in 1984 (having returned from Kunming in 1979) and has three departments: Forestry; Economics and Management; and Forest Industries. In 1963 there were three departments: Forestry (with four sections including economics); Engineering (three sections including wood technology and wood chemistry); and Landscape Gardening. The Beijing Forestry University is one of three in China; the others are at Nanjing and the Northeast University of Harbin (Heilongjiang). In addition, there are three regional forestry colleges under the ministry, and the five provincial colleges noted above. The Forestry University in Beijing is one of sixty-eight universities in the capital, of which four come under the Ministry of Education. Student numbers in forestry are the same as in 1963, but there are rather more graduate students and the undergraduate intake of 500 compares with 300–400 in 1963. (If the landscape gardening option is removed, the numbers are roughly the same.) Although the number of teachers appears to have doubled, there is no evidence of significantly greater research output. Two-thirds of the staff do no research whereas in 1963 research was a requirement of all teaching staff.

Students majoring in forestry spend a month in the field in the first year, one and a half months in the second, two months in the third, and six months in the fourth year. This is about the same as in 1963. English is compulsory at the postgraduate level but not for undergraduates, although a foreign language of some sort is required. (In 1963 there were 240 hours of language per year and mainly English or Russian was selected.) There are 113 courses in nine subject matter groups: forestry, forest protection, water and soil conservation, forest economics, forest finance, landscape design, landscape species, wood processing, and forest machinery. Forestry majors need twenty-seven courses including compulsory mathematics, a foreign language, history, politics, and physical training. The mandatory ratio of two-thirds male to one-third female students is observed. There is a teaching forest in the west mountain suburb of Beijing that includes 750 ha for experimental trials and the like. The university is well equipped, certainly much better so than in 1963, but much of the equipment is new and its purpose is not always clear. Much of it is designed for research, not teaching.

Changes in curriculum give greater emphasis to economics and finance and perhaps less to landscape design and gardening. The Beijing Forestry University is said to be the best in China for general forestry; Nanjing specializes in forest products and the Northeast University in logging and transport. There are exchange relationships with other universities in China and overseas, but clearly there needs to be considerable expansion in this area. There also appears to have been a shift since 1963 from the European examination system to North American test methods.

The Northeast Forestry University at Harbin has an enrollment of 3500 students, of whom 200 are postgraduates; the faculty comprises more than 800. It has ten departments (in addition to social science and what is described as liberal arts): Forestry (forestry, forest protection, forestry inventory and planning, urban forestry); Wildlife (wildlife, biology); Forest Products (wood products manufacturing, forest products chemical industry, chemistry); Forest Civil Engineering (forest road and bridge engineering); Logging and Log Transportation; Forest Machinery (forest machinery design and manufacture, wood products machinery design and manufacture, vehicle operation and engineering, forest industry automation); Forest Economics; Mathematics; Physics; and Foreign Languages (English).

Political education appears to occupy a less prominent part in the curriculum, as does physical training, than was the case in 1963. The Northeast Forestry University is better served than Beijing for experimental forests (totaling 26,000 ha, as well as 6400 ha of supposedly virgin *Pinus koraiensis* forest), and it is also much better equipped than was the case in 1963.

The provincial colleges are smaller in staff numbers and student enrollment (see Table 29) and have few postgraduate students. They are said to have a more practical orientation and to cater to students from the provinces in which they are located, as distinct from the six "regional" institutions. Similarly, the technical training schools are not as well equipped—nor do they attempt to teach the whole field of forestry subjects. They offer one-year or two-year courses with a practical orientation. Cadre schools, on the other hand, set out to provide a working understanding of forestry, not a comprehensive coverage. Apart from the Nanjing Institute, there are no facilities geared to forest industries.

We have no data on success rates in the forestry schools, but apparently no one leaves without some form of recognition. The University of Tianjin (sometimes used as a sounding board for educational reform) in 1987 announced five new education guidelines: The best students, with an interest in scientific research, will be assigned to postgraduate work or recommended for study abroad; those who complete credits ahead of schedule will graduate in advance; those with wide interests may take extra subjects, extend their period at the university by up to twelve months, and receive "double degrees"; those who need to retake examinations frequently before passing will be given a diploma instead of a degree (so will those who cheat!); and those who fail will receive a certificate and, if they can pass within a year of leaving university, will receive a diploma.

It is difficult for a foreigner to make any useful comment on China's education system, or even its forestry components, based on short visits to institutions. As in all countries, it can be criticized. (Jack Westoby has observed that "we have all had some education; and therefore regard ourselves as experts in the field.") But the tasks facing China today are so great as to daunt the imagination. Now, more than ever before, China needs to "walk on two legs"—the slogan that epitomized the policy of using all possible means of development: traditional and modern, theoretical and empirical.

A criticism leveled at the system of forestry education in 1963 related to the use of trained personnel. In field forestry, the trained forester seldom had executive control; he was, in effect, a technical consultant to nontechnical administrators. The only training available to the administrators was through correspondence courses and most of them elected not to take them. Nowadays the influence of technically trained foresters is greater and they have management roles (and responsibilities).

RESEARCH

China has long recognized a central role for science and technology in economic development—during the eighteenth-century "modernization" as well as in the post-Mao period of "adjustment and reform." The series of volumes published by Joseph Needham under the title *Science and Civilization in China* documents that earlier progress. And the fact that science and technology is one of the Four Modernizations (along with agriculture, industry, and the PLA) is evidence of continuing acknowledgment. On 13 March 1985, the Party Central Committee issued a "Decision on the Reform of the Science and Technology Management System" (*Beijing Review* of 8 April 1985) confirming commitment to a new industrial revolution based on science, technology, research, and development. More detailed statements of policy and research management reform have since been published. Sweeping changes are taking place as China, like other countries, searches for cost effectiveness in research.

The Reorganization of Science

As with its industrialization model, China in the 1950s adopted the Soviet system of organizing science and technology—leading to the separation of industrial research from production and from higher education. Moreover, research funding came to be considered a matter of right—irrespective of the researcher's merit or the relevance of his work. Five separate research systems evolved: the academies, the university sector, the economic ministries, the national defense (PLA), and the "local" sector (which covers over 80 percent of the 4300 research institutions in China, according to Suttmeier, 1986). The Academia Sinica was designated the leading organization—under the State Science and Technology Commission (SSTC)—but it was never amenable to much control. Nor was it effective in coordinating the activities of the rash of institutions that broke out under various decentralization policies during the 1960s and 1970s. Over the years, the universities were downgraded (the separation of research from education began long before the Cultural Revolution) and the strong elements of the bureaucracy emerged as the Academia Sinica, the economic ministries, and (presumably) the PLA. Various planning bureaus and coordinating committees have subsequently been set up and reforms of what had become a frustrating and cost-ineffective system are in train.

The Academia Sinica has been renamed the Chinese Academy of Sciences (CAS); the SSTC has been given a more limited policy role within its capacity; and a new think tank, the National Research Center for Science and Technology for Development (NRCSTD), has been established. The Chinese Association of Science and Technology (CAST) serves the interests of scientific societies in popularizing science and its application in production. There is also an Academy of Social Sciences—a much smaller organization than the CAS, with independent status dating from 1976—that directs the work of seven institutes.

The CAS consists of some 400 academicians who elect a presidium on a four-year membership. It has five academic departments (physics, chemistry, biology, earth sciences, and technology) to administer the 122 scientific research institutes under its control. It also has a policy and management science group. In 1982 the Science and Technology Leading Group, a "supracommission" formally headed by the premier, was set up to provide a forum for national policy coordination and intervention in interministerial demarcation disputes.

These developments have led to significant semipublic discussion of science policy and organization, some of it translated into English and published in issues of the JPRS (1982a; 1982b; 1983a; 1983b; and JPRS-CST 1984). In November 1986, the joint science "white paper"—a 323-page volume outlining the nation's science and technology policies—was released.

The need for reform came from five main deficiencies in the research and development system. First, the CAS was intended to be an applied research organization (90 percent of its budget was for applied research)—yet it had no formal links with the users of research (the production departments). Second, because of the virtual separation of research and teaching, the universities were in no position to undertake the basic research needed to support the academy (especially since the Cultural Revolution). Third, the ministerial research organizations had become virtually autonomous, paying little regard to the needs of production and preaching egalitarianism with respect to research budgets and rewards—thus funds were allocated annually, irrespective of performance, to institutes and to individual members. Fourth, the "hierarchies and horizontal cleavages" that characterize the administration encouraged excessive duplication and what a vice-premier has described as the "repetition of antiques." Finally, the lack of job mobility in China has severely restricted work incentives and professional development.

The major reforms designed to mitigate these shortcomings relate to funding, commercialization of research results, recruitment, and increased mobility for scientists. Seventy percent of CAS research is now to be funded through contracts (in some cases linked with "responsibility contracts" with research workers). And a special fund—along the lines of the U.S. National Science Foundation—has been set up from which researchers (whether in the CAS system, universities, or even the ministries) can be awarded grants according to the merits of their proposals as judged by their peers. The white paper includes a section on "environment and resources" and emphasizes the transfer of technology from the military to civilian sectors, as well as high-technology development (space and nuclear programs).

Ultimately, 100 percent of research will be contracted, though it is recognized that some areas (astronomy and certain agricultural research, for example) cannot immediately make profits and will require allocations. Such research will need scientific justification, and it is acknowledged that perhaps as much as one-third of CAS research may be curtailed. A period of five years has been set to accomplish the reform, but the CAS vice-president thinks this period may be too short. The CAS controls 122 research institutes, 40,000 research workers, and a 1985 budget of 700 million yuan (US$220 million). The principle of "let the user pay" is to apply to all research, not merely that under the CAS.

Guidelines for in-house contracts with researchers suggest a distribution of "net profits" from research of 50 percent to further research (to be allocated by the institute director), 30 percent to "welfare" (of all staff, not only those involved in the contract), and 20 percent as individual incentive. (Income above one and a half times the average salary—about 150 yuan a month—is taxed. Despite some fears, it seems unlikely that the reform will lead to undue concentration of personal wealth.) The introduction of contract research has led to acceptance of the patent concept, though not yet universally. If a market system is to work, however, knowledge becomes a salable commodity and the discoverer has a claim to the ownership of his discoveries. The patent law, which took effect on 1 April 1985, acknowledges rights to intellectual property but was not introduced without ideological misgivings (see Suttmeier, 1986); it will at least give China easier and cheaper access to technology and will protect China's exports on some world markets. Finally, the rehabilitation of intellectuals and commitment to (limited) mobility of scientific manpower allows job advertising for research contracts.

China's reorganization of science and technology has been prompted by its growing awareness of waste, ineffectiveness, and low productivity. The movement owes much to overseas practice and foreign influences on Chinese scientists. In turn, the outside world is watching China's reforms with great interest—especially its application of the user-pays principle and its attempts to achieve horizontal as well as hierarchical integration.

FORESTRY-RELATED RESEARCH

In the pecking order of scientific research, forestry does not rate very highly. And "horizontal cleavage" is even more evident than in field operations. The following paragraphs outline the institutions involved in research and some of their programs; overseas evaluations are noted and comment offered on the problems in research administration that China shares with other countries.

INSTITUTIONS Forestry research is carried out in seventeen laboratories of earth science and biology institutes of the CAS, in eight research institutions of the Academy of Forestry (under the Ministry of Forestry), and in over 200 provincial institutions (varying greatly in scope and status). The Academy of Forestry also runs three forestry experiment stations. Provincial institutes concentrate on particular problems (sand fixation and shelter planting at Zhangwu, Liaoning, for example) or groups of species (such as the Poplar Research Institute, Gaixian); and forest products institutes serve particular economic ministries (pulp and paper under the Ministry of Light Industry; wood preservation under the Ministry of Railways; materials testing in MOURCEP). Some institutes are classified as provincial but supported by central funds. (The Yunnan Institute of Environmental Science gets significant funds from MOURCEP, for example, and the He Shan Experimental Ecofarm, Guangzhou, integrating forestry, agriculture, animal husbandry, and fisheries in a severely degraded area, is supported by the CAS.) Other institutes have grown from China's involvement in international projects. (The Ding Hu Forest Ecosystem Station, which was based on the Man and the Biosphere program of UNESCO, now has an interdisciplinary program involving the South China Institute of Botany of CAS, the Guangzhou Institutes of Geography, Soil Science, Microbiology, and Entomology, and the Zhongshan University Departments of Biology and Meteorology.)

Apart from centers classified as primarily involved with forestry research, literally hundreds more are engaged in studies of interest to foresters. They include some thirty botanical gardens (sixteen of which are annotated in Dale, 1984); earth science and natural resource commissions of the CAS; and horticultural and agricultural research institutes of the Chinese Academy of Agronomy—the Ministry of Agriculture's counterpart to the Academy of Forestry. Given the magnitude of the task, it is perhaps not surprising that research coordination leaves much to be desired.

Forestry and related research under the CAS touch upon three aspects. The first component, ecological systems, is studied at the Mountain Research Station of Forest Ecosystems, Changbaishan (Jilin province), and the Ding Hu Forest Ecosystem Station, Guangdong. The second component, an umbrella category, the "nature of change," covers a range of work—from the use of trees as windbreaks at the Desert Research Institute at Lanzhou (Gansu province) to catchment research in South China and forest soils at the Soils Research Institute in Shenyang (Liaoning). The third component, species introduction, is undertaken in botanical gardens in Beijing, Wuhan (Hubei), Kunming (Yunnan), Guangzhou (Guangdong), and Xishuangbanna (Yunnan).

The Academy of Forestry comprises:

Forestry Research Institute, Beijing
Forest Economics Research Institute, Beijing
Forest Science and Technology Research Institute, Beijing
Wood Industry Research Institute, Beijing
Tropical Forestry Research Institute, Guangzhou
Subtropical Forestry Research Institute, Fuyang
Chemical Processing Research Institute, Nanjing
Shellac Research Institute, Jingdong (Yunnan)
Mount Daoging Experiment Station, Guangxi Zhuang
Deng Kou Experimental Station, Nei Monggol
Mount Daqing Experimental Bureau, Jiangxi

The World Bank lists fifty-three "main forestry research institutes" in China (see Appendix E), while FAO (1986b) has published an annotated list of seventeen institutes in its *World Compendium of Forestry and Forest Products Research Institutions.* The FAO list is a particularly useful outline including notes on fields of activity, special facilities, and publications. It covers seven of the Academy of Forestry institutes, as well as provincial institutes in Chengdu (Sichuan), Harbin (Heilongjiang), Nanning (Guangxi Zhuang), Shanghai, Wuchang (Hubei), and Yangling (Shaanxi). The CAS Institute of Desert Research at Lanzhou (Gansu) is included, together with two university departments (Nanjing in Jiangsu province and the Central South College of Forestry, Zhuzhou, Hunan). Although many important centers have been omitted (notably most CAS institutes), the FAO list provides a promising beginning. (The entire compendium covers 600 research organizations in FAO member countries.)

RESEARCH PROGRAMS As in most countries, judgments about the relative importance of different research institutions vary with the interests of the assessor. The Chinese forestry delegation to the Ninth World Forestry Congress presented a paper on "Forestry Research and Technology," but it makes no reference to work done by anyone outside the Ministry of Forestry. Similarly, reports from the CAS concentrate on work within CAS institutes. Visitors are usually no better placed to make judgments, since they are sponsored in some way by one component of a governmental system. Their itineraries are largely determined by their hosts and the hierarchical system militates against interaction between organizations.

More serious, perhaps, is the lack of contact between scientists of different affiliations in China—even when located in adjacent institutes. The Tropical Botanical Institute at Xishuangbanna of the CAS, for example, is concerned with identifying potential new crops. It is in a remote part of Yunnan four and a half hours by road from an airport; in the next village there is a provincial Tropical Crops Research Institute; despite their scientific isolation, the institutes have no contact, formal or informal—there is not even any sharing of library facilities, yet the excellent periodicals section of the Botanical Institute is obviously underutilized. Even within the CAS there would appear to be obvious scope for rationalizing library services, yet each institute wants self-containment. Although there is considerable duplication of research and a waste of scarce scientific manpower, the underutilization of capital investment in research facilities is just as important.

The unnatural divisions between institutions in China may be encouraged by the involvement of international agencies and other funding bodies. The United Nations family of FAO, UNIDO, UNESCO, and so forth have their own hierarchical system and there is perhaps as much cleavage between them as there is in China. Similarly, bilateral aid works through particular ministries and seldom funds multidisciplinary and multiagency projects.

Apart from the Ninth World Forestry Congress review of MOF research priorities, there have been reports from international agency visitors (World Bank, FAO, and others) and specialist review teams. (The Forestry Study Exchange Program between the United States and the PRC embodies four teams: integrated pest management; forest genetics and tree improvement; processing of forest products; and forest tree seed testing.) Of particular interest in the Asia-Pacific region is a detailed assessment of four forest research institutes (Guangdong, Guanxi, Sichuan, and Shaanxi) by the Australian Center for International Agricultural Research (ACIAR)—under cofinancing by the World Bank and the Australian Development Assistance Bureau.

The World Forestry Congress paper enumerates ten research components of *practical* significance:

Genetic improvement of poplars
Establishment of seed orchards (Chinese fir, slash pine, larch, and others)
Exotic introductions (poplars and southern pines)
Fast-growing plantations (mainly native species)
Farm shelterbelts
Sand fixation
Diseases and pests (especially human control and chemical prevention)
Silvicultural machinery (for ground preparation and pest control)
Logging
Forest products (aimed at reducing wood consumption)

The priorities to the end of the century are fast-growing plantations, sustained forest utilization, and economic timber consumption. The duplication in research and deficiencies in coordination are acknowledged, and five proposals are advanced to meet future needs:

1. Establishment of four research centers (for *Paulownia*, *Eucalyptus*, bamboo, and poplar) to "concentrate research work and avoid repetitive work." The centers are being established with bilateral assistance (Canada, Australia, West Germany), but there is no evidence

yet of any concentration of existing research—rather, the work appears to be supplementary.

2. Strengthening of research and management. The objectives here are to involve the users of research in the selection of projects and to introduce the user-pays principle but with a quid pro quo of scientific responsibility. This goal is highly commendable and challenging.
3. Improvement of forest extension. This objective will involve setting up a vertical extension system from central government to county level (the details are not clear) and the encouragement of "specialized households" to undertake extension. It is a particularly important experiment that is to be patterned on the highly successful "mass scientific network" in agriculture. The "Sangie Experience" used to popularize scientific farming is a contract system providing advice, plant materials, and so forth to peasant farmers; excluding natural disasters, reductions in yield resulting from inappropriate advice are fully compensated, while increases in income are shared between the extension technicians and the contracting farmer. The technicians belong to specialized households who are production contractors as well as field experimenters. It remains to be seen how effectively this system can be applied in forestry. It will be particularly difficult on poor sites, requiring rotations beyond ten years, but it is the most innovative and exciting development in forestry extension worldwide in decades. (It has been suggested that the principle of financial compensation to be paid by extension personnel for inappropriate advice might be extended to international consultants.)
4. Increasing research training in China and overseas.
5. Improvement in information servicing. This objective involves centralized data collection and processing and is desirable if only to reduce the duplication that now exists. More important, however, is the need to communicate with other forestry-related institutions.

COMMENT The impressions of visitors have much in common and confirm what China's policymakers already know. In the first place, forestry research institutions—with notable exceptions—are not very dynamic. There is a lack of well-trained, committed researchers and too much unproductive competition between them. Second, the system does not facilitate coordination and there is excessive and expensive duplication of research. Third, there is a lack of clear policies relating research to the needs of users—both in forestry and in forest products. Research programs are drawn up—and priorities established—by the producers rather than the users of research.

Fourth, many projects are ad hoc; planning is unsystematic; design and analysis lack statistical validity. If forestry research is to become cost-effective, there is a need to establish problem-related, long-term programs using interdisciplinary research teams. Fifth, program review procedures are inadequate and generally limited to in-house evaluation. Although procedures for setting up projects are improving, the review system leaves much to be desired and there is inadequate understanding of why work is being done. (A much-quoted example is that of tissue culture—of species easy to propagate conventionally and in situations where 95 percent of "production" derives from unselected seed of unknown provenance.) And, sixth, the lack of job mobility in China militates against efficiency in the use of scientific manpower.

The ACIAR (1986) evaluation makes a number of recommendations that provide a basis for technical assistance to the four institutes. It stresses the need for more clearly defined (and reasoned) national research priorities and proposes that each institute develop a core research program (replicated rather than duplicated), together with a specialization in areas where there is comparative advantage. Thus it is suggested that Guangdong specialize in fast-growing plantations and the properties of timber grown in them; Guangxi should focus on improving the yield and quality of forest chemicals; Sichuan might develop expertise in ecology and conservation; and Shaanxi should give particular attention to semiarid forestry and shelterbelt plantings.

The advice is sound. But the most urgent problem in forestry research in China is one that short-term visitors are not in a strong position to address—wastage of scarce resources resulting from needless repetition and duplication. It is perhaps of interest that in 1963, as director of a forestry research institute, I diagnosed three major deficiencies in China's forestry research: a shortage of trained manpower; a dichotomy between research programs and practical requirements; and duplication of research. I assumed that problems of coordination arose because of the "many widely scattered organizations having at least a fringe interest in forestry matters." Also, China's self-reliance policies of the 1960s encouraged institutes to seek self-sufficiency—even to the extent of manufacturing their own chemicals and equipment. During the Cultural Revolution, indeed, scientists were urged to ignore published research and "learn from the rich-experienced peasants." It is scarcely surprising that they became not merely self-sufficient but self-contained. The legacy remains.

But duplication of research is nowadays not always unintentional: It may be deliberate and competitive (though, in China, the motives are not financial). It can arise in research communities where imagination is lacking, scientists are secretive, and communication with non-researchers is inadequate. It can arise from the reluctance of researchers to move beyond the confines of their training environment. And it can be the result of bad advice—and similar reluctance—on the part of aid donors. Paradoxically, duplication of research can also result from over-aggressive salesmanship on the part of donors. Certainly there is inadequate dialogue between representatives of the agencies themselves, and demarcation disputes between international agencies are not unknown.

Visitors who travel extensively in China looking at research have a responsibility to do more than identify gaps in program coverage: They should also indicate areas that are receiving *too much* attention. Tissue culture is an obvious candidate throughout China, for example, and Krugman et al. (1983:64) point out that "vast orchards are being established with trees selected at low intensity and in quantities having no bearing to annual regeneration programs." In forest products, timber testing procedures have become ossified, pulping trials (nowadays irrelevant since most paper properties can be predicted from simple studies of wood anatomy) are being repeated ad nauseam, and kiln drying schedules have been developed in all provinces that have facilities—despite the fact that kiln drying remains a virtually unused technique. My observations on forest products research in 1963 could be repeated here verbatim; except for the search for wood substitutes, the program appears unchanged.

Similarly, research programs on multiple-purpose species appear little changed since 1963. Species trials are somewhat improved in design, but there is little extension of the species range in most of the country. The U.S. team involved in cooperative studies in forest genetics and tree improvement, however, has recorded its satisfaction with the maintenance and care of joint trials in China.

China can ill afford unproductive investment in equipment or manpower. Its most urgent need is to review nationwide forestry and related research with a view to substantially *reducing* such investment—perhaps by closing some institutes altogether and concentrating *relevant* programs in areas where there are specific needs. The division of China into three development areas acknowledges the country's separate economies: It might provide an appropriate basis as well for the reorientation of forestry and forest products research. The concept of regional institutes—each with basic research capabilities plus local specializations—is being adopted in India and has much to offer China.

CONCLUSIONS

It is too soon to pass judgment on China's new education system, though differences in the general knowledge of university graduates between 1963 and 1986 have been noted. China is well on the way to reestablishing a stratified system and, perhaps, the revival of elitism. The important task is to achieve and maintain equality of opportunity—a difficult challenge in a relatively immobile population. Another challenge will be the accommodation within the system of graduates returning from overseas. As sometimes happens in other countries, there may be a temptation to channel them into research—yet China needs them in management.

The picture of forestry and related research in China painted here is gloomy. One reason is that forestry is not a very glamorous profession; nor is it in the front ranks of science and technology. (This is true in most countries, but perhaps particularly so in China.) There are, however, grounds for optimism.

First, deficiencies in the organization and administration of forestry research are recognized and acknowledged. Second, reforms in other areas of science and technology appear to be working. An article by Zhou (1986)—a deputy director of the Science and Technology Policy and Management Research Institute (CAS)—reports that contract research has increased from 20 to 60 percent of the CAS budget; that interdisciplinary problem-oriented research has increased (he cites the involvement of more than 280 scientists and technicians in six disciplines in a Huang River Basin project); that technology transfer is improving (twelve research institutes in Shanghai have technical links with 460 enterprises and have established sixty-four joint research–production bodies); and that, in 1985, thirty-five research institutes established relations with universities. One hopes that forestry may follow the same pattern.

Third, the close relation between economic reform and the reorganization of science and technology is appreciated. Economic improvement began in rural China and moved to the urban fringes; the first successes were in small- to medium-sized enterprises. The transfer of research findings to production is reportedly following the same route, and the "best-selling" technologies at technical trade fairs are those promising quick returns. This trend augurs well for the forest industries in the southern provinces of China (where there is a free log market), but it poses problems for the state sector and for long-term forest management projects.

Fourth, excellent research is in fact being done in China: The CAS basic research institutes (Plant Physiology and Biochemistry in Shanghai; Botany in Beijing) stand comparison with similar institutions anywhere in the world; there are outstanding Chinese taxonomists and soil scientists; there is doubtless good work being done in other fields. There is no reason why forestry should prove to be exceptional—given that it is an applied science requiring multidisciplinary involvement and a taskforce (rather than individual) approach to practical problems.

Finally, China's innovative approach to extension through the "mass scientific network" system exemplifies traditional aptitude and adaptability that are part of the Chinese genius. If the success in agriculture can be repeated in forestry, the Asia-Pacific region—indeed, the entire developing world—will be greatly beholden. The effort should be closely monitored, therefore, and a feedback system to research established.

9

LEARNING FROM CHINA

THE ANCIENT HISTORY of China presents a pattern of frequent political change, military conquest, foreign invasion, and insurrection—turmoil that undoubtedly contributed significantly to forest destruction and accounts for early reports of industrial wood shortages (Adshead, 1979). But it can be argued that a more pervasive determinant of the present-day forest economy has been agricultural production. Land clearance began in the third millennium B.C., and the first floods were recorded shortly afterward. By 1400 B.C. it was necessary to move the imperial capital because of disastrous flooding along the Huang River. The so-called Golden Period (roughly 1100–250 B.C.) saw the first attempts at systematic land-use management—through Commissions of Mountain Forests and of Swamps and Marshes, Police of the Foothill Forests, and Police of Rivers and Streams. Decrees protecting trees date from this period, and there were controls on the marketing of immature trees. People who failed to replant trees were not allowed to have coffins. And the first articulation of a sustained yield policy came from Mencius himself (372–289 B.C.). Centralized government declined, however, with the end of the Qin dynasty. There followed a succession of rulers and the familiar cycle of peace and prosperity alternating with agricultural depression and political upset. Agricultural development in peacetime involved forest clearance, which led to flood and drought, then famine and insurrection. The interdependence of agriculture and forestry and the importance of timber resources in the development of communications and industry (which provide the base for agricultural progress beyond subsistence) are unchanging features of China's rural economy.

My purpose in this final chapter is to discuss recent changes that have influenced forestry policies and practices in China. As already noted, such perceptions reflect change in the observer as well as in what he sees. The viewer with hindsight is less tolerant of incompetence, but more aware of the constraining effects of bureaucracies on life-styles and personal relationships. He is also influenced by changing global perceptions of the functions of forests and the role of foresters.

The chapter highlights several aspects of change in China's forest policies and practices and cites lessons that might be applied more generally to Third World development. There is intense interest worldwide in China's political and social experiments as well as a professional interest in resource uses, concepts, and technology. It stems in part from the mystery that has for so long surrounded China and in part from the observation of Chinese communities in other countries. There is admiration—perhaps subconscious envy—of the cultural unity of the Chinese. And in the West there is fear, too, of possible domination by what is assumed to be a huge, disciplined, and hardworking labor force dedicated to catching up, and presumably overtaking, the industrialized world within a few decades. The reality of China is rather different.

IDEOLOGY AND INFRASTRUCTURE

Visitors to China in the 1960s were, inevitably, overwhelmed by the size of the country and the astonishing fact of political unity in so vast and diverse a nation. The scale of operations was staggering—in labor-intensive construction of earth dams, mass reforestation campaigns (planting gangs of more than 50,000 people were regularly mobilized by communes), recycling of wastes, or insect pest control (by hordes of "red pioneer" schoolchildren). The notion that manpower could substitute for inputs of capital (or technology) anywhere in China was promulgated as an article of faith; Mao Zedong had declared that man was "the greatest of earthly treasures" and "the masses" were China's most glorious resource. To "learn from Dazhai"—the commune in Shanxi province supposedly transformed from poverty to relative wealth through labor—was the battle cry.

Upon my return in the 1980s, my impressions of vastness were unchanged but appeared irrelevant to economic development. China is now more clearly seen to be a series of disparate economies in a state of (relatively peaceful) coexistence. They may be geographic, demographic, topographic, or commercial. Thus differences between the coastal economies and those of the hinterlands are acknowledged in the establishment of the Special Economic Zones and in the economic regions established under the most recent five-year plan. Differences between the northeast and south, southwest and north, and the unique isolation of west and northwest are reflected in attempts to regulate the movement of raw materials and to base development on "natural advantage." Recognition of these differences is reflected in forest policies with respect to industrial development (and the movement of logs) as well as in priorities for the establishment of fast-growing plantations and the accommodation of sideline production.

Differences between urban and rural economies are apparent in most of China (except the poorest counties)—but they are not the same as in other Third World countries. Relative affluence in the rural areas (and extending into the rural/urban fringes) is reflected especially in housing standards and stems from the continued acceptance of private property in rural areas (and the opportunities it provides for personal or family investment) in contrast to the deterioration of living conditions in the urban centers, where property is state or collectively owned.

Earlier reference has been made to Klatt's (1983) perception of differences between agricultural and industrial production with respect to economies of scale and the practicability of command planning. These features have played an important role in China's recent development—in that decision-making has come more easily to peasant farmers than to the urban proletariat nurtured on the comfort of the "great iron ricebowl." Rural entrepreneurs have responded to the new reforms, and the vigor of sideline production is an astonishing contrast to earlier times when even the retention of small private plots was castigated (though never to the point of confiscation). Forest products now provided from the state forest resource as sidelines range from ginseng in nurseries to edible wood-rotting fungi in timber yards and endangered animal species from forests. The products marketed—and the eagerness with which they are offered—testify to the survival of peasant mercantilism.

Nor is this difference limited to rural production. Everywhere that people congregate there is a wide range of trading activities—in contrast to the drab, almost apologetic, offerings of noodles and water-ices that featured in Mao's China. There are ornamental shrubs raised in window boxes, artistic renderings on sticks of toffee strands, homemade clothing, and all kinds of artifacts for sale; services provided for a fee include on-the-spot blood-pressure readings and horoscopes—and there are traveling discotheques! Even the formal markets of the rural communities (and the urban fringes) are more diverse and extrovert than the dark covered halls of the cities.

There have always been differences between lowland and high-country economies. They exist primarily because of constraints imposed by topography and climate on the kind of agriculture that can be practiced, but they are accentuated by problems of transport and energy. The best examples are in autonomous regions and minority areas of the border provinces visited in the 1980s but not in 1963.

In the Tibetan autonomous region, the settled and poor agriculture of the valleys contrasts with the hard—but apparently healthy—life-style of nomadic herdsmen. And both are very different from the floodplain agriculture of Inner China. Equally striking is the relative insensitivity of the high-country dwellers to policy change. There is an evident time lag between policy enactment in Inner China and the border provinces (particularly, minority areas). Tibet escaped the excesses of the Great Leap and the Cultural Revolution. The "contract responsibility" system has had little formal impact—perhaps because households have always made their own economic decisions anyway; or because "real deprivation is too close for the population to view the present move from state controls to market forces as a stable development" (Clarke, 1987:45).

The tradition of independence dies hard and poses problems for forestry. Pressures on the forest resources have increased beyond the levels of sustainable timber and fuelwood production. And the new plantations of the Lhasa Valley have to be fenced—at a high cost of imported materials—and irrigated. These clonal "high-tech" plantations provide a sharp contrast with the natural scrub forests of the hills.

The multiple economies of China make scale more important than size. The success of small-scale agroforestry in the North China Plain is longstanding, and smaller scale in the wood-processing industry is rapidly coming to be appreciated. It is obvious in the cost effectiveness of stand-alone village sawmills, compared with the waste, poor quality, and high costs of processing in the integrated wood complexes. And it fits neatly into the "reform" concepts of contract responsibility and specialized households. Perhaps the first lesson to be learned from China, then, is this: There are diseconomies of scale as well as economies. Where transport and energy are major constraints—and there is an underemployed labor force—the diseconomies dominate the economies.

Related to the question of appropriate operational scale and the efficient allocation of resources are the decentralization policies that have featured (on and off) in China over the past forty years. Decentralization and "local self-sufficiency" were very much part of Mao Zedong's philosophy—though, as Myrdal (1984) has observed, implementation of the policies involved cycles of rigid collective control focusing on social structures, mass campaigns of central accumulation and disposition of resources, followed by periods of decentralization and increased production and distribution. The difference between the present and earlier "open" periods—and perhaps the key to success—is the abrogation of centralized decision-making and reliance on market signals to guide development. Given the multiple economies of China, however, universal success in market driving is by no means certain; nor is the continuity of the mercantile philosophy assured. There is considerable discussion among Chinese about whether the reform policies will last, and it is a particular concern among foresters. Uncertainty of continued rights to forest use and land tenure has already led to illegal forest destruction; to the extension of the tenure period in attempts to mitigate short-term exploitation; and to growing discussion of a possible free market in rural land-lease rights.

In December 1987, the first ever public auction of land-use rights in socialist China took place (in the Shenzhen Special Economic Zone). Land nationalization is an emotional aspect of socialist ideology; even long-term leasing has been criticized as infringing a basic Marxist principle and has led to "bitter struggle" (Morgan, 1988). The notion of a free market in lease rights raises the specter of wealthy landlords exploiting the landless peasants—the very situation that fomented revolution in the first place. Moreover, policies can change very quickly. The 1982 constitution specifically prohibits the appropriation of land or property "by any organization by whatever means"; the constitution was amended in April 1988 to enable the transfer of land-use rights—four months after the Shenzhen transfer (Morgan, 1988). The constant amendment of the forest law from 1979 to 1986 reflects policy changes with respect to land tenure and the adoption of prescriptions (and penalties) to replace the customary exhortation.

Shifting from a system of command to guidance planning puts a burden on the market—and its ability to receive and generate signals—that it may not yet be able to support. Markets are imperfect everywhere: In the industrialized world the absence of a "price" that effectively controls pollution is a classic example of market failure. In China, such failure is exacerbated by the lack of commercial infrastructure. Foreigners who have entered into joint ventures complain of deficiencies in law and in concepts of accountability. Langston (1985) notes a traditional antipathy toward written laws in Chinese business, and one businessman has observed that, in the Chinese view, access to law is a privilege—not a right—and subject to negotiation. Since "important questions" have to be resolved through consultation, not by law, each party effectively holds a veto. And the application of foreign law—despite evident deficiencies in indigenous law—may be regarded as an infringement of sovereignty.

Nor are there realistic sanctions. Until recently, there was no law on bankruptcy; liability is ill-defined; insurance is rudimentary. China's banking system lags behind the needs of an economy increasing in complexity. Banks have little experience of credit control and have not been held responsible for profit and loss; they have hitherto served merely as distribution channels for funds "ordered" by the borrowers and released through the rigid, vertical hierarchy according to political clout rather than creditworthiness. The banks have been looked on in the past as another "iron ricebowl." But in the new climate, they have to undertake studies and make judgments; they are part of the system of checks and balances that every market economy must have; they have to learn new roles.

Reference has been made to Prybla's complaint that when bureaucrats are exposed to a market environment, they behave more like black marketeers than capitalists. One of the less pleasant aspects of revisiting China is the awareness of mercenary and exploitative attitudes that were virtually absent earlier. They are exemplified by the new slogan "to get rich is glorious" (and by a much older one from another part of the world—"make hay while the sun shines"). Many examples could be cited. Certainly there is nostalgia for the puritanism of the old days—when schoolteachers brought their classes to provincial guesthouses in the early morning "to see the foreigner" and to sing a welcome; now they demand payment (in foreign currency). In all countries the loss of innocence can be ugly, and nostalgia reflects change in the observer perhaps more than in the observed. Nonetheless, duplicity with respect to land and resource use, the lack of commercial infrastructure and accountability, and the stumbling search for the checks and balances needed in a market economy must call into question the durability of the new reforms. Reformed poachers are said to make the best gamekeepers; China in some ways is demonstrating the converse.

The second lesson from China, then, is this: Free markets need an appropriate infrastructure as well as responsible operators. The participants may also need the protection of the state. Even in mature economies the market system has failed to cope adequately with environmental protection and with the needs of conservation—or, indeed, with the time frame required by reforestation and sustained-yield resource management. In China, the eventual outcome of forest land contracting will depend upon the ability of the state to defend rights of usufruct: Landholders need certainty, continuity, and common sense. China has some way to go yet before reaching "that combination of modern science with local inventiveness and local responsibility that is at the core of the only really effective and sustainable ecological balance" (Ward and Dubos, 1972:236).

FOREST POLICY AND PRACTICES

Chapters 5, 6, and 7 have documented recent developments in forest policy and practice. Forest policy is more detailed now than in the 1960s, but the objectives are little changed and manifestos remain predominantly exhortatory rather than prescriptive. Planning is undoubtedly more soundly based, but reforestation has always been its major focus—and the aim is to increase the area under trees to the point where China can be self-sufficient in industrial timber and fuelwood. Somewhat naively, targets have been set as percentages of the land surface area to be covered, without much regard to the *distribution* of forest production. To some extent this approach was a result of emphasis given to protection forests (and their location according to the needs of agriculture) and to continuing reliance on the remaining natural forests for industrial timber. Only recently have questions of supply and demand and the location of industrial raw material supplies been debated. Changes of greater interest to the outside world relate to methods of policy implementation—in particular, the replacement of mass spring and autumn "campaign" plantings by contract responsibility systems and the involvement of specialized households.

Planning

The long gestation period in reforestation makes it particularly vulnerable to haphazard development. Despite the emphasis on planning in socialist economies, it is doubtful whether the five-year module has conferred much benefit upon forestry. It is too short a period to judge the success of planting and too long to arouse and maintain the evangelical fervor needed in a campaign system. Moreover, targets have been established in terms of percentage increases over past performance—frequently with the base data changed or redefined.

Planning in Chinese forestry is becoming more prescriptive and more openly discussed among academics and the executive hierarchy. In particular, long-term supply and demand forecasting is now being undertaken and the implications of shortfall are debated. In the 1960s, any suggestion that the planting program would be inadequate to provide all the raw materials needed for China's forest industries by the year 2000 would have been firmly rejected: The possibility of China importing nearly 10 million cubic meters of logs annually (as in 1985), or purchasing cutting rights in North American forests, would have been unimaginable. Now there is serious discussion of the date when supply and demand should be in balance and talk of the pros and cons of importing logs or lumber—to the point where the processing of imported logs for reexport is readily accepted—and argument about the realism of "designing timber out of the economy." The reality of China's timber shortage has stimulated interest in the Asia-Pacific forest economy—indeed the global economy—which would not have been acknowledged in the 1960s. These are signs of China's maturity.

The supply/demand crisis has aroused interest in the remoter regions of China, increasing the time and money devoted to forest inventory. At the same time, the shift from command to guidance planning has stimulated arguments between the center and the regions over the control of resources. The fact that both Yunnan and Tibet, for example, have not formally revised their inventory statistics since 1973 does not facilitate rational planning, and the switch in resources noted in Chapter 3 between Heilongjiang and Nei Monggol (Yakashih) is evidence that utilization planning leaves something to be desired.

In all countries of the world where there is regional variation in the distribution of resources (even if it is only rainwater), there are arguments about valuation and the control of distribution. Particular problems for forestry result from the history of "cost-plus" pricing of timber from natural forests. In North America there is growing criticism of "below-cost" or "free good" timber sales from the national forests; in China, with the massive increase in prices that free markets have triggered, there is outrage at state-fixed procurement prices in the timber-rich provinces. The provincial administration of Heilongjiang, for example, in the context of a World Bank loan, has argued that for decades the state has taken timber at well below world prices and that in equity the state (not the province) should now accept responsibility for servicing reforestation loans. Similar altercations are taking place between county administrations and provincial governments; they are exacerbated by emerging boundary disputes. There is no evidence that China has solved these problems any more effectively than other countries, whatever their ideology.

The planning of industrial timber harvesting ought to be simpler than much-longer-term reforestation planning—but, again, there are gaps between the policy of the center and what happens in the counties. Though, overall, the central plan may call for a 10 percent cut in timber production, at enterprise levels where there is a free market, the aim is to increase production rather than cut back. There is also an air of unreality about utilization plans—due, apparently, to a lack of integration between production and allocation of product and between the availability of finance and equipment. The evidence documented in Chapter 4 regarding waste and incompetence in the timber industry indicates that China has some way to go to get planning mechanisms into gear and responsive to market forces.

Thus, while long-term forestry planning is more evident in China today than in the 1960s, it is also a bit academic in the face of the realities in the regions and counties. There can be no more tragic example of failure to mobilize and coordinate resources than the 1987 forest fires. In this case the loss of hundreds of lives and huge timber stocks resulted from the inability of individuals within their hierarchies to mobilize resources outside them. Despite ample time, and access to a multi-million-dollar international forest fire management project, no attempt was made to mobilize firefighting facilities outside the region—let alone outside China.

China has been slow to adopt modern planning techniques. Planning is still not taught in the forestry departments of universities, and the Forest Economics Research Institute is more concerned with long-term trends than the immediate need for hard data. With many years' experience of the work-point system to determine the distribution of income within the commune, it is perhaps surprising that there is so little operations research carried out in forestry. All provinces supposedly use books of work norms—but, in fact, some are out of date and some have never been completed. (This is true in the case of the Tibet autonomous region, which, for the WFP-assisted project in 1988, was basing "food for work" requirements on norms for Sichuan modified according to topography and altitude.) There is an urgent need for operations research and, in particular, for case studies on which to base negotiations of contract responsibility more realistically. Realism in the establishment of work targets was, perhaps, less crucial in the days of the communes—which provided a safety net for families that were significantly below average in accumulating work points. Under the *Baogan Daohu* (contract responsibility system) financial incentives have improved for the successful, but the mesh size of the safety net has also increased.

Implementation

Changes over the last two decades in forestry practices, like so much else in China, present contrasts. The conservatism of silvicultural treatments—ultra-close plantation spacings, reluctant and light thinning, inadequate culling in nurseries—stand alongside boldness in planting species well outside their natural range. The quality of urban planting and, in particular, the care devoted to environmental forestry stand in contrast to the almost cavalier management of industrial plantations; tending is certainly better than it used to be, but there is a long way to go. The gap between research and forestry practice has been noted. And despite meticulous study of ecological minutiae in the natural forests, there is an apparent lack of concern that their specific composition is changing radically as a result of "creaming" and artificial enrichment—sometimes with exotic species—in secondary forests. As noted earlier—and not limited to forestry in China—there is a similar contradiction between concern about the maintenance of biological diversity and the enthusiastic espousal of clonal plantations.

In protection forestry, China has made considerable technical progress but continues to face enormous bureaucratic problems. Administrative streamlining in Beijing has yet to have a visible effect in the regions, while devolution of control from communes to counties has led to irrational decision-making on water conservation projects, neglect of maintenance, and the abandonment of protective plantations. It is no coincidence that—generally throughout China—where there are fewer opportunities for sectional wrangling—as in the PLA—silviculture and plantation management are of a higher quality than under collective control.

In Chapter 5 I suggested that the specialized households could become as powerful a lever for forestry development as the original *Baogan Daohu* (contract responsibility) system. The lack of adequate data bases in forestry—particularly in the remoter and more intractable areas that are earmarked for plantations—makes it difficult to devise equitable contracts under responsibility systems. In agriculture, it is relatively easy to set a target for production of an annual crop and allow the contractor to market the surplus. This is not the case in forestry—hence the wide spectrum of shareholder and contract systems that are developing.

Motivation in tree planting is seldom straightforward and, in peasant economies, plantation forestry is not easily market-driven. On virtually all sites there are alternative crops offering quicker and higher returns. (On the poorest of sites in many countries, *Cannabis* and the opium poppy are classic examples.) In some countries, under various systems of "*taungya*" (raising tree seedlings together with food crops), the farmer's need for land to grow subsistence crops is used to enforce tree planting (new land is allocated only after successful establishment of trees); in others, landless itinerant farmers are given tenure on condition that they grow tree crops for the landowner. In China, many of the sharecropping and contract tree-planting schemes involve short-term, high-value tree crops (spices, fruits) as well as industrial timber and fuelwood. Where such combinations are not possible, "fringe" benefits and infrastructural provisions (roads, schools, and the like) will be important, as will perceptions of security.

There can be problems of success as well as failure. Chinese foresters emphasize tailoring contracts to individual needs; but success can breed envy and lead to demands for change from the less successful contractors, and change may reintroduce insecurity. For these reasons, the specialized households may prove more durable and acceptable.

Though it is too early to judge, China's land reform—and associated *Baogan Daohu* systems—seem to offer the most promise of lessons that the rest of the developing world might appropriately learn. The challenge for the administration is to maintain flexibility without compromising security—both in time and with respect to the basic needs of households that must be provided from outside if people are to be persuaded to settle in "difficult" areas.

The use that is made of the PLA—though difficult for a visitor to evaluate—may be of interest in other Third World countries that need to maintain an underemployed defense capacity. It provides a disciplined work force, amenable to training, deployed in areas not popular for settlement by Han Chinese. There is no evidence that military organizations are less "bureaucratic" than civilian systems, but they are probably less prone to sectional squabbles (pecking orders are more clear-cut) and inefficiencies can be mitigated more rapidly. Dreyer (1986) documents the huge extent to which the PLA is involved in economic production: In industry its role is controversial, but in farming and reforestation it is noncompetitive. (Its farm network is aimed at feeding its own members, and its tree-planting is generally in remote areas.) It may in fact be more cost-effective to use the PLA in these roles than to accommodate reluctant settlers.

Another aspect of China's policies that other nations will follow closely is the application of the "user pays" principle and "accountability." In particular, the attempt to apply *Baogan Daohu* in research and extension is of interest outside China. The reasons that have prompted it—the rift between research and production, excessive duplication of research, and the fact that research priorities have been overwhelmingly established by the producers rather than the users of research—are of concern in many countries. The implications of contract research for the marketability of knowledge—and the acceptance of responsibility under laws of patents—may set precedents for transfer of technology from west to east and north to south. Similarly, the involvement of specialized households in extension—with income-sharing between the extension workers and the contracting landholder (and compensation for the latter in the event of bad advice)—is an innovation that has wider application.

CONCLUSIONS

There is perhaps a tendency to exaggerate change in China—by visitors and the Chinese themselves. China was so little known to the outside world for so long that (especially in the ancient arts of agriculture) visitors may see innovation where there is none. At the same time, the use of political jargon, the substitution of slogans for directives, and the limited interpretive ability of cadres and managers can confuse even the Chinese. It has been argued that agricultural collectivization was only a small step from the mutual aid systems that existed before the revolution. There have been few major changes in agricultural and forestry technology—those acclaimed as such (agroforestry, for instance) in fact have a long history. The Chinese people the world over have a reputation for pragmatism; the returning visitor to China becomes aware of the extent to which it is grafted onto conservatism. While policies may be pragmatic, practices are conservative.

Nursery practice, silviculture, choice of species, research, and education are imitative and reveal little by way of innovation. Even in areas where China is acknowledged a world leader (the culture of bamboo, China's pharmacopoeia, the development and use of locally produced fertilizers), techniques have long been documented. There is no doubt that, in their context, they are successful. But the crucial problems of forestry in China do not relate to traditional forest areas and practices. There is little evidence that problems of intractable sites—with poor eroding soils, inadequate moisture, and a harsh climate—have yet been solved. There are still unresolved political and economic problems of "adjustment and reform"; and there is an enormous need of trained manpower able and willing to make sensible management decisions.

China's greatest achievement is undoubtedly with respect to land reform and the apparent provision of security of tenure without ownership. It remains to be seen whether the leadership will succumb to pressures to allow the inheritance of land (as distinct from tenancy) and to permit a free market in lease rights. If it does, the door will be opened to the accumulation and inheritance of wealth and the "great socialist dream" will be at an end. At the same time, if the barren hills and desert fringes are to be brought into production, greater incentives are needed.

Sustainable development in the remoter parts of China will involve massive investment in infrastructure (roads, schools, security, marketing and banking services, and the like) as well as production incentives. The returns will not come quickly, nor will the investment demonstrate a high benefit/cost ratio by conventional economic analysis. Such problems are familiar.

In all countries, too, there is a conflict between market forces and "externalities." In the West, economists have argued the case for the control of pollution through taxation for many years. The debate is being taken up in China in the context of pulp and paper mills and China's dilemma with respect to plant size. In no country have traditional "command and control" systems proved very efficient as instruments of environmental policy, and taxes provide incentives both for governments to enforce pollution controls and for the polluters to develop and install effective abatement techniques. In a cellular economy such as that of China, the appeal is evident. At the same time, China faces particular problems of communicating such ideas to management because of the poor caliber and training of personnel. Responsibility at this level often rests with the lost generation of the Cultural Revolution—uneducated, insecure, and resentful, the "jokers in the pack" (according to the late China specialist David Bonavia). To communicate radical economic change, to reawaken enthusiasm, and to impose discipline will take time and considerable political skill. The next generation of managers will no doubt be more attuned to new ideas and technologies; even so, their training has of necessity been naive and lacking in breadth. Young cadres are numerous but of limited literacy and poorly endowed with the capacity of reason that derives from balance; high technology enables—perhaps encourages—them to make mistakes with total confidence. Restricted vision is seldom realized by those who suffer from it and, in China, there are as yet inadequate yardsticks against which to measure experience.

In conclusion, then, there can be no conclusions. China is still learning to define its needs and to formulate policies based on the accumulated knowledge of history. Experience in the 1950s with Soviet models taught China that there are no ready-made socialist recipes for successful economic development. And experience now is revealing the costs as well as the benefits of a free market philosophy. What China has to teach is not revolution but the ancient Confucian virtues. The country with the most to learn from China is China.

It is always easier to learn from other people's history than one's own. Whether China can weave the lessons of history into the fabric of its new society must remain an open question. Because of changing global perceptions of the roles of forests (and foresters), it is a question of universal interest.

APPENDIX A

INDUSTRIAL WOOD USE IN CHINA

Analyses of wood use in China in my earlier commodity reports (Richardson, 1965, 1972) were needed as much to measure total industrial timber consumption as to provide a basis for the projection of future needs. Now that end-use allocations have been published and, for the first time, accurate population statistics are available, global projections of demand can be based on demographic parameters. It is of interest, however, to identify qualitative as well as quantitative determinants of consumption. In this appendix, therefore, I examine the major wood-using sectors.

CHINA'S POPULATION

No aspect of China's economy has preoccupied analysts more than its population. It has been said that demographers in the seventies spent more time wrestling over the problem of whether China's population was 740 million or 750 million than it took China to produce the extra 10 million bodies. To the Chinese it has not until recently been a statistic of overriding importance. Indeed, Zhao Enlai in 1971 said: "We tend to think that we are more than 700 million but not yet near the 800 million mark." One of the earliest of the present reforms, however, was population growth control—written into the 1978 constitution and involving somewhat draconian measures to contain the birth rate (see Aird, 1986). The census data show an astonishing reduction in growth from 3 percent in 1971 to 1.8 percent (Orleans, 1986) and a rate of natural increase in 1985 of 1.2 percent. The excitement outside China with the results of the 1982 census is perhaps not surprising.

Traditionally, China had a system of population registration (based on ten-family groups) that began in 1644 and survived until the end of the nineteenth century. Modern estimates, however, derived from a census in 1953 published by the State Statistical Bureau in 1958. The growth rate was then around 2 percent with a target rate of 1 percent. Birth control measures—including incentives for late marriage and oral contraception ("the medicine of the twenty-one remembrances")—have been applied in varying degrees of intensity since 1963, but with four-fifths of the population living in rural areas, success was limited.

The 1982 census was the most ambitious ever undertaken and has been described as a "monumental achievement . . . that is certain to increase China's statistical credibility throughout the world" (Aird, 1983:640). Thirty-two volumes comprising data down to county level are being published.

According to preliminary analyses, the current net population increase is 1.2 percent, but this has come about within the last five years through birth control and one-child-family policies. Life expectancy is rising. (It reached age 68 in 1985, compared with 35 in 1949; the death rate dropped to 6.6 per 1000 from 25 in 1949 and infant mortality to 34.68 per 1000 from 200 in 1949—all figures from the *China Daily* of 21 July 1986 citing "Economic Information.") The total fertility rate (roughly the number of children per completed family) was still 2.6 overall in 1981 (Vlassoff, 1986).

A summary tabulation of population by provinces is presented in Table A-1. For present purposes, the important features of the census are the division between rural and urban dwellers, the age structure (as a basis for projecting marriage and household formation), mortality, and literacy. These are all factors that affect the demand for wood and wood products. Timber is needed to build houses, furniture, tools, implements—even coffins. And there is an obvious relation between literacy and the requirements for paper.

TABLE A–1

PROVINCIAL DATA FROM THE 1982 CENSUS

Province, Region, or Municipality	Population	Density per km²	% Annual Average Growth Rate Since 1964	% Urban	% Illiterate or Semiliterate
China	1,003,937,078	105	2.1	20.6	23.5
Beijing	9,230,687	549	1.1	64.7	12.4
Tianjin	7,764,141	687	1.2	68.7	13.9
Hebei	53,005,875	282	1.7	13.7	22.3
Shanxi	25,291,389	162	1.9	21.0	18.0
Nei Monggol	19,274,279	16	2.5	28.9	22.4
Liaoning	35,721,693	245	1.6	42.4	12.9
Jilin	22,560,053	120	2.0	39.6	16.2
Heilongjiang	32,665,546	69	2.7	40.5	16.1
Shanghai	11,859,748	1913	0.5	58.8	14.3
Jiangsu	60,521,114	590	1.7	15.8	27.2
Zhejiang	38,884,603	382	1.8	25.7	24.2
Anhui	49,665,724	356	2.6	14.3	33.4
Fujian	25,873,259	213	2.4	21.2	26.3
Jiangxi	33,184,827	199	2.6	19.4	22.0
Shandong	74,419,054	486	1.6	19.1	28.0
Henan	74,422,739	446	2.2	14.1	26.9
Hubei	47,804,150	255	2.0	17.3	23.3
Hunan	54,008,851	257	2.1	14.4	17.6
Guangdong	59,299,220	280	2.1	18.7	16.4
Guangxi	36,420,960	158	2.5	11.8	17.2
Sichuan	99,713,310	176	2.1	14.3	23.4
Guizhou	28,552,997	162	2.9	19.7	32.1
Yunnan	32,553,817	83	2.6	12.9	33.7
Xizang	1,892,393	1.6	2.3	9.5	51.8
Shaanxi	28,904,423	141	1.9	19.0	24.7
Gansu	19,569,261	43	2.5	15.3	34.9
Qinghai	3,895,706	5	3.4	20.5	31.9
Ningxia	3,895,578	59	3.5	22.5	28.9
Xinjiang	13,081,681	8	3.3	28.4	20.9

SOURCE: Aird (1983).

NOTES: The total population statistic has since been revised to include military personnel at 1,008,175,288. At a growth figure of 1.2 percent net, the 1989 midyear population was nearly 1,070,000,000.

RESIDENTIAL CONSTRUCTION

Since the late 1970s, housing has greatly concerned China's leaders. Vice-Premier Yu Chiuli focused on its problems at the Fourth NPC (*Beijing Review,* 4 November 1977) as did the premier in a report to the Fifth NPC (*China Daily,* 15 December 1981) and again in a review of the Seventh FYP to the Sixth NPC on 25 March 1986. Table A-2 indicates investment in residential building (as a percentage of total construction). State housing is overwhelmingly urban, reflecting the realization that city growth is inevitable under modernization. (Some 200 towns produced 74 percent of China's total GNP according to Kraus, 1979.) The urban/rural ratio grew from about 16 percent in 1964 (Koshizawa, 1978) to the 1982 census figure of 20.6 percent. The State Statistical Bureau's figure for 1983 (including the military population) was 241,260,000 (23.5 percent according to Banister, 1986). The percentage is not out of line with other underdeveloped countries, but it has to be realized that only three countries in the world have a *total* population in excess of China's *urban* population (Orleans, 1982). Moreover, migration to urban areas is by young people of marriageable age needing accommodation and likely to add disproportionately to the birth rate.

TABLE A–2
INVESTMENT IN CONSTRUCTION:
1975–1985

Year	Productive Construction (%)	Nonproductive Construction (%)	Residential Building (%)
1975	85.7	14.3	5.9
1976	85.1	14.9	6.1
1977	83.3	16.7	6.9
1978	82.6	17.4	7.8
1979	73.0	27.0	14.8
1980	66.1	33.7	20.8
1981	58.7	41.3	25.5
1982	54.5	45.5	25.4
1983	58.3	41.7	21.1
1985[a]	61.1	38.9	17.5
1986[a]	61.7	38.3	16.7

SOURCE: Statistical Yearbook of China (1984).
[a] First six months only.

A recent report (Chan and Xu, 1985) notes a shift in the official definition of urban population in 1963–1964 and provides a reclassification for the period 1949–1982. It demonstrates that, in fact, urbanization in China is close to that of other large developing countries—such as India—and that although there was a massive urban exodus during the period 1966–1976, it was countered by in-migration, especially after 1977.

During the Fifth FYP, according to Premier Zhao Ziyang, 630 million square meters of floor space was added to the stock of urban residential construction (and 3.2 billion square meters in rural areas). One source (USDA/FAS, 1986) puts current construction at "roughly 1.1 billion square meters" annually, of which urban housing is about 200 million square meters. This is more than double the rate at the beginning of the Fifth FYP (CTR, 1985).

A carefully documented study by Taubman (1983) estimates the 1982 urban housing stock at 978 million square meters and ratios of construction/living space of 1:0.08 for old housing (pre-1949) and 1:0.55 for new housing,[1] giving an available living area per capita of 4.5 m^2. He derives values (per capita living space in m^2) for various Chinese cities in 1981 as follows:

Beijing	4.91
Shanghai	4.00
Shenyang	2.47
Wuhan	3.61
Guangzhou	3.21
Harbin	2.71
Chongqing	2.38
Chengdu	2.22

The absolute poverty rating is 2 m^2 and the target for the year 2000 is 8 m^2 (*People's Daily* of 5 August 1980). A recent broadcast report (FE, 1986) claimed an average per capita living space in cities of 6.32 m^2. A report in the *China Daily* (17 July 1986), however, notes that despite a doubling of Guangzhou's housing stock in eight years, 11,600 families live in houses averaging less than 2 m^2. To put these values in perspective, it should be noted that in most working urban families, both husband and wife are employed—often at the same workplace—and the children attend day-care centers. To some extent family life takes place outside the home. Rural living areas are rather bigger (5.5 m^2) and construction areas considerably so—averaging 11.0 m^2.

The 1982 census showed an urban population of 20.8 percent (reconstructed by the SSB—see Banister, 1986) growing to 23.5 percent in 1983. (Figures for both years include military personnel.) The sudden jump is probably a statistical artifact, but there is no doubt that the urban fringes are expanding much faster than the country overall. If a net population growth rate of 1.2 percent is assumed (implying highly successful family planning policies)—together with an annual increase in urban population equal to the average over the period 1978–1983 (20.1 percent)—China will need to build 2 billion square meters of additional urban residential accommodations merely to maintain the existing standard at the end of the current FYP (that is, more than three times the area built during the Sixth FYP). To achieve the target of 8 m^2 per capita by the year 2000 for an urban population that stabilizes at the 1990 level of 450 million—assuming an optimistic 2 percent annual renewal—would require an *additional* 1.6 billion square meters of construction during the 1990 decade.

The premier's comments on "nonproductive construction" are not encouraging: "Housing conditions should be steadily improved, but residential building standards should not be too high and rooms should not be too large. . . . In urban development, priority should be given to infrastructural projects that support production and facilitate daily life" (Zhao Ziyang, 25 March 1986). Two factors may bring change. First, as noted in a World Bank staff working paper (Gaag, 1984), low income elasticity for rents (0.535) reflects urban housing subsidies and may severely underestimate the future latent demand. Second, the precedents set by rural developments may force a reordering of priorities. Lardy (1986b) emphasizes the sanctity of private housing in rural areas and its focus for owner investment; following the success of reform, private housing investment rose to 15.7 billion yuan in 1982 and 21.4 billion yuan in 1983—many times the state investment in the entire agricultural sector. It is doubtful whether urban dwellers will be satisfied with their present cramped conditions in the face of such growing rural prosperity.

The World Bank (1985a) puts the urban rental housing stock at 600 million square meters, urban owner-occupied at 300 million square meters, and rural at 10 billion square meters. In an "international comparison" table, China's percentage of GDP invested in housing was the lowest (of India, Indonesia, Korea, Brazil). According to Taubman

(1983), citing the *People's Daily* of 5 August 1980, one-third of all urban households in 1980 had no housing and were forced to share with other families. In Shanghai, 25 percent of urban dwellers lived in buildings ripe for demolition. Moreover, even "new" housing areas do not provide much more living space. Taubman's field survey of twenty-five new housing areas (completed mainly between 1974 and 1978) indicates an increase of only 1 m^2 over the mean urban value. (See Table A-3.)

Housing availability in urban China is closely linked to working units that supply accommodations at token rentals. (This practice not only discourages private construction but has also prompted many owners to hand over their houses to the state in return for rental accommodations.)[2] There are evident differences in housing standards according to working unit (and, of course, position within the unit). Thus in 1979 in Beijing (Lin, 1983), with an overall average space per resident of 4.57 m^2, employees of Central Party and government organizations averaged 7.65 m^2, municipal government personnel rated 5.5 m^2, district government staff 4.67 m^2, and other residents (including primary-school teachers) only 3.6 m^2. Housing thus serves as a fringe benefit; it is also used as an incentive to family size restriction (see Aird, 1986). Some three-fourths of urban populations are state-employed and have housing allocated through work units; the rest belong to the collective system and rely on municipal housing administrations. The latter are responsible for housing the elderly and unemployed and for significant new housing (using state funds).

There are constraints upon urban renewal and development in addition to costs. The association of housing with industry has created environmental problems, but relocation of factories would require worker relocation too. The fringes of the cities are intensively cultivated for town-supply horticulture, which China can ill afford to abandon. (Between 1957 and 1980, some 33 million hectares of cultivable land—one-third of the present agricultural area—was forfeited to construction, according to the *China Daily* of 14 January 1982.) Expansion has, therefore, to be upward. In the major cities, high-rise living (over six stories) is presently preferred by the bureaucracy (but is far from universally liked by the inhabitants). In the smaller conurbations, multistoried building is more often restricted to five or six stories. This is the limit for construction without using powered hoists and for buildings without elevators; the unit construction cost is about half that of ten- to twelve-story blocks.

TABLE A–3

LIVING SPACE IN SELECTED NEW HOUSING AREAS: 1983

Category	Year of Construction	Inhabitants (in 1000's)	Number of Floors	Living Space per Head (m^2)	Floor Space per Household (m^2)
"Workers' Villages"					
Tongling–Zianqiao Village (Anhui)	1973/74	5.95	4.0	4.6	39.5
RA of Vinyl Factory (Guangxi)	1972/74	4.11	3.0	5.7	43.3
RA of Chemical Fertilizer Factory (Hubei)	1975/77	5.48	3.7	4.1	37.2
RA of Tower Station Xindian (Shandong)	1973/77	2.57	3.2	4.6	45.9
Residential Areas in Suburbs (or Satellite Towns)					
Yingfeng Xincun—RA No. 1 of Petrochemical Plant (Beijing)	1975/78	3.50	5.2	5.4	43.0
RA of new Huangpu Port (Guangzhou)	1975/80	9.75	5.0	5.2	39.0
RA of new Port (Dalien)	1974/76	4.60	3.1	4.3	39.0
Wanshan Xin Village of Depot, Xiangyang (Hubei)	1973/78	11.34	4.0	4.3	38.1
RA of Shanghaiguan Shipyard (Qinhuangdao)	1974/79	12.80	4.2	4.7	43.0
RA of Petrochemical Factory (Guangzhou)	1974/78	7.01	4.8	5.3	43.1
Residential Areas in Cities					
Tuanjiehu (Beijing)	1976/79	21.26	6.9	6.2	53.3
Jinsong (Beijing)	1976/79	25.57	7.5	6.2	53.5
Chaoyang, Village No. 2 in Changsha (Hunan)	1976/78	14.00	5.5	5.8	45.0
Tiantannanxiao (Beijing)	1972/78	7.80	4.4	5.4	49.7
Bingcaogang, Neighborhood Quarter No. 2 (Dukou)	1975/78	3.50	5.1	5.2	39.0
Daban on Nanlu Street (Nanning)	1975/78	9.44	6.0	6.5	48.0

Redevelopment of Old Residential Quarters					
Mingyuan Xincun (Shanghai)	1973/78	4.96	5.6	5.5	39.5
Guiyang Street (Tianjin)	1976/79	16.83	5.0	5.8	43.5
Binjiang Street, Neighborhood Quarters No. 1 + 2 (Nanning)	1976/78	2.08	5.8	6.2	42.1
Dongfang Xincun (Changzhou)	1977/79	2.12	5.8	6.8	51.1
Lane 991 of Xinzhaozhou Street (Shanghai)	1974/78	4.15	5.8	5.0	35.7
Residential block of Wuhan Theater	1964/66	2.45	4.4	5.2	41.5
Hebei Main Street (Tianjin)	1974/79	4.30	5.1	5.3	47.9
Residential block in Binjiang—North Street (Guilin)	1977/80	1.28	6.1	4.9	42.7
Qiansanmen—Main Street (Beijing)	1976/78	32.50	11.3	5.9	55.0

SOURCE: Taubman (1983).

NOTE: RA = residential area.

Despite the commendable efforts of the Sixth FYP, living conditions for China's urban population will not improve without substantially increased investments. Costs have risen significantly, even since the Seventh FYP was drafted. Delfs (1986b) reports current building costs in Tianjin—China's third-biggest city—at 200–300 yuan m^2. As a result, attempts have been made to encourage private construction and financing. Housing sales to collectives and individuals have been attempted; government mortgages—for fifteen years at 1 to 2 percent—are available; increasing rents in line with quality has been adopted as a formal strategy; foreigners—especially overseas Chinese—are being urged to invest in private building projects and even to purchase and restore old courtyard houses in Beijing (Taubman, 1983). Enterprises that have benefited from "reform" are encouraged to invest welfare funds in housing.

Attempts are being made to improve efficiency in the construction industry. MOURCEP—serviced by bureaus of building design, science and technology (with subsidiary Building Technology Centers), and building administration—is responsible for setting standards and monitoring codes of practice. Because materials (including timber) are allocated by the state to the client enterprise (not to the building unit, which simply provides labor and equipment), responsibility for economic efficiency is ill-defined (*People's Daily,* 5 August 1980); there is considerable wastage and deterioration from poor materials handling on building sites. An article in the *Beijing Review* (11 August 1986) complained of "truckloads of perfect timber" being buried or burned on building sites in Beijing and urged the extension of total contract building.

Some attempts to rationalize residential construction have been made through the Municipal Construction and Development Corporation (*China Daily,* 22 January 1982). This organization builds satellite towns and new living quarters, as well as carrying out development projects—using funds supplied by institutions. All units—whether employing state funds or their own—must contract with the corporation. It provides a focus for the transfer of technology and, in particular, the development of construction systems.

Historically, brick and wood were the major urban construction materials in China. Shortages have led to substitution of wood by concrete (applied with reusable steel or plank shuttering) and steel or aluminum joinery. Despite price advantages of wooden joinery, designers specify metal wherever possible—partly because it is government policy to do so and partly because of low-quality wooden products. Joinery is poorly finished and built from semigreen and knotty timber subject to on-site deterioration through exposure to the extremes of a generally continental climate. Panel products made in China are usually not structural grades or exterior-glued. Given better-quality products and wood-oriented construction systems, there is no doubt that cost effectiveness could be much improved. The NFPA (1986) report reproduces "Regulations for Economical and Rational Applications of Wood and Wood Substitutes." They prohibit the use of wood in traditional building applications (floors, stairs, walls, piling) as well as in shipbuilding, railways, and mining. They are difficult to enforce and outside the state sectors they are widely disregarded. Urban population pressures may force reconsideration of present substitution policies.

In rural construction, wood is traditionally used in the round for roof trusses and tile supports constructed on the job. Increasing affluence—and access to village sawmills—has enabled the use of prefabricated squared trusses and purlins, while the envied "10,000 yuan households" are building houses with attics floored with wood.

There were few signs in 1986 of economic downturn in the rural economy—except in the northern and western border provinces, which are traditionally the poorest and the farthest from markets. Even if the government wanted to apply brakes, it is doubtful whether it would succeed in view of the dismantling of the allocation systems and timber price controls. The macroeconomic levers still under central control influence primarily the state enterprises—which have contributed little to the growth of the rural economy. It is likely, therefore, that private-sector residential construction in the present FYP will at least match that of the last one.

There are several implications for wood consumption in housing. For planning purposes, the Chinese use roundwood input values of 0.058 m^3/m^2 for rural housing and 0.039 m^3/m^2 for urban construction. (These values differ slightly from MOURCEP guidelines.) On this basis, and assuming an annual program during the current FYP of 1 billion square meters, the roundwood requirement would be 42.8 million cubic meters. The urban sawnwood component would be 11.6 million cubic meters, which compares with allocations of 15 million cubic meters of roundwood under the State Plan (the average of 1983–1986) for *total* construction. (There are no allocations for the private sector.) Making the calculation on another basis—using a 1979 index of 0.14 m^3 sawnwood/1000 yuan of investment in construction (Shiraishi, 1983) and Zhao Ziyang's budget of 500 billion yuan during the next FYP—wood use in official construction will average 14 million cubic meters annually. For this kind of exercise, the agreement is acceptably close.

NONRESIDENTIAL CONSTRUCTION

In recent years nonresidential construction in the major cities of China has been dominated by hotel and office building. Although the hotel boom has now lost momentum, there are few signs yet of any downturn in other nonresidential construction.

According to one of China's senior planners (*South China Morning Post,* 8 September 1986), investment in construction was still running free in 1986, up 14 percent on the same period in 1985. Outside the State Plan, the rise may be much higher since investors no longer need to rely on the state for capital. Nonfiscal controls can be avoided by disguising expenditure as "technical renovation," and it seems likely that any underused capacity left from the hotel slump may be absorbed by other subsectors—at least in the short term. The premier's discouraging remarks in 1986 on overinvestment in building were no doubt carefully calculated—but again it is difficult to see what the state could do without abrogating the principles of reform. In fact, under new policies (SCMP, 8 September 1986) the People's Bank of China in five cities no longer has to restrict lending to cash reserves allocated by the central government. The specialized sector banks (including the People's Construction Bank) are expected to expand both deposit taking and lending.

FURNITURE

The demand for furniture is linked with the rate of household formation and in general follows a similar pattern to construction. In 1982, the Chinese industry reportedly produced 63.83 million pieces of furniture

(NFPA, 1986) of which 61.5 percent was of wood. Zhong's estimate of increasing output was 1.6 percent annually, consuming 2.3 million cubic meters of solid wood a year.

The NFPA report considers wood consumption in furniture manufacture to be grossly underrecorded and says that perhaps five to ten times more wood is used than is acknowledged by the Ministry of Light Industry. Certainly furniture manufacture is a growth sector and is under consideration as an export industry, based perhaps on the use of imported hardwoods. A recent study by Araman (1987) highlights the increasing demand in Asia for the blander close-grained hardwoods for use in furniture manufacture for reexport to North America. China uses similar species (poplar, lime, birch) and is importing some from Canada. The establishment of joint ventures in furniture production for export would be a logical development.

TRANSPORT

Transport, including railways, accounts for 5 percent of roundwood allocated under the State Plan. Most of this demand may be expected to grow more or less in accord with GNP, but railways have long been recognized as a weak point in the economy. Japanese commentators (Nakajima, 1985) and the World Bank (1985b) have analyzed the sector and envisage substantial line expansion (to 60,000 km by 1990 and 75,000 to 80,000 km by 2000 from 52,000 km in 1985) and double-tracking (to 25 percent by 2000 from a present 18 percent). The World Bank report points out that in the USSR between 1955 and 1975, rail freight densities tripled as a result of the replacement of steam locomotives by diesel and electric power, combined with some double-tracking (hence high returns from a relatively modest investment). China, too, is following a policy of increasing the existing capacity rather than expanding the network. The scenario recommended by the World Bank estimates annual investments to the year 2000 of 5.5 billion yuan for infrastructure and 1.2 billion yuan for freight cars. Although double-tracking may involve mostly concrete sleepers and freight cars are predominantly steel, wood is still needed for sleepers on bridges and in stations, as well as for freight car repairs. Moreover, it is difficult to mix concrete and wooden sleepers (the ballast for concrete sleepers has to be tamped rather than packed), so maintenance and some double-tracking will demand wooden sleepers. According to the NFPA report, the Railways Ministry envisages an increasing annual requirement for wood to 1.5 million cubic meters (sawn) by 1990.

MINING

The mining industry is a major user of wood, both for pit props and as sawn timber. Of particular interest—because of its scale and relatively primitive technology—is the coal industry. China's enormous reserves of good quality coal are well documented, as is the historical dominance of coal in China's energy balance. (See, for example, Keidel, 1986; *China Coal Industry Yearbook,* 1983.) Production rose from 30 million tons in 1950 to over 600 million tons in 1980 when the industry faced an investment and "capacity to deliver" crisis. The current development strategy emphasizes mechanization and the application of foreign technology to increase production and exports. With the development of petroleum came the decision to substitute coal for oil in domestic industry and to export the petroleum saved. This strategy has been implemented and the role of coal in easing the "energy-constrained economy" (Field, 1982) is expected to grow.

The coal industry's production target for the year 2000 is 1.2 billion tons (raw coal)—and is believed to be conservative, representing a growth rate already overtaken. Given a modest 4.5 percent annual growth, the final output target will be closer to 1.5 than 1.2 billion tons. Other mine production is historically about one-fourth that of coal (CIA, 1975), giving a total mine production of some 1.8 billion tons.

The roundwood allocation to mining in 1983 was 6.5 million cubic meters (used to produce about 715 million tons of coal). Assuming no change, the roundwood requirement by the year 2000 would be some 15 million cubic meters for the state allocation. There could, of course, be some reduction in timber requirement per unit of mine production in the future. Certainly the mining industry has a high priority for wood savings through preservation. (At present, according to Zhong [pers. comm., 1986], only 20 percent of the timber used in mining is treated with preservative.) It is not impossible, however, that the increased use of preservative-treated timber could extend rather than reduce the use of timber. Wooden props are preferred to steel (or to such devices as concrete-filled bamboo). They provide an early warning of failure—particularly in locally operated (county or village) mines, which are expected to increase production at a higher rate than state mines (Keidel, 1986). China has more than 120,000 collective mines, compared with 6000 in the state sector (*South China Morning Post*, 8 September 1986).

PLYWOOD

The use of plywood is growing: The NFPA report estimates increasing consumption from 796,000 m^3 in 1982 to 1,227,000 m^3 in 1985. Evident problems associated with fiberboard and particleboard have given reconstituted boards a poor image, but the outlook for plywood is bright. The NFPA report discusses plywood in some detail to discover possible areas of softwood substitutes for what is in China a predominantly hardwood product. In the 1972 Tuolumne report (Richardson, 1972:108), I argued that "China might well set her sights on development as an in-transit processing country" for plywood. This eventuality now seems unlikely, but China has obvious intentions of exporting veneers. New panel-product technology is of interest and some 4 million cubic meters of total production is envisaged by the end of the century. At present, China imports some 800,000 m^3 of plywood—mostly thin lauan.

OTHER SECTORS

Apart from the uses already discussed, there are in addition pulp and paper, packaging, and the multitude of end uses included in Table 18 as "miscellaneous" and "other." Pulp and paper is a priority area of light industry (EIU, 1986), which currently uses an allocation of nearly 4 million cubic meters of roundwood. This supply is acknowledged to be inadequate (it produces only 1 million tons of paper), and there have been growing environmental and other problems associated with the use of agricultural residues. The MOF estimates the requirement of wood for the paper industry in the year 2000 at 30 million cubic meters (NFPA, 1986).

Paper production in 1982 was 5.9 million metric tons for a population (Table A-1) of 1003.9 million—indicating a per capita consumption of 5.68 kg. According to Zhong (1985), 25 to 27 percent of the total raw material used in paper manufacture in China is wood; Shiraishi (1983) reports that 4.7 m^3 is used per metric ton of chemical pulp and 2.5 m^3 per metric ton of mechanical pulp; from these data, along with production and import statistics, he deduces the 1980 roundwood volume input to paper manufacture to be 3.8 million cubic meters. (This compares with a 1985 MOF *allocation* to pulp and paper of 3.9 million cubic meters.) If per capita consumption increased to 20 kg (roughly half the present world average), the wood requirement would be 35 million cubic meters a year. A Chinese projection (Zhong, pers. comm., 1986) puts per capita consumption at 13 kg by the year 2000.

Projections of this sort have little significance. Demand for paper is selective. China's increasing needs are for printing and writing papers and for improved packaging media. Moreover, the Seventh FYP increase in paper production (which will rely heavily on imported machinery) is likely to be based on wood rather than agricultural residue. In view of China's energy constraints, increased imports of wood pulp to the country's eastern seaboard mills—rather than the establishment of new pulp mills—would be a logical development.

The relation between education—in particular, literacy—and paper consumption was noted earlier. The 1982 census revealed some surprises with respect to literacy. Despite previous literacy campaigns, there was no change in the absolute level of literacy between 1964 and 1982 (233.3 million in 1964 and 235.8 million in 1982)—though the percentage of illiterates in the total population did decline (Banister, 1986). In no province was the illiteracy rate below 15 percent of the population aged 12 and above, and in seven provinces it exceeded 40 percent. In countries without a phonetic script, the definition of literacy is problematic: The minimum standard is set in China at 1500 characters, but Lardy (1986b) emphasizes the difficulties of making comparisons with other countries where definitions vary. The fact remains that China still has a huge task ahead to bring literacy to acceptable standards and to raise female literacy to the same level as that of males. The main weapon in this campaign is education and its ammunition is paper.

Primary school enrollment in 1980 was 146 million—more than double the 1960 levels—but not all registered pupils attend regularly, especially in rural areas where there are competing demands for their energies. Poor performance is also ascribed to poor facilities—including constant shortages of books and paper (Yahuda, 1986).

Elementary education defines the potential for higher education. The Cultural Revolution did incalculable harm at the higher levels resulting in a virtual gap in professional manpower. To mitigate its effects, China has concentrated on ad hoc vocational middle schools to train technicians and managers, while the university system is being rehabilitated. Thus middle-school enrollment in 1985 (7.5 million) included a sixfold increase in agricultural middle-school enrollment to 2.25 million. Higher education enrollments rose from 564,000 in 1976 to 1 million in 1980 and 1.4 million in 1984. World Bank experts argue that three to four times the present level of vocational/technical trainees will be needed to meet China's needs. The Seventh FYP forecasts vocational and technical enrollment by 1990 of 3.6 million and an increase of 21 percent in tertiary education entrants. Adult education institutions are to produce 2.1 million specialists over the five-year period.

It is not possible to relate educational needs directly to the demand for paper, but that demand can only increase. Moreover, the need is for printing and writing papers, rather than newsprint, and China's recycling potential is limited. (Nearly 2200 newspapers are published in China, but most of them are single sheets intended for wall posting and communal reading.) Since the quality of the newsprint is inadequate for de-inking, it can be recycled only for cardboard.

Imports of woodpulp for cardboard manufacture are increasing. The NFPA report notes that the Packaging Import/Export Corporation used its 1986 foreign exchange allocation to buy paper pulp rather than wood; and, wherever possible, substitutes for wood are to be used. There are frequent press reports of dissatisfaction with packaging standards, and the China Packaging Technology Association is charged with reducing the waste of both materials and products. The import of wastepaper is also increasing, recently from North America. Recycling Chinese-made paper, however, has reached only 20 percent—the constraints are technical (de-inking) and logistic.

Packaging is a significant and growing user of both sawnwood and plywood. The NFPA (1986) indicates 7.7 million cubic meters of roundwood consumption in 1985, of which 0.5 million was imported softwood—mainly Radiata pine from Chile, larch from the USSR, and U.S. hemlock. Some 160,000 m^3 of plywood was also used. The standard of packaging—especially for exports—is causing concern, and losses resulting from poor materials and practices are reported to have exceeded 10 billion yuan in 1985. (This is probably an exaggeration—it is an impossible statistic to measure.) The Seventh FYP, however, specifies "packaging and materials handling" as requiring particular attention and MOFERT has established a promotion group for export packaging.

"Miscellaneous" and "other" end uses for wood include agriculture, shipbuilding, exploration, textiles (there are more than 500,000 wooden looms in China), footwear, musical instruments, sporting goods, matches, coffins (despite policies favoring cremation, it is still very much a family tradition in rural China to acquire a coffin early in life and to display it), vehicles, implements, tools, and handicrafts.

CONCLUSION

The visitor from an industrialized country often has difficulty imagining the myriad uses for wood (including bamboo) in rural China. In a market in Chengdu (Sichuan), in the course of two hours' strolling, I noted more than 250 separate items with wooden components. Often one forgets how many wooden household and rural industry implements, toys, games, ornaments, and the like, are still in use. Indeed, I was much chastened in 1963—on describing what I took to be innovative woodworking tools to an elderly lady born of missionary parents in rural China—to learn that they were familiar from her childhood. What in the United States and Western Europe would be put in a folk museum will be in use in China for a long time to come. There is little point, therefore, in trying to forecast the demand for wood in the "miscellaneous" and "other" categories.

NOTES

1. Construction space is defined as the total area of all floors, including partitions and exterior walls as well as secondary rooms (kitchen, bath, lavatory, hall, and the like), which are often shared. Living space does not include secondary rooms.
2. *Hsinhua* (9 April 1982) cites repair and maintenance costs in Beijing as high as 13.88 yuan/m^2, while rents were only 1.48 yuan/m^2.

APPENDIX B

TIMBER SPECIES OFFERED FOR EXPORT SALE IN 1986

This appendix lists the species of Chinese timber that were advertised as available for export in 1986. The first list is compiled from English-language publications of the China National Timber Import and Export Corporation (TIMEX), a subsidiary of the China National Native Produce and Animal By-products Import and Export Corporation (TUHSU), operating under the Ministry of Foreign Economic Relations and Trade (MOFERT). The second list of species is from a trade publication from Yunnan. (Sichuan has also issued a list of fifty-one species that are available as logs and lumber; see TUHSU/TIMEX, 1986.) Obvious reexports from imported logs have been ignored here: They include (for the record) clear grades of Douglas fir, hemlock, and Radiata pine.

TUHSU LIST

Branch	Product
SHANTUNG	
Paulownia	logs, boards, planks
Birch	blocks, turnery, etc.
Aspen	turnery, splints
Teak	sawnwood, flooring
Ash	logs, sawnwood, veneer, flooring
Basswood	veneer
Phellodendron	cork products, logs
HEILONGJIANG	
Birch	logs, small goods

Branch	Product
Oak	logs, sawnwood, veneer
Walnut	logs, sawnwood, veneer
Maackia	logs
Poplar	various products
INNER MONGOLIA	
Birch	sawnwood
LIAONING	
Ash	veneer, sawnwood
Lime (basswood)	veneer, sawnwood
Oak	veneer, sawnwood
SHENYANG	
Birch	veneer, plywood
Ash	veneer, plywood, planks
Schima superba	veneer, plywood
Lime	veneer, plywood
Cedar	sawnwood
Camphor pine	sawnwood
Elm	planks
Catalpa	planks
DALIEN	
Ash	logs, veneer, plywood
Oak	logs, veneer, plywood
Walnut	logs
Phellodendron	logs
Birch	logs, veneer, plywood
Elm	logs, veneer, plywood
Aspen	logs, turnery
Maple	logs
Lime	plywood
TIANJIN	
Red pine	sawnwood
Birch	sawnwood
Oak	sawnwood
Lime	plywood
Ash	veneer, plywood
HENAN	
Paulownia	logs, boards, sawnwood, lacquer
JIANGSI	
Paulownia	logs, shooks, plywood
Mulberry	logs

TUHSU LIST (*Continued*)

Branch	Product
FUKIEN	
Masson's pine	plywood, sawnwood
Chinese fir	coffin planks
KWANGSI CHUANG AUTONOMOUS REGION	
Chinese fir	logs, sawnwood, coffins
Eucalyptus	chips
GUIZHOU	
Chinese fir	logs, sawnwood
Cedar	logs, sawnwood
Camphor wood	logs, sawnwood
Phoebe nanmu	logs, sawnwood
Chinese sweetgum	logs, sawnwood
Chinese sassafras	logs, sawnwood
HUNAN	
Masson's pine	plywood
Schima superba	plywood
ZHEJIANG	
Torreya (T. grandis)	logs 300–400 years old
Chestnut	logs
KIRIN (JILIN)	
Ash	poles
Walnut	poles
Oak	poles
Huai (*Sophora japonica*)	logs
Maple	logs
Elm	logs
Phellodendron	logs

YUNNAN LIST

Cyclobalanopsis glaucoides
Ehretia corylifolia
Cupressus duclouxiana
Biota orientalis
Cyclobalanopsis delavayi
Populus yunnanesis
Catalpa duclouxii
Dalbergia fusca
Symingtonia tonkinensis
Chukrassia tabularis
Paramanglietia aromatica
Schima crenata
Castanopsis fargesii
Pseudosassafras laxiflora
Paulownia fortunei
Carya tonkinensis

YUNNAN LIST (*Continued*)

Pometia tomentosa	*Albizzia odoratissima*
Toona surenii	*Pseudotsuga sinensis*
Lithocarpus grandifolius	*Betula alnoides*
Choerospondias axillaris	*Cyclobalanopsis kerii*
Parmichelia baillonii	*Abies ferreana*
Cinnamomum glanduliferum	*Mesua ferrea*
Cassia siamea	*Dalbergia szemaoensis*
Picea likiangensis	*Dalbergia burmanica*
Keteleeria evelyniana	*Quercus senescens*
Betula utilis var. *sinensis*	*Quercus acutissima*
Larix potanini	*Fagus longipetiolata*
Pinus densata	*Castanopsis calathiformis*
Pinus armandi	*Lithocarpus echinotholus*
Alnus nepalensis	*Cinnamomum parthenoxylon*
Populus rotundifolia var. *duclouxiana*	*Pinus khasya* var. *longbianensis* (*P. kesiya*)
Prunus conradinae	*Pinus yunnanensis*
Acer franchetii	*Quercus pannosa*
Juglans regia	*Neonauclea griffithii*
Burretiodendron hsienmu	*Tectona grandis*

SOURCE: Cotchell Pacific (1987).

APPENDIX C

THE FOREST LAW

Adopted by the Seventh Session of the Standing Committee of the Sixth National People's Congress on 20 September 1984.

Chapter I. General Provisions

Article 1. This Law is hereby enacted for the purpose of protecting, cultivating, and rational utilizing of forest resources, speeding up the drive to afforest the land and ensuring that the forest plays its role of conserving water and soil, adjusting the climate, improving the environment, and providing forest products, so as to meet needs arising from the socialist construction and people's lives.

Article 2. This Law must be observed in all activities relating to forest cutting, utilization, cultivation, and forest management and administration in the territory of the People's Republic of China.

Article 3. The forest resources, with the exception of those owned by the collective as provided for by law, are owned by the whole people.

The forest, forest trees, and forest land owned by the whole people and by the collective, as well as the forest trees owned and forest land used by the individual, shall be registered at local people's governments at and above the county level, which shall issue certificates to confirm such ownership or the right of use after verification has been made.

The legitimate rights of the owner and user of the forest, forest trees, and forest land are protected by law, which no institution or individual should infringe upon.

Article 4. The forest falls into the following five categories:

(1) Shelter forest: the forest, forest trees, and groves that are mainly used as shelter. They consist of forest for protecting headwaters of rivers, forest for conserving water and soil, windbreak and sand-fixation forest, farm and

pastureland shelter forest, embankment protective belts, and road protection belts.

(2) Timber forest: it comprises timber-producing forest and forest trees as well as bamboo forest producing bamboo products.

(3) Economic forest: the forest trees that mainly produce fruits, edible oil, ingredients of drinks, condiments, industrial materials, medicinal herbs, etc.

(4) Fuel forest: the forest trees that are mainly used as fuel.

(5) Forest for special uses: the forest and forest trees that are mainly for defense, environmental protection, and scientific experiment purposes. They include forest for defense, forest for experiment, maternal forest, forest for environmental protection, scenic forest, as well as the forest trees in scenic and historical spots and places with historic significance in the Chinese revolution, and the forest in natural preservation zones.

Article 5. The forestry development is guided by the policy of taking cultivation as the basis, protecting the forest in a comprehensive way, vigorously promoting afforestation, integrating cutting with cultivation, and ensuring the sustained utilization of forest resources.

The state encourages research in forest science so as to raise the level of forest science and technology.

Article 6. The state applies the following measures of protection with regard to forest resources:

(1) Quota is imposed on forest cutting, and efforts to plant trees, close hillsides to facilitate afforestation, and enlarge forest coverage are encouraged.

(2) Economic assistance of long-term loans shall be provided to collectives and individuals engaged in tree planting and forest cultivation in accordance with the relevant regulations of the state and people's governments at local levels.

(3) Forest cultivation funds shall be collected for the exclusive use of tree planting and forest cultivation.

(4) Coal mines and paper mills should set aside certain amounts of funds in proportion to the output of coal, wood pulp, and paper for the exclusive use of cultivating timber forest that will be used to make pit props and paper.

(5) The forestry fund system shall be established.

Article 7. With regard to the development of forest production in areas of national autonomy, the state and people's governments at the level of province and autonomous region, in accordance with the state regulation concerning the exercise of autonomy for regions of national autonomy, grant these areas greater decision-making power and more economic benefits than other areas in forest development, timber distribution, and use of the forest fund.

Article 8. The competent forestry department under the State Council is responsible for the work of national forestry. The competent forestry departments of the people's governments at and above the county level are responsible for the work of forestry in their respective regions. Full-time posts or concurrent posts shall be set up in people's governments at the township level to take charge of forestry work.

Article 9. To plant trees and protect forest is the bounden duty of every citizen. People's governments at various levels should organize people to take part in voluntary tree-planting activities.

Article 10. Commendation and material rewards shall be given by people's governments at various levels to institutions and individuals with outstanding achievements in tree planting, forest protection, and administration.

Chapter II. Forest Management and Administration

Article 11. Competent forestry departments at various levels administer and supervise the protection, utilization, and renewal of forest resources in accordance with the provisions of this Law.

Article 12. Competent forestry departments at various levels are responsible for conducting surveys and keeping files of record on forest resources so as to have a good grasp of changes in forest resources.

Article 13. People's governments at various levels should work out long-term forestry development programs. State-owned forestry enterprises and institutions as well as natural preservation zones should, in accordance with the long-term forestry development programs, draw up forest management plans to be implemented following approval by the competent departments at the higher level.

The competent forestry departments should guide the rural collective economic organizations as well as state-owned farms, pasture farms, industrial enterprises, and mines in drawing up forest management plans.

Article 14. Disputes over ownership and right of use of forest trees and forest land arising among institutions owned by the whole people, among institutions owned by the collective, and between institutions owned by the whole people and institutions owned by the collective shall be handled by people's governments at and above the county level.

Disputes over ownership and right of use of forest trees and forest land arising among individuals and between individuals on the one hand and institutions owned by the whole people or institutions owned by the collective on the other shall be handled by local people's governments at the level of county or township.

Parties to a dispute in disagreement with the decision of the people's government may file complaints with the people's court within one month from the date of being notified of the decision.

Pending the settlement of the dispute over the ownership of the forest trees and forest land, no party is allowed to cut forest trees under dispute.

Article 15. No forest land should be used or no more forest land than necessary as required should be used in carrying out prospecting, designing, and engineering construction as well as in mining. In case it becomes necessary to use or requisition forest land, procedures should be completed in accordance with the relevant legal provisions. Application should be submitted to the State Council for approval for use or requisition of forest land exceeding 2000 *mu* in area.

CHAPTER III. FOREST PROTECTION

Article 16. Local people's governments at various levels should see to it that departments concerned under them set up agencies for protecting forest. With a view to strengthening forest protection, they should increase forest protection facilities in areas with large forests in light of the actual need. They should see to it that grass-roots institutions owning forest land or located in the forest region formulate the forest protection pledge, mobilize people to protect forest, define areas of protection responsibility in the forest, and appoint full-time or part-time forest guards.

The forest guards shall be appointed by people's governments at the county or township level. The main duties of the forest guard are: Inspect the forest and prevent the destruction of forest resources. The forest guard has the right to bring for punishment those committing acts of destroying forest resources to local departments concerned.

Article 17. Local people's governments at various levels should adopt practical measures to prevent and extinguish forest fires:

(1) Fix a fire prevention period. During this period, it is forbidden to light fires in the forest area. In case it is necessary to use fire as required under special circumstances, prior approval must be secured from people's governments at the county level or departments with power of approval as granted by them.

(2) Install fire prevention facilities in the forest area.

(3) They are responsible for mobilizing local military units and people as well as departments concerned to extinguish forest fires when they occur.

(4) With regard to people who are wounded, disabled, or caused to die in the course of putting out forest fires, those employed by state-owned institutions shall be given medical treatment and the bereaved families given compensation by such institutions; those who are not employed by state-owned institutions shall be given medical treatment and the bereaved families given compensation by institutions causing the outbreak of the fire in accordance with the provisions laid down by the relevant competent departments under the State Council. In case the institutions involved are not responsible for the outbreak of the fire or are financially incapable of bearing such expenses, these

people shall be given medical treatment and the bereaved families given compensation by the local people's government.

Article 18. The competent forestry departments at various levels are responsible for preventing and controlling plant disease and insect pests in the forest.

The competent forestry departments are responsible for determining categories of forest tree seedlings to be quarantined, setting up quarantine zones and protection zones, and quarantining tree seedlings.

Article 19. It is forbidden to cut trees for the purpose of opening up farmland and to engage in stone-cutting and sand- and earth-digging as well as other activities of deforestation.

It is forbidden to cut firewood and graze in the young-growth area and in the special-purpose forests.

People entering the forest and fringe area of the forest are forbidden to move or damage marks in the service of forestry.

Article 20. The competent forestry department under the State Council as well as people's governments at the level of province, autonomous region, and municipality directly under the State Council should set up natural preservation zones in areas with typical forest ecology in different natural regions, in forest areas where valuable plants and animals propagate, and in forest areas calling for special protection such as natural tropical rain forest so as to better protect and administer these forest areas.

The administrative regulations for natural preservation zones shall be formulated by the competent forestry department under the State Council to be implemented following the approval of the State Council.

The valuable trees growing outside the natural preservation zone as well as plant species of special value in the forests area should be put under protection, whose cutting and gathering are forbidden unless with permission from the competent forestry departments of provinces, autonomous regions, and municipalities directly under the State Council.

Article 21. It is forbidden to hunt wild animals that have been put under state protection in the forest area. In case it is necessary to hunt such animals for special uses, the relevant laws and regulations of the state should be observed.

Chapter IV. Afforestation

Article 22. People's governments at various levels should, in the light of specific conditions of their regions, work out afforestation plans and set forth targets for increasing the forest coverage of their respective regions.

People's governments at various levels should mobilize people of all walks of life in the urban and rural areas and various institutions to fulfill the task provided for in the afforestation plans.

The competent forestry departments and other competent departments are responsible for carrying out afforestation on the barren hills and uncultivated land suitable for afforestation owned by the whole people, and the collective economic organizations are responsible for carrying out afforestation on those owned by the collective.

Afforestation of areas along railways, highways, and rivers and around lakes and reservoirs should be carried out by competent departments concerned in the light of actual conditions of these areas. Afforestation of factories, mines, ground occupied by government departments and schools, and army barracks, as well as farms, pasture farms, and fish farms, should be carried out by those institutions concerned.

Both the barren hills and uncultivated land suitable for afforestation owned by the whole people and by the collective may be contracted to the collective or individual for tree planting.

Article 23. The forest trees planted by state-owned enterprises and institutions are under the cultivation of these bodies, which may use earnings from the forest trees according to state regulations.

Forest trees cultivated by institutions of collective ownership belong to such institutions.

Trees planted by rural inhabitants around their houses and on the private plots and hills under their management belong to themselves.

In the case of barren hills and uncultivated land suitable for afforestation owned by the whole people and by the collective that are contracted by the collective or individual for planting trees, the forest trees planted by the contracting collective or individual belong to themselves, unless otherwise provided for in the contract. In the latter case, provisions of the contract should be followed.

Article 24. Local people's governments are responsible for closing the newly cultivated young-growth land and other forest land that should be closed to facilitate afforestation.

CHAPTER V. FOREST CUTTING

Article 25. The state, acting on the principle that the consumption of the timber forest should be lower than its growth, imposes strict control on the annual forest cut. Annual cutting quotas should be fixed at the level of state-owned enterprises and institutions, farms, factories, and mines for cutting forest and forest trees owned by the whole people and at the county level for cutting forest and forest trees owned by the collective. These quotas should then be reported by the competent forestry departments at the level of province, autonomous region, and municipality directly under the State Council for verification by people's governments at the same level, before they are finally submitted to the State Council for approval.

Article 26. The state adopts a unified annual timber production plan, which should not exceed the approved annual cutting quota. The scope that the plan covers shall be stipulated by the State Council.

Article 27. The following provisions must be observed in cutting forest and forest trees:

(1) Selective cutting, clear-cutting, and gradual cutting may be carried out in cutting the grown timber forests in light of their different conditions. Clear-cutting should be under strict control, and reforestation should be completed during the same year or the following year after such cutting has been carried out.

(2) Only cultivation and regeneration cutting are allowed in shelter forest and in forest for defense uses, maternal forest, forest for protecting the environment, and scenic forest in the category of forest for special uses.

(3) It is strictly forbidden to cut the forest trees in scenic and historical spots and places with historical significance in the Chinese revolution in the category of forest for special uses, as well as forest trees in natural preservation zones.

Article 28. A cutting license should be applied for and its provisions followed in cutting forest trees. However, no cutting license is required for rural inhabitants in cutting scattered trees belonging to themselves on their private plots and around their houses.

A cutting license is required for cutting forest trees by state-owned forestry enterprises and institutions, government agencies, organizations, army units, and schools, as well as other state-owned enterprises and institutions, which is to be issued, after verification has been made, by the competent forestry departments at and above the county level in areas where they are located.

A cutting license is required for regeneration cutting of road belts along railways and highways as well as trees in cities and towns, which is to be issued by the relevant competent departments after verification has been made.

A cutting license is required for cutting forest by rural economic collective organizations, which is to be issued by the competent forestry departments at the county level after verification has been made.

A cutting license is required for rural inhabitants to cut forest trees on hills contracted to them and forest trees of the collective that are contracted to individuals for cultivation, which is to be issued by the competent forestry department at the county level or people's government at the township or town level authorized by it after verification has been made.

Provisions in this article apply to cutting bamboo forest mainly for producing bamboo products.

Article 29. The departments in charge of issuing cutting licenses upon verification should not exceed the approved annual cutting quotas in issuing cutting licenses.

Article 30. The state-owned forestry enterprises and institutions must present documents of survey and design of the cutting area when applying for cutting licenses. Other institutions must present documents concerning the purpose of cutting, the location, types of trees, state of the forest, area, and reserve of the cutting area concerned, as well as ways of cutting and regeneration measures, when applying for cutting licenses.

The departments issuing cutting licenses have the right to revoke cutting licenses to the institution that violates regulations in carrying out cutting in the cutting area and stop its cutting activities until such violations are rectified.

Article 31. Institutions or individuals engaged in cutting forest trees must complete reforestation within the time limit specified and meet requirements in terms of area, number of trees, and types of trees as provided for in the cutting license. The area of reforestation and the number of trees to be planted must be larger than the cutting area and the number of trees cut.

Article 32. Regulation concerning the use and supervision of timber from the forest area will be separately formulated by the State Council.

Article 33. With the exception of timber under unified distribution of the state, a transport certificate issued by the competent forestry departments is required for transporting timber from the forest area.

With the approval of people's governments at the level of province, autonomous region, and municipality directly under the State Council, timber checking stations may be set up to supervise timber transportation in forest areas. The timber checking station has the right to prohibit transporting timber without a transport certificate or allocation notice issued by the competent goods and material departments.

Chapter VI. Legal Responsibilities

Article 34. Whoever illegally cuts forest or other woods, when the circumstances are not serious, is to compensate for losses arising therefrom as ordered by the competent forestry department, plant trees several dozen times the number of trees illegally cut, and pay a fine three to ten times the illegal earning. Whoever denudes forest or other woods, when the circumstances are not serious, is to plant trees five times the number of trees denuded as ordered by the competent forestry department and pay a fine two to five times the illegal earning.

Whoever illegally cuts and denudes forest and other woods, when the circumstances are serious, is to bear criminal responsibility as stipulated in Article 128 of the Criminal Law.

Whoever illegally cuts and keeps possession of forest trees of a huge amount is to bear criminal responsibility as stipulated in Article 152 of the Criminal Law.

Article 35. In the case of issuing a forest tree-cutting license in excess of the approved annual cutting quota or issuing a forest tree-cutting license beyond one's power in violation of this Law, whoever is directly responsible is to receive administrative sanction. When the circumstances are serious and cause major destruction to the forest, those directly responsible are to bear criminal responsibility as stipulated in Article 187 of the Criminal Law.

Article 36. Whoever counterfeits or sells a forest tree-cutting license is to be fined and illegal earnings confiscated by the competent forestry department and, when the circumstances are serious, bear criminal responsibility in the light of Article 120 of the Criminal Law.

Article 37. Whoever opens up farmland, cuts stones, digs sand and earth, gathers seeds and resin, and engages in other activities in violation of this Law and causes destruction to forest and forest trees is to compensate for losses arising therefrom as ordered by the competent forestry department and plant trees one to three times the number of trees destroyed.

Article 38. The cutting-license-issuing departments have the right to stop issuing cutting licenses to institutions or individuals that fail to fulfill reforestation targets as required until they have met such targets. When the circumstances are serious, they are to be fined by the competent forestry department and those directly responsible are to receive administrative sanction from the institutions to which they belong or the competent department at the higher level.

Article 39. Parties in disagreement with the decision of fine by the competent forestry department may file a complaint with the people's court within one month from the date of receiving the fine notice. If the party fails either to pay the fine or to file complaint upon the expiration of the specified time, the competent forestry department may request the people's court to enforce the decision of fine by coercive measures.

CHAPTER VII. SUPPLEMENTARY PROVISIONS

Article 40. The competent forestry department under the State Council shall formulate provisions of implementation in accordance with this Law, which are to be put into effect following approval by the State Council.

Article 41. In areas of national autonomy where some of the provisions of this Law do not suit local conditions, the organs of autonomy in these areas may, in accordance with the principle of this Law and in the light of actual conditions of areas of national autonomy, formulate alternative or additional provisions to be submitted through due legal procedure to the standing committee of the provincial people's congress and that of the autonomous region or the Standing Committee of the National People's Congress for approval and implementation.

Article 42. This Law is to become effective as of 1 January 1985.

APPENDIX D

RELEVANT ARTICLES OF THE CRIMINAL LAW QUOTED IN THE FOREST LAW

Article 128. Whoever violates the laws and regulations on forestry protection, illegally cuts trees, or denudes forest or other woods, when the circumstances are serious, is to be sentenced to not more than three years of fixed-term imprisonment or criminal detention and may in addition or exclusively be sentenced to a fine.

Article 152. Whoever habitually steals or habitually swindles or steals, swindles, or forcibly seizes articles of public or private property of a huge amount is to be sentenced to not less than five years and not more than ten years of fixed-term imprisonment; when the circumstances are especially serious, the sentence is to be not less than ten years of fixed-term imprisonment or life imprisonment, and the offender may in addition be sentenced to confiscation of property.

Article 187. State personnel who, because of neglect of duty, cause public property or the interests of the state and the people to suffer major losses are to be sentenced to not more than five years of fixed-term imprisonment or criminal detention.

Article 120. Whoever, for the purpose of reaping profits, counterfeits or resells planned supply coupons is, if the circumstances are serious, to be sentenced to not more than three years of fixed-term imprisonment or criminal detention and may in addition or exclusively be sentenced to a fine or confiscation of property.

In the case of a ringleader committing the crime in the preceding paragraph or especially serious circumstances, the sentence is to be not less than three years and not more than seven years of fixed-term imprisonment, and the offender may in addition be sentenced to confiscation of property.

APPENDIX E

MAJOR FORESTRY RESEARCH INSTITUTES IN CHINA

Forestry Research Institute*
Beijing

Research Institute of Forest Economics*
Beijing

Research Institute of Forest Science and Technology Information*
Beijing

Research Institute of the Wood Industry*
Beijing

Forestry Research Institute
Academy of Agriculture
Beijing

Beijing Research Institute of Forest Machinery
Beijing

Forestry Research Institute of Anhui
Hefei, Anhui Province

Forestry Research Institute of Fujian
Fuzhou, Fujian Province

Forestry Research Institute of Gansu
Lanzhou, Gansu Province

Research Institute of Sand Dune Stabilization of Gansu
Waxian, Gansu Province

Forestry Research Institute of Guangdong
Guangzhou, Guangdong Province

Research Institute of Tropical Forestry*
Guangzhou, Guangdong Province

Forestry Research Institute of Guangxi*
Nanning, Guangxi Zhuang Autonomous Region

Mount Dagin Experimental Bureau
Pingxiang, Guangxi Zhuang Autonomous Region

Forestry Research Institute of Guizhou
Guiyang, Guizhou Province

Forestry Research Institute of Hebei
Shijiazhuang, Hebei Province

Forestry Research Institute of Heilongjiang
Harbin, Heilongjiang Province

* Listed in FAO (1986b).

SOURCE: World Bank (1983).

Research Institute of Forest Protection of Heilongjiang
Harbin, Heilongjiang Province

Research Institute of Logging and Transportation of Heilongjiang
Harbin, Heilongjiang Province

Wood Industry Research Institute of Heilongjiang*
Harbin, Heilongjiang Province

Forestry Research Institute of the Forest Management Bureau of Da Hinggan
Jiagdaqi, Nei Monggol Autonomous Region

Research Institute of Subsidiary and Special Forest Products of Heilongjiang
Mudanjiang, Heilongjiang Province

Forestry Research Institute of Henan
Zhengzhou, Henan Province

Forestry Research Institute of Hubei*
Wuchang, Hubei Province

Forestry Research Institute of Hunan
Changsha, Hunan Province

Research Institute of the Forest Products Industry in Hunan
Changsha, Hunan Province

Experimental Bureau of Denkou
Bayannaoer, Nei Monggol Autonomous Region

Forestry Research Institute of Nei Monggol
Hohhot, Nei Monggol Autonomous Region

Forestry Research Institute of Jiangsu*
Nanjing, Jiangsu Province

Research Institute of Chemical Processing and Utilization of Forest Products*
Nanjing, Jiangsu Province

* Listed in FAO (1986b).

Mount Daqing Experimental Bureau
Fenyi, Jiangxi Province

Forestry Research Institute of Jiangxi
Nanchang, Jiangxi Province

Southern Institute of Forest Plant Quarantine
Yiyang, Jiangxi Province

Forestry Research Institute of Jilin
Changchun, Jilin Province

Research Institute of Forest Management of Liaoning
Benxi, Liaoning Province

Research Institute of Economic Forestry of Liaoning
Dalian, Liaoning Province

Poplar Research Institute of Liaoning
Gaixian, Liaoning Province

Research Institute of Afforestation in Arid Zones of Liaoning
Jianpin, Liaoning Province

Northern Institute of Forest Plant Quarantine
Shenyang, Liaoning Province

Forestry Research Institute of Liaoning
Xinmin, Liaoning Province

Research Institute of Sand Dune Afforestation of Liaoning
Zhangwu, Liaoning Province

Forestry Research Institute of Ningxia Academy of Agriculture and Forestry
Yinchuan, Ningxia Autonomous Region

Forestry Research Institute of Qinghai Academy of Agriculture and Forestry
Xining, Qinghai Province

Forestry Research Institute of Shandong
Jinan, Shandong Province

Forestry Research Institute of Shanxi
Taiyuan, Shanxi Province

Wood Industry Research Institute of Shanghai*
Shanghai

Forestry Research Institute of Shaanxi
Xian, Shaanxi Province

Forestry Research Institute of Sichuan*
Chengdu, Sichuan Province

Forestry Research Institute of Xijiang
Wulumuqi, Xinjiang Autonomous Region

* Listed in FAO (1986b).

Shellac Research Institute*
Jingdong, Yunnan Province

Forestry Research Institute of Yunnan
Kunming, Yunnan Province

Research Institute of Subtropical Forestry*
Fuyang, Zhejiang Province

Forestry Research Institute of Zhejiang
Hangzhou, Zhejiang Province

REFERENCES

ACIAR (Australian Center for International Agricultural Research). *China Forestry Development Project.* Canberra: 1986.

Adshead, S. A. M. "Timber as a Factor in Chinese History: Problems, Sources and Hypotheses." *Proceedings of the 1st Symposium on Asian Studies.* Hong Kong, 1979.

Aird, J. S. "The Preliminary Results of China's 1982 Census." *China Quarterly* 96(1983):613–640.

———. "Coercion in Family Planning: Causes, Methods and Consequences." In *CELT 2000,* vol. 1, pp. 184–221. Washington, D.C., 1986.

Anon. "A New Landmark in Pasture Building—The Grass Kulun." *Acta Botanica Sinica* 18(1). Wushenchao, Nei Monggol, 1976.

———. "Combating Desertification." Institute of Glaciology, Cryopedology, and Desert, CAS. Lanzhou, 1977.

———. *China Facts and Figures.* Beijing, 1982.

———. *A Brief Account of China's Forestry.* Ministry of Forestry. Beijing, 1984.

APFC (Asia Pacific Forestry Commission). "Forestry Development in China." Country Report. Beijing, 1987.

Araman, P. A. "Pacific Rim Demands for U.S. Hardwoods." Typescript. 1987.

Banister, J. "Implications of China's 1982 Census Results." In *CELT 2000,* vol. 1, pp. 160–183. Washington, D.C., 1986.

Betke, D., and J. Kuchler. "Shortage of Land Resources as a Factor in Development: The Example of the PRC." In B. Glaeser (ed.), *Learning from China?* London, 1987.

Biswas, M. R., and A. K. Biswas. *U.N. Conference on Desertification.* Nairobi, 1977.

Blandon, P. *Soviet Forest Industries.* Boulder, 1983.

Bretschneider, E. V. *History of European Botanical Discoveries in China.* 2 vols. St. Petersburg, 1898.

Brown, J. *Project China 2606.* Typescript, World Food Program. Beijing, 1985.

Buchanan, K. "The Changing Landscape of Rural China." *Pacific Viewpoint* 1(1)(1960):11.

CBR (*China Business Review*). "China's Papermakers." July/August 1981.

CELT 2000 (*China's Economy Looks Towards the Year 2000*). Vol. 1: *The Four Modernizations;* vol. 2: *Economic Openness in Modernizing China.* Washington, D.C., 1986.

Chan, K. W., and Xueqiang Xu. "Urbanization in China Since 1949." *China Quarterly* 104(1985):548–613.

Chang, C. T. (ed.). *National Atlas of China.* Vol. 5: *General Maps of China.* Taipei, 1962.

Chang Hsin-shi. "On the Eco-geographical Characters and the Problems of Classification of the Wild Fruit Tree in the Ili Valley of Xinjiang." *Acta Botanica Sinica* 15(2)(1973):239–253. (English summary.)

Chao Xinshen. *Farmland Protection Forestry.* Beijing, 1983.

Chen, H. C. "Forest Flora of Lu-shan (Kiangsi)." *Journal of Agriculture and Forestry* 13(35)(1936):974–985.

Chin, S. Y., and G. Toomey. "Paulownia, China's Wonder Tree." *IDRC Reports* (April 1986):11–12.

China Coal Industry Yearbook. Ministry of Coal. Beijing, 1983.

China Daily (Beijing). Issue of 3 June 1985.

Chu, K. C. "An Expedition to Inquire into the Feasibility of Transmitting Water from the Upper Course of the Yangtze to the Yellow River." *Ti-lih-Chih-shih* [Geographic knowledge] 10(4)(1959a):145.

_______. "The Fight Against Deserts." Typescript, FAO. Rome, 1959b.

Chun, W. Y., and K. Z. Kuang. "A New Genus of Pinaceae—*Cathaya* Chun et Kuang, gen. nov., from Southern and Western China [*Genus novum pinacearum ex Sina australi et occidentali*]. *Botanicheskii Zhurnal SSSR* (Leningrad) 43(1958):461–470.

CIA (Central Intelligence Agency). *PRC Handbook of Economic Indicators,* Washington, D.C., 1975.

_______. *China: Economic Performance in 1985.* Report to Joint Economic Committee, 17 March 1986.

Clarke, C. M. "Reorganization and Modernization in Post-Mao China." In *CELT 2000,* vol. 1, pp. 90–109. Washington, D.C., 1986.

Clarke, G. E. "China's Reforms of Tibet and Their Effects on Pastoralism." Institute of Development Studies, DP 237. London, 1987.

Cotchell Pacific. *The Cotchell Report on China.* Hong Kong, 1987.

Cox, E. H. M. *Plant Hunting in China.* London, 1945.

Cressey, G. B. *Land of the 500 Million . . . A Geography of China.* New York, 1955.

Crook, F. W. "The *Baogan Daohu* Incentive System." *China Quarterly* (1985):102, 291–303.

________. "The Reform of the Commune System and the Rise of the Township-Collective-Household System." In *CELT 2000*, vol. 1, pp. 354–375. Washington, D.C., 1986.

CTR (*China Trade Report*). Issue of November 1985.

Dale, V. R. *Science and Technology in China.* Wellington, N.Z., 1984.

David, A. "Journal d'un Voyage dans le Centre de la Chine et dans le Tibet Oriental." *Nouvelles Archives de la Musée d'Histoire Naturelle Paris,* bulletins 8(3)(1872); 9(15)(1873); 10(3)(1874).

Davie, J. L. "China's International Trade and Finance." In *CELT 2000,* vol. 2, pp. 311–334. Washington, D.C., 1986.

Delfs, R. "Disconnected Panic Button." *Far Eastern Economic Review* (Hong Kong), issue of 5 June 1986a.

________. "Rising from the Ashes." *Far Eastern Economic Review* (Hong Kong), issue of 31 July 1986b.

________. "Policy Advice Helps the Huge Country." *Far Eastern Economic Review* (Hong Kong), issue of 1 October 1987.

________. "Fiscal Feudalism." *Far Eastern Economic Review* (Hong Kong), issue of 6 April 1989.

Deng [Teng] Suchun. "The Early History of Forestry in China." *Journal of Forestry* 25(1927):564–570.

Dickerman, M. B. *Forestry in the People's Republic of China.* Society of American Foresters. Bethesda, Md., 1980.

Dowdle, S. "Seeking Yields from Fewer Fields." *Far Eastern Economic Review* (19 March 1987):78–80.

Dreyer, J. T. "The Role of the Military in the Chinese Economy." In *CELT 2000,* vol. 2, pp. 186–198. Washington, D.C., 1986.

DTS. *Direction of Trade Statistical Yearbook.* Washington, 1986.

Eberstadt, N. "Material Poverty in the People's Republic of China in International Perspective." In *CELT 2000,* vol. 1, pp. 263–324. Washington, D.C., 1986.

EIU (Economist Intelligence Unit). *China: Country Report No. 2.* Hong Kong, 1986.

Enderton, C. S. "Nature Preserves and Protected Wildlife in the People's Republic of China." *China Geographer* 12(1985):117–140.

FAO (Food and Agriculture Organization). *China: Forestry Support for Agriculture.* Forestry Paper no. 12. Rome, 1978.

________. *China: Integrated Wood Processing Industries.* Forestry Paper no. 16. Rome, 1979.

________. *Forestry in China.* Forestry Paper no. 35. Rome, 1982.

________. *Forest Resources, 1980.* Rome, 1985.

———. *Forest Products: World Outlook Projections.* Forestry Paper no. 73. Rome, 1986a.

———. *World Compendium of Forestry and Forest Products Research Institutions.* Forestry Paper no. 71. Rome, 1986b.

FBIS (U.S. Foreign Broadcast Information Service). Report, p. Q/3, Chinese, 8 February 1982.

———. Report—Agriculture, no. 29, Chinese, 1 November 1984.

FE (Far East). BBC World Broadcasts. FE/6069/BII/14 of 17 March 1979.

———. FE/6376/C/1 of 21 March 1980.

———. FE/6530/C/10 of 23 September 1980.

———. FE/W115/A/13 of 7 January 1981.

———. FE/6661/C/5 of 28 February 1981.

———. FE/W1123/A/13 of 4 March 1981.

———. FE/6673/C1/1 of 14 March 1981.

———. FE/W1160/A/20 of 18 November 1981.

———. FE/6982/BII/147 of 23 March 1982.

———. FE/7455/BII/17 of 4 October 1983.

———. FE/8079 of 12 October 1986.

FEER (*Far Eastern Economic Review,* Hong Kong). Issue of 16 January 1986.

———. Issue of 20 March 1986.

———. Issue of 26 March 1986.

———. Issue of 15 May 1986.

———. Issue of 19 March 1987.

Fewsmith, J. "Rural Reform in China: Stage 2." *Problems of Communism* (July/August 1985):48–55.

Field, R. M. "Growth and Structural Change in Chinese Industry: 1952–79." *China Under the Four Modernizations, Part I.* U.S. Congress Joint Economic Committee. Washington, D.C., 1982.

Field, R. M., and H. L. Noyes. "Prospects for Chinese Industry in 1981." *China Quarterly* 85(1981):96–106.

Fingar, T. "Overview: Energy in China." In *CELT 2000,* vol. 2, pp. 1–21. Washington, D.C., 1986.

Fischer, W. A. "Chinese Industrial Management: Outlook for the Eighties." In *CELT 2000,* vol. 1, pp. 548–570. Washington, D.C., 1986.

Franchet, A. "Plantae Davidianae ex Sinarum Imperio Pt. I." *Nouvelles Archives de la Musée d'Histoire Naturelle Paris,* bulletins 2(5)(1883):153; 2(6)(1883):1; 2(7)(1884):55. Pt. II: 2(8)(1885):183; 2(10)(1888):33.

Freeberne, M. "The People's Republic of China." In *The Changing Map of Asia.* London, 1971.

Gaag, J. v.d. "Private Household Consumption in China." World Bank Staff Working Paper no. 701. Washington, D.C., 1984.

Garside, R. *Coming Alive: China After Mao.* London, 1981.

Gibson, J. M., and D. M. Johnston. *A Century of Struggle.* Canadian Institute of International Affairs. Ottawa, 1971.

Ginsburg, N., and B. A. Lalor. *China: The 80's Era.* Boulder, 1984.

Glaeser, B. (ed.). *Learning from China?* London, 1987.

Gold, T. B. "Personal Relations in China Since the Cultural Revolution." *China Quarterly* 104(1985):654–675.

Grainger, A. "China's Special Approach to Forestry." *World Wood* 21(4)(1980):18–19.

Guan, Junwei, and Wang Lixian. "The Problem of Land Use in Soil Erosion Zones of China." Cited in Holder et al. (1985).

Guangdong (Guangdong Institute of Botany). *Guangdong Province's Vegetation.* Beijing, 1976.

He Kang. "Interview." *China Reconstructs* (Beijing). Issue of October 1985.

Holder, F. G., et al. "Erosion Control and Development Through Forestry and Pasture in Xiji County." Interim Evaluation of Project China 2605. UN/FAO World Food Program. Rome, 1985.

Ho Ping Ti. *Studies on the Population of China, 1368–1953.* Cambridge, Mass., 1959.

Hou, H. Y. "Vegetation of China with Reference to Its Geographical Distribution." *Annals of the Missouri Botanical Gardens* 70(1983):509–548.

Hou, H. Y. [Hou Xue Yu], C. T. Cheng, and H. P. Wang. "The Vegetation of China with Special Reference to the Main Soil Types." *Report for VIth International Congress of Soil Science.* Beijing, 1956.

Hou, Zhizeng, and Yuchen Wang. "A Study in the Afforestation of Barren Hills and Wastelands by Peasants in China." Typescript, FAO. Bangkok, 1986.

Hsu, C. "China Aims to Turn Its Land Green." Typescript, FAO. Rome, 1959.

Hsu, Y. C. "A Preliminary Study of the Forest Ecology of the Area About Kunming." *Contributions of the Dudley Herbarium* 4(1)(1950).

Hsuing, W. Y. *Forests and Forestry in China.* H. R. MacMillan Lectureship in Forestry. Vancouver, Canada, 1980.

Hughes, T. J., and D. E. T. Luard. *The Economic Development of Communist China, 1949–1958.* London, 1959.

Hwang, T. T. "On the Turpan Depression." *Science* 27(3)(1944):7.

IDRC (International Development Research Center). *Paulownia in China: Cultivation and Utilization.* Asian Network for Biological Sciences. Singapore, 1986.

________. *Recent Research on Bamboos.* Ottawa, 1987.

Ishikawa, S. "China's Economic Growth Since 1948—An Assessment," *China Quarterly* 94(1983): 242–281.

Jiang Yuxu. "On the Characteristics and Management Strategies of Sub-alpine

Coniferous Forests in Southwestern China." Typescript, Institute of Forestry, Academy of Forestry. Beijing, 1986.

JPRS (Joint Publications Research Service). *China Examines Science Policy.* Washington, D.C., 1982a.

———. No. 81620 of 24 August 1982b.

———. No. 83240 of 12 April 1983a.

———. No. 84524 of 13 October 1983b.

JPRS-CST. No. 84011 of 17 April 1984.

Kalish, J. "Report from China, 1–4." *Pulp and Paper International* 24(3)(1982):54–57; 24(4)(1982):55–57; 24(5)(1982):59–62.

Kang Chao. *The Construction Industry in China.* Edinburgh, 1968.

Keidel, A. "China's Coal Industry." In *CELT 2000,* vol. 2, pp. 60–86. Washington, D.C., 1986.

Keswick, M. *The Chinese Garden.* New York, 1978.

Klatt, W. "The Staff of Life: Living Standards in China, 1977–81." *China Quarterly* 93(1983):17–50.

Koshizawa, A. "China's Urban Planning: Toward Development Without Urbanization." *Developing Economies* 16(1)(1978):3–33.

Kraus, W. *Wirtschaftliche Entwicklung und Sozialer Wandel in der Volksrepublik China.* Berlin, 1979.

Krugman, S. L., and R. C. Kellison. "Administrative Report on Forest Genetics and Tree Improvement Programs in the People's Republic of China: 1986." USDA, Forest Service, 15 December 1986.

Krugman, S. L., et al. *Forest Genetics and Tree Improvement in the People's Republic of China.* USDA and SAF. Washington, D.C., 1983.

Kuo, C. C., and Y. C. Cheo. "A Preliminary Survey of the Forests in Western China." *Sinensia* 12(1941):81–133.

Lampton, D. M. "Water Politics and Economic Change in China." In *CELT 2000,* vol. 1, pp. 387–406. Washington, D.C., 1986.

Langston, O. "Laying Down the Law." *Far Eastern Economic Review,* 24 January 1985.

Lardy, N. R. *Agriculture in China's Modern Economic Development.* Cambridge, 1984.

———. "Consumption and Living Standards in China, 1978–83." *China Quarterly* 100(1986a):849–865.

———. "Overview: Agricultural Reform and the Rural Economy." In *CELT 2000,* vol. 1, pp. 325–335. Washington, D.C., 1986b.

Lawrence, A. "Guest Travellers Tales." *Far Eastern Economic Review,* 4 September 1986.

Lee, P. N. "Enterprise Autonomy in Post-Mao China: A Case-Study of Policy Making, 1978–83." *China Quarterly* 105(1986):45–71.

Leung, Stanley. "Growth Threatened by Ecological Neglect." *China Daily,* 18 August 1986.

Li Tang. "Combination of Forestry with Industry and Commerce." In *China Agriculture*, pp. 53–54. Beijing, 1987.

Li Wenhua and Zhang Mingtao. "Watershed Management in Mountain Regions of S.W. China." International Workshop on Watershed Management in the Hindu Kush–Himalaya Regions. ICIMOD/CAS. Beijing, 1985.

Liang Heng and J. Shapiro. *Son of the Revolution.* London, 1983.

Lin Jung. *Handbook of Water and Soil Conservation in the Middle Yellow River Loess Regions.* Beijing, 1959.

Lin, P. T. K. "The Educational Revolution." In J. M. Gibson and D. M. Johnston, *A Century of Struggle.* Ottawa, 1971.

Lin, Z. Q. "Problems and Proposals on Housing Development." *Architectural Journal* 1(1983):40–44.

Liu Peihua. *A Short History of Modern China.* Beijing, 1954.

Lowdermilk, W. C. "Erosion and Floods in the Yellow River Watershed." *Journal of Forestry* 22(6)(1924):11–18.

_______. "Forest Destruction and Slope Denudation in the Province of Shanxi." *China Journal of Science and Arts* 4(3)(1926):127–136.

_______. "Forestry in Denuded China." *Annals of the American Political Science Society* 152(1932):98.

Luukkanen, L. "Notes on the Forests of Northeastern China and Their Utilization." *Silva Fennica* 14(4)(1980):332–341.

MacFarlane, I. H. "Construction Trends in China." In *China: A Re-assessment of the Economy.* Joint Economic Committee of the U.S. Congress. Washington, D.C., 1975.

McFadden, M. W., et al. "Integrated Pest Management in China's Forests." *Journal of Forestry* 79(11)(1981):714–726.

Malraux, A. *Anti-Memoires.* London, 1968.

Matthews, J. D. "Observations on Trees and Forests in China." U.K. Forestry Commission Occasional Paper no. 8, 1980, pp. 32–55.

Menzies, N. *The History of Forestry in China.* Cambridge (in press).

Merrill, E. D. *Plant Life of the Pacific World.* New York, 1945.

MOF (Ministry of Forestry). *Forestry Development in China.* Beijing, 1984.

_______. *Statistics of China's Forestry.* Beijing, 1985.

Morgan, S. "Ideology on the Block." *Far Eastern Economic Review,* 14 July 1988.

Morris, R. C. "Report of Insect Damage to Poplar Trees in the PRC." Typescript, FAO. Beijing, 1985.

Moss, L. "Space and Direction in the Chinese Garden." *Landscape* 14(3)(1965):29–33.

Myrdal, J. *Return to a Chinese Village.* New York, 1984.

Nakajima, S. "Reform of China's Transport System." *JETRO China Newsletter,* March/April 1985.

Nations, R. "Deng Xiaopeng's Reforms Worry Kremlin Bosses." *Far Eastern Economic Review* (Hong Kong), 14 August 1986.

Naughton, B. "The Profit System." *China Business Review* (U.S.), November/December 1983.

_______. "Finance and Planning Reforms in Industry." In *CELT 2000*, vol. 1, pp. 604–629. Washington, D.C., 1986.

NCNA (New China News Agency, Beijing). News release of 1 June 1961.

_______. News release of 1 December 1961.

_______. News release of 8 August 1962.

_______. News release of 21 November 1962.

Needham, J. "Science and China's Influence on the World." In *The Legacy of China*, pp. 234–308. Oxford, 1964.

New Zealand Mission Report. Ministry of Foreign Affairs. Wellington, 1987.

NFPA (National Forest Products Association). *The Market for Softwood Lumber and Plywood in the People's Republic of China.* 2 vols. Washington, D.C., 1986.

Nien Cheng. *Life and Death in Shanghai.* London, 1985.

Noyes, H. L. "United States–China Trade." In *CELT 2000*, vol. 2, pp. 335–347. Washington, D.C., 1986.

Oates, W. E. "Taxing Pollution: An Idea Whose Time Has Come?" *Resources for the Future* 91(1988):5–7.

Orleans, L. A. "China's Urban Population: Concepts, Conglomerates and Concerns." In *China Under the Four Modernizations*, pt. I, U.S. Congress Joint Economic Committee. Washington, D.C., 1982.

_______. "Overview: China's Human Resources." In *CELT 2000*, vol. 1, pp. 147–159. Washington, D.C., 1986.

Ovington, J. D., et al. "A Report on Gardens, Parks, and Open Spaces in China." Department of National Parks. Canberra, 1975.

Pei, Sheng-ji. "Some Effects of the Dai People's Cultural Beliefs and Practices on the Plant Environment of Xishuangbanna, Yunnan Province, Southwest China." In *Cultural Values and Human Ecology in Southeast Asia.* Michigan Papers on South and Southeast Asia, no. 27. Ann Arbor, 1985.

Petrov, M. P. "Land-use of Semi-deserts in Central Asia." In *Management of Semi-arid Ecosystems.* Amsterdam, 1979.

Pollard, D. F. W. "Impressions of Urban Forestry in China." *Forestry Chronicle* 53(5)(1977):294–297.

Potanin, G. N. "Sketch of a Journey to Szechuan and the Eastern Frontier of Tibet in 1892–3." *Isv. Russk. Geogr. Obshch* 35(1899):363–418.

Prybla, J. S. "Contribution to 'Economic and Political Developments in China.'" *Heritage Round Table* 27(1986a):27–46.

_______. "China's Economy from Mao to Market." *Problems of Communism* (January/February 1986b):21–38.

Qu Geping and Li Jinchan. "Environmental Management in China." *Unasylva* 33(134)(1981):2–18.

Ren Mei'e, Yang Renzhang, and Bao Haosheng. *An Outline of China's Physical Geography.* Beijing, 1985.

Richardson, S. D. "Commodity Report: Production and Consumption of Forest Products in China (Mainland)." *Unasylva* 19(1)(1965):24–31.

_______. *Forestry in Communist China.* Baltimore, 1966.

_______. *The Production and Consumption of Forest Products in Mainland China: Future Requirements and Trade Prospects.* Report to the Tuolumne Corp. San Francisco, 1972.

_______. "Letter from Dailing." *Far Eastern Economic Review* (Hong Kong), 27 January 1987a, p. 74.

_______. "Forestry in Outer Mongolia." *Commonwealth Forestry Review* 66(3)(1987b):265–272.

_______. "Forestry Beyond the Rim." *New Zealand Forestry* 33(4)(1989): 17–19.

Richardson, S. D., and E. Salem. "A Policy in the Ashes." *Far Eastern Economic Review* (Hong Kong), 4 June 1987, 63–64.

Ross, L. "Obligatory Tree Planting: How Great an Innovation in Implementation in Post-Mao China?" Typescript, Purdue University, 1983.

_______. *Environmental Policy in China.* Bloomington, 1988.

SAF (Society of American Foresters). "Forestry in the People's Republic of China." Washington, D.C., 1980.

SCMP (*South China Morning Post*). "Giant Shelter Belt Project Transforms Formerly Sand Ravaged Area of Northeastern China." *Survey of the China Mainland Press,* no. 2512, 1 June 1960.

Shiraishi, K. "Chugoku Nogyo Chiri Soron." *Bezai Uikuri,* nos. 470–474, pts. 1–5 (1983).

Shirk, S. "The Evolution of Chinese Education: Stratification and Meritocracy in the 1980's." In Ginsburg and Lalor, *China: The 80's Era.* Boulder, 1984.

Sichuan (Editorial Board of Sichuan's Vegetation). *Sichuan Province's Vegetation.* Chengdu, 1980.

Siddiqi, T. A., Jin Xiaoming, and Shi Minghao. "China–USA Governmental Cooperation in Science and Technology." Occasional Papers of the East-West Environmental and Policy Institute, no. 1. Honolulu, 1987.

Smil, V. "Energy Development in China: The Need for a Coherent Policy." *Energy Policy* 9(1981):113–126.

_______. "The People's Republic of China: Environmental Aspects of Economic Development." World Bank. Washington, D.C., 1982.

_______. *The Bad Earth: Environmental Degradation in China.* Armonk, 1984.

Smith, S. H. "A Traditional Decade: Higher Education in China in the 1980's." In Ginsburg and Lalor, *China: The 80's Era.* Boulder, 1984.

South China Morning Post (Hong Kong). Issue of 8 September 1986.

Sowerby, A. de C. *A Naturalist in Manchuria.* 3 vols. Tientsin, 1923.

———. "Approaching Desert Conditions in North China." *China Journal of Science and Arts* 2(3)(1924a).

———. "Forestry in China." *China Journal of Science and Arts* 2(4)(1924b).

SSB. *Statistical Yearbook of China, 1984.* Beijing, 1984.

Su Wanju. "News Report." *Beijing Review,* issue of 21 July 1986.

Surls, F. M. "Agricultural Policy and Growth." *China Quarterly* 100(1984):866–869.

Suttmeier, R. P. "Overview: Science and Technology Under Reform." In *CELT 2000,* vol. 2, pp. 199–215. Washington, D.C., 1986.

SWB (Summary of World Broadcasts). Hong Kong statement of Wan Li. Transcript of 24 January 1984.

Taubman, W. "Problems of Urban Housing in the PRC." Typescript, Center of Asian Studies, University of Hong Kong, 1983.

Thompson, M., M. Warburton, and T. Hatley. *Uncertainty on a Himalayan Scale.* Milton Ash, England, 1986.

Tong, H. K. *China Yearbook, 1937–45.* Chongqing, 1947.

Travers, L. "Peasant Non-agricultural Production in the People's Republic of China." In *CELT 2000,* vol. 1, pp. 376–386. Washington, D.C., 1986.

TUHSU/TIMEX. *Timbers and Timber Products of Sichuan, China.* China National Native Produce and Animal By-products Import/Export Corp. Sichuan, 1986.

Turnbull, J. W. "Eucalypts in China." *Australian Forests* 44(4)(1981):222–234.

UNEP (UN Environmental Program). *Combating Desertification in China.* Nairobi, 1982.

USDA (U.S. Department of Agriculture). "PRC Forestry." American Consulate-General, Report no. HK2041. Hong Kong, 1983.

———. "China: Outlook and Situation Report." Economic Research Service. Washington, D.C., 1985.

USDA/FAS. "China—Forest Products Annual: Global Economic Data Exchange System AGR." Report Ch6043. Washington, D.C., 1986.

Vlassoff, C. "The One-Child Solution." *IDRC Reports* 15(3)(1986): 22–23.

Wagner, R. G. "Agriculture and Environmental Protection in China." In B. Glaeser (ed.), *Learning from China?* London, 1987.

Walder, A. G. "Rice Bowl Reforms." *China Business Review* (U.S.), November/December 1983.

———. "The Informal Dimension of Enterprise Financial Reforms." In *CELT 2000,* vol. 1, pp. 630–645. Washington, D.C., 1986.

———. "Wage Reform and the Web of Factory Interests." *China Quarterly* 109(1987):22–41.

Walker, K. R. *Food Grain Procurement and Consumption in China.* Cambridge, 1984.

Wang, C. W. *The Forests of China, with a Survey of Grassland and Desert Vegetation.* Maria Moors Cabot Foundation. Cambridge, Mass., 1961.

Wang Huenpu. "Nature Conservation in China: The Present Situation." *Parks* 5(1)(1980):1–9.

_______. "The Perspective of Natural Conservation on Western Uplands in China." Typescript, Academia Sinica. Beijing, 1984.

Wang, H. Y., Y. H. Li, D. H. Li, and X. J. Yu. "A Study on the Oil Bearing Plants in Tropical and Sub-tropical Areas in Yunnan Province." *Collected Research Papers, Yunnan Institute of Tropical Botany* (1982):80–100.

Wang, J., N. Hua, et al. "Morphology and Ultra-structure of Some Non-wood Paper Making Materials." Paper prepared for 40th Appita Conference, Auckland, 1986.

Wang, Yifeng. "Characteristics of the Vegetational Zones in the Nei Monggol Autonomous Region." *Acta Botanica Sinica* 21(3)(1977):274–284. (English summary.)

Ward, B., and R. Dubos. *Only One Earth: The Care and Maintenance of a Small Planet.* London, 1972.

Westoby, J. C. "Forestry in China." *Unasylva* 108(27)(1975):20–28.

Wijkman, A., and L. Timberlake. "Natural Disasters: Acts of God or Acts of Man?" International Institute for Environment and Development (Earthscan). London, 1984.

Wilson, E. H. *A Naturalist in Western China.* 2 vols. London, 1913.

Wong, C. P. W. "Ownership and Control in Chinese Industry: The Maoist Legacy and Prospects for the 1980's." In *CELT 2000,* vol. 1, pp. 571–603. Washington, D.C., 1986.

World Bank. "Country Study: China." Annex 5 to *China: Long Term Development Issues and Options.* Washington, D.C., 1985a.

_______. "China: The Transport Sector." Annex 6 to *China: Long Term Development Issues and Options.* Washington, D.C., 1985b.

_______. "Country Study: China." Annex 2 to *China: Long Term Development Issues and Options.* Washington, D.C., 1985c.

_______. *China: Forestry Development Project.* Staff Appraisal Report. Washington, D.C., 1985d.

Wu, Chengyih. "The Regionalization of the Chinese Flora." *Acta Botanica Yunnanica* 1(1)(1979):1–22. (English summary.)

Xinjiang (Integrated Survey Team of Xinjiang and Institute of Botany). *Vegetation of Xinjiang and Its Utilization.* Academia Sinica. Beijing, 1978. (In Chinese.)

Xu Shige. "China's Cooperative Forestry Economy." Typescript, Ministry of Forestry. Beijing, 1986.

Xu Zhenbang, Dai Hongcai, and Li Xin. "Rational Management of Broadleaved Korean Pine Forest and Improvement of Its Woodland Productivity in N.E. China." Institute of Forestry and Soil Science, Academia Sinica. Beijing, 1986.

Yahuda, M. "China." *Asia and Pacific Review* (1986):73–88.

Yang Chiang. *Six Chapters of Life in a Cadre School: Memoirs from China's Cultural Revolution.* Boulder, 1986.

Yang, C. Y. "The Forest Vegetation of Si-shan and Hsiao-wu-tai-shan." *Bulletin of the Chinese Botanical Society* 3(1937):1–97.

Yang, M. T. "Landscape Characteristics of Southern Guizhou and Some Problems of Demarcation of Natural Districts." Ti-li no. 4. Cited in *Joint Publications Research Service* 20(1962):119.

Yang Yupo. "Great Importance Should Be Attached to the Ecological Balance of the Subalpine Forest of Western Sichuan." Typescript, Forest Research Institute of Sichuan. Chengdu, 1986.

Yeh, K. C. "Macroeconomic Changes in the Chinese Economy During the Readjustment." *China Quarterly* 100(1984):691–716.

Yong Wentao. Article in *Nonggye Tingji Wenti* 9(1982).

Yu, P. H., Z. E. Xu, and Y. L. Huang. "A Research on the Ethnic Timber Utilizations in the District of Xishuangbanna." *Collected Research Papers, Yunnan Institute of Tropical Botany* (1982):108–115.

Yu, Y. "Reed Pulping in China." Progress Report no. 14. Atlanta, Ga., 1983.

Zhong, X. "Modernization of the Chinese Paper Industry." *Das Papier* 39(10A) (1985):V60–V64.

Zhou, Chengkui. "Revamping Science and Technology System." *Beijing Review* 24(1986):21–27.

Zhu Zhisong. "The Green Great Wall of China." *American Forests* May(1981):1–4.

Zweig, D. "Prosperity and Conflict in Post-Mao China." *China Quarterly* 105(1986):1–18.

BOTANICAL INDEX

SUBJECT INDEX

ABOUT THE AUTHOR

A graduate of Oxford University, S. D. (Dennis) Richardson has been director of research for the New Zealand Forest Service, senior forestry specialist with the Asian Development Bank, and professor and head of the Department of Forestry in the University of Wales and the University of Technology, Papua New Guinea. Dr. Richardson was visiting professor of forestry at the University of Wisconsin in 1964–65, and has been a research fellow with the East-West Center, a nonprofit educational institution that examines issues related to population, resources and development, the environment, culture, and communication in Asia, the Pacific, and the United States. Dr. Richardson's first book, *Forestry in Communist China,* was published in 1966. The author of more than two hundred papers and articles, Dr. Richardson is currently an independent forestry adviser and a frequent consultant to international agencies. He is a director-at-large of the International Society of Tropical Foresters (U.S.A.), a governing councillor of the Commonwealth Forestry Association (U.K.), and an honorary life member of the New Zealand Institute of Foresters. He lives with his wife in New Zealand.